PRACTICE MASTERS
LEVELS A, B, AND C

HOLT, RINEHART AND WINSTON

A Harcourt Classroom Education Company

Austin • New York • Orlando • Atlanta • San Francisco • Boston • Dallas • Toronto • London

To the Student

Practice Masters Levels A, B, and C consist of three levels of exercises graded by level of difficulty. There is a full page for each of the three levels of exercises for each lesson in the *Pupil's Edition*. Level A exercises are the least difficult, designed for students to practice the lesson objectives through the use of examples. Level B exercises are middle-range exercises that students can handle with the use of current examples together with some prior knowledge. Level C exercises are the most challenging exercises, relevant to the lesson and appropriate for the student who has mastered the lesson.

Copyright © by Holt, Rinehart and Winston

All rights reserved. No part of this publication may be reproduced or transmitted in any form or by any means, electronic or mechanical, including photocopy, recording, or any information storage and retrieval system, without permission in writing from the publisher.

Teachers using ALGEBRA 2 may photocopy complete pages in sufficient quantities for classroom use only and not for resale.

Photo Credit
Front Cover: Tom Paiva/FPG International

Printed in the United States of America

ISBN 0-03-064817-3

2 3 4 5 6 7 066 05 04 03 02 01

Table of Contents

Chapter 1	Data and Linear Representations	1
Chapter 2	Numbers and Functions	25
Chapter 3	Systems of Linear Equations and Inequalities	46
Chapter 4	Matrices	64
Chapter 5	Quadratic Functions	79
Chapter 6	Exponential and Logarithmic Functions	103
Chapter 7	Polynomial Functions	124
Chapter 8	Rational Functions and Radical Functions	139
Chapter 9	Conic Sections	163
Chapter 10	Discrete Mathematics: Counting Principles and Probability	181
Chapter 11	Discrete Mathematics: Series and Patterns	202
Chapter 12	Discrete Mathematics: Statistics	226
Chapter 13	Trigonometric Functions	244
Chapter 14	Further Topics in Trigonometry	262
Answers		280

NAME _____ CLASS _____ DATE _____

Practice Masters Level A
1.1 Tables and Graphs of Linear Equations

State whether each equation is a linear equation.

1. $y = 4x + 1$ _____
2. $y = x - 7$ _____
3. $y = 2x^2 + 3x$ _____
4. $y = -5x$ _____
5. $y = 1 - x^2$ _____
6. $y = 4 + 2.5x$ _____
7. $y = \frac{1}{2}x^2 + 3$ _____
8. $y = \frac{3}{4}x - 1$ _____
9. $y = \frac{-2}{3}x - x^2$ _____

Graph each linear equation.

10. $y = 3x + 2$

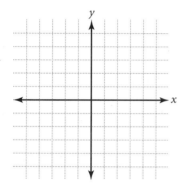

11. $y = 4x - 1$

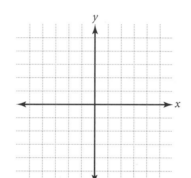

12. $y = \frac{1}{2}x + 5$

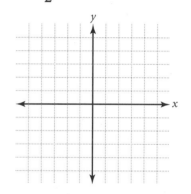

13. $y = 2x + 1$

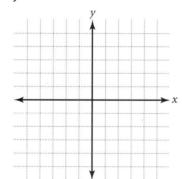

14. $y = 2x$

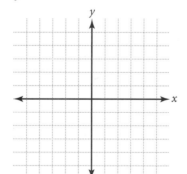

15. $y = -x - 4$

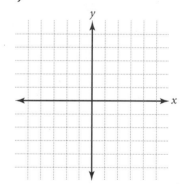

Determine whether each table represents a linear relationship between x and y. If the relationship is linear, write the next ordered pair that would appear in the table.

16.

x	0	1	2	3
y	4	5	6	7

17.

x	2	3	4	5
y	5	7	8	11

18.

x	-2	-1	0	1
y	-7	-4	-1	2

_____ _____ _____

Algebra 2

NAME _____ CLASS _____ DATE _____

Practice Masters Level B
1.1 Tables and Graphs of Linear Equations

State whether each equation is a linear equation.

1. $y = \frac{2}{3}x + \frac{1}{2}$ _____

2. $y = 4 - \frac{1}{2}x$ _____

3. $y = \frac{x^2}{2} - 4$ _____

4. $y = -2 - 3x$ _____

5. $y = x - \frac{3}{4}x^2$ _____

6. $y = 5.5$ _____

7. $y = \frac{x^2}{2} - 4$ _____

8. $y = 2x + 4 - 3x$ _____

9. $y = \frac{1}{x} - 7.4$ _____

Graph each linear equation.

10. $y = 2x - \frac{21}{3}$

11. $y = \frac{-2}{3}x$

12. $y = -(x - 7)$

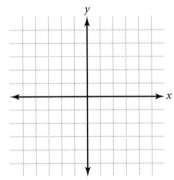 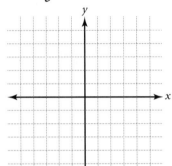 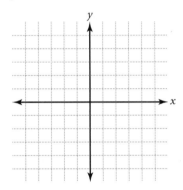

13. $y = -4 - x + 1$

14. $y - 5 = x + 1$

15. $y + 4 = x - 2$

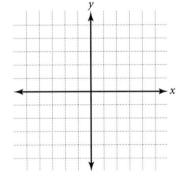 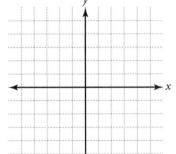 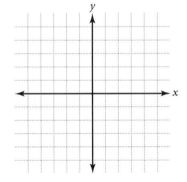

Determine whether each table represents a linear relationship between x and y. If the relationship is linear, write the next ordered pair that would appear in the table.

16.
x	−2	−4	−6	−8
y	4	5	6	7

17.
x	2	3	4	5
y	0	−3	−4	−5

18.
x	−4	−3	−2	−1
y	7	5	3	1

Practice Masters Level C

1.1 Tables and Graphs of Linear Equations

State whether each equation is linear. If so, then graph each equation.

1. $y = \frac{3}{4} - \frac{1}{4}x^2$ _____

2. $y - 3x = \frac{1}{2}$ _____

3. $y = \frac{x}{6} - \frac{1}{3}$ _____

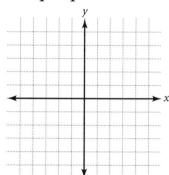
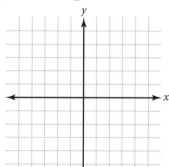
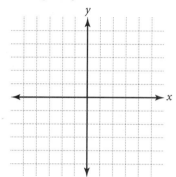

4. Describe the graph of $y = mx + b$ when $m = 0$. _____

5. Describe the graph of $y = mx + b$ when $b = 0$. _____

6. A teen night club charges $5.00 for admission and $1.50 for a soft drink.

   a. Write a linear equation that relates the total cost, C, in terms of the number of soft drinks, d. _____

   b. What is the total cost if 3 soft drinks are purchased? _____

   c. How many soft drinks are purchased if the total cost is $12.50? _____

7. A ball is rolled down an inclined plane. The results from several experiments are shown in the table. Determine whether the data are linearly related. If so, write a linear equation that relates these data.

Height, h	Distance rolled, d (ft)
3	45
4	60
5	75

_____

8. At 6:00 A.M., the temperature was 58°F. At 9:30 A.M., the temperature was 79°F. Assume that the temperature increased at a steady rate per hour.

   a. Write a linear equation relating elapsed time from 6:00 A.M. to temperature. _____

   b. What will the temperature be at 11:00 A.M.? _____

Practice Masters Level A

1.2 Slopes and Intercepts

Write the equation in slope-intercept form for the line that has the indicated slope, *m*, and *y*-intercept, *b*.

1. $m = 2, b = -3$ _____
2. $m = \dfrac{1}{2}, b = 3$ _____
3. $m = -1, b = 4$ _____
4. $m = \dfrac{-2}{3}, b = 0$ _____
5. $m = 0, b = -7$ _____
6. $m = \dfrac{-1}{2}, b = \dfrac{3}{4}$ _____

Find the slope of the line containing the indicated points.

7. $(2, 3)$ and $(4, 5)$ _____
8. $(-2, 4)$ and $(2, -3)$ _____
9. $(5, 0)$ and $(0, -5)$ _____
10. $(5, 4)$ and $(2, 1)$ _____
11. $(0, 0)$ and $(6, -2)$ _____
12. $(4, 6)$ and $(1, -3)$ _____

Identify the slope, *m*, and *y*-intercept, *b*, for each line. Then graph.

13. $y = 5x - 3$

14. $4x + y = 1$

15. $\dfrac{-1}{2}x + y = 3$

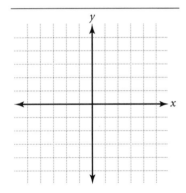
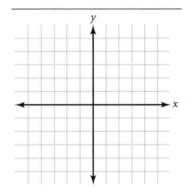
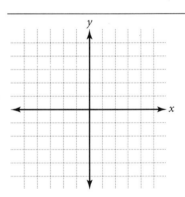

16. $y = 3$

17. $x = -4$

18. $4x - 2y = 6$

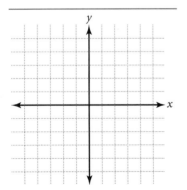
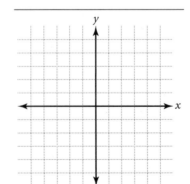
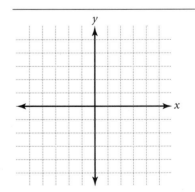

Practice Masters Level B
1.2 Slopes and Intercepts

Write the equation in slope-intercept form for the line that has the indicated slope, m, and y-intercept, b.

1. $m = \dfrac{-2}{3}, b = \dfrac{-1}{3}$ _____ 2. $m = 0, b = \dfrac{-6}{7}$ _____

3. $m = \dfrac{-3}{5}, b = 0$ _____ 4. $m = -1, b = 0$ _____

Find the slope of the line containing the indicated points.

5. $(-3.5, 2)$ and $(4.2, 7)$ _____

6. $(-4.1, -2)$ and $(-2, -6.2)$ _____

7. $\left(\dfrac{-1}{2}, 3\right)$ and $\left(\dfrac{-3}{2}, -3\right)$ _____

Identify the slope and y-intercept for each line.

8. $2x + 3y = 8$ _____ 9. $-2x + 5y = -5$ _____ 10. $2x - 2y = 0$ _____

11. $y - 4 = 6(x - 8)$ _____ 12. $6x = 4y$ _____ 13. $-5y - 10x = 10$ _____

Write an equation in slope-intercept form for each line.

14. A line passing through $(2, 7)$ with a slope of 3. _____

15. A line with a slope of 0 passing through $(-3, -4)$. _____

16. A vertical line through $(-1, 0)$. _____

Use the intercepts to graph each equation.

17. $x + 2y = 10$

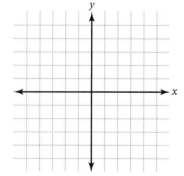

18. $x + 3y = -2$

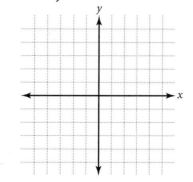

Algebra 2

Practice Masters Levels A, B, and C 5

NAME _____ CLASS _____ DATE _____

Practice Masters Level C

1.2 Slopes and Intercepts

Find the slope of the line containing the indicated points.

1. $\left(\frac{2}{3}, \frac{4}{9}\right)$ and $\left(\frac{4}{9}, \frac{2}{9}\right)$ _____

2. $(-1.2, -6.4)$ and $(-3.4, -5.3)$ _____

Graph each equation.

3. $5x - y = 2x + 4$

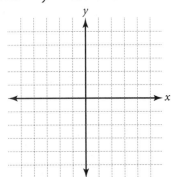

4. $4x + 3y = 5$

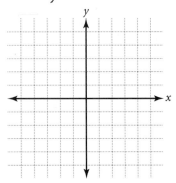

5. $2 - 2y = -x$

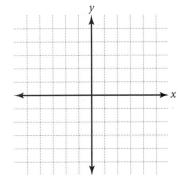

6. $5(y - 1) = -2(x + 3)$

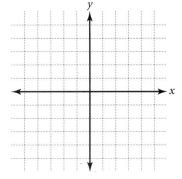

7. The points $(1, -3)$, $(4, 0)$, and $(10, a)$ are on one line. Find a. _____

8. An extreme ride at an amusement park drops its riders 220 feet in 4.8 seconds. What slope models this situation? _____

9. A chicken weighs 4.2 pounds on Day 36. On Day 60, the chicken weighs 7.4 pounds.

   a. What is the increase in weight per day? _____

   b. When will the chicken weigh 8 pounds? _____

   c. Write a linear equation to model the weight, w, in terms of the number of days, d. _____

   d. What does the y-intercept represent? _____

NAME _____ CLASS _____ DATE _____

Practice Masters Level A
1.3 Linear Equations In Two Variables

Write an equation for the line containing the indicated points.

1. $(4, 2)$ and $(5, 4)$ _____
2. $(0, 3)$ and $(2, 10)$ _____
3. $(-3, 2)$ and $(2, 1)$ _____
4. $\left(-4, \dfrac{1}{2}\right)$ and $(-2, 6)$ _____
5. $(3, -5)$ and $(6, -5)$ _____
6. $(2, -3)$ and $(2, 0)$ _____

Write an equation in slope-intercept form for the line that has the indicated slope, m, and contains the given point.

7. $m = 2$ and $(4, 4)$ _____
8. $m = -3$ and $(-1, 6)$ _____
9. $m = \dfrac{1}{2}$ and $(6, -3)$ _____
10. $m = \dfrac{-3}{4}$ and $(8, 0)$ _____
11. $m = 4$ and $(0, 0)$ _____
12. $m = 0$ and $(2, 2)$ _____

Write an equation in slope-intercept form for the line that contains the given point and is parallel to the given line.

13. $(2, 5); y = x + 6$ _____
14. $(-3, 1); y = -5x + 2$ _____
15. $(3, -4); y = \dfrac{5}{3}x - 2$ _____
16. $(2, 5); y = x + 6$ _____
17. $(4, 1); y = \dfrac{-1}{2}x + \dfrac{1}{2}$ _____
18. $(0, -3); y = 5$ _____
19. $(2, -2); y = -3x + 6$ _____
20. $(4, 6); y = 2x - 1$ _____
21. $(6, -3); y = -9x - 18$ _____
22. $\left(-3, \dfrac{3}{2}\right); y = x + 1$ _____

Write an equation in slope-intercept form for the line that contains the given point and is perpendicular to the given line.

23. $(0, -5); y = \dfrac{3}{4}x + 7$ _____
24. $(6, 2); y = -3x - 1$ _____
25. $(-4, 0); x + y = 5$ _____
26. $(-6, -2); 3x - y = 1$ _____
27. $(-2, 2); y = \dfrac{-1}{4}x$ _____
28. $(7, -1); y = 3$ _____
29. $(6, -2); y = 6x + \dfrac{6}{8}$ _____
30. $(4, -10); y = 4x - 8$ _____
31. $(8, 16); y = 2x + 12$ _____
32. $(3, -2); y = \dfrac{3}{2}x - \dfrac{6}{8}$ _____

Algebra 2   Practice Masters Levels A, B, and C

Practice Masters Level B

1.3 Linear Equations In Two Variables

Write an equation for the line containing the indicated points.

1. $\left(4, \frac{1}{2}\right)$ and $\left(0, \frac{5}{2}\right)$ _____

2. $(-2, -4)$ and $(-1, -8)$ _____

3. $\left(\frac{3}{4}, \frac{-2}{3}\right)$ and $\left(\frac{1}{2}, \frac{4}{3}\right)$ _____

4. $\left(\frac{-1}{5}, 0\right)$ and $(-6, 0)$ _____

Write an equation in slope-intercept form for the line that has the indicated slope, m, and contains the given point.

5. $m = -5$ and $(-2, -1)$ _____

6. $m = \frac{3}{4}$ and $(-15, 5)$ _____

7. $m = \frac{-1}{2}$ and $\left(\frac{2}{3}, 0\right)$ _____

8. $m = 0$ and $\left(\frac{7}{4}, \frac{-11}{4}\right)$ _____

9. $m =$ undefined and $(-2, 1)$ _____

10. $m =$ undefined and $(0, 0)$ _____

Write the equations in slope-intercept form for the line that contains the given point and is (a) parallel and (b) perpendicular to the given line.

11. $(5, 0); x - 3y = 6$ _____ _____

12. $(3, -1); x - y = 4$ _____ _____

13. $(2, 1); 2x + 2y = 5$ _____ _____

14. $\left(0, \frac{-1}{2}\right); 2x - 3y = 12$ _____ _____

Write an equation for the line that is perpendicular to each equation.

15. $3x + 2y = 8$ at the y-intercept

16. $-2x + 5y = -20$ at the x-intercept

_____ _____

Use the slope to determine whether the following lines are parallel, perpendicular or neither.

17. l_1 contains $(2, 1)$ and $(-1, 4)$
    l_2 contains $(3, -2)$ and $(1, 0)$

18. l_1 contains and $\left(\frac{1}{2}, 2\right)$ and $\left(\frac{1}{2}, -3\right)$
    l_2 contains $(0, -4)$ and $(1, 1)$

NAME _____ CLASS _____ DATE _____

Practice Masters Level C
1.3 Linear Equations In Two Variables

For Exercises 1–6, write an equation in slope-intercept form for each line.

1. a line with a slope of $\frac{4}{5}$ and containing the point $(10, 6)$ _____

2. a line with no slope containing the point $(2, 0)$ _____

3. a line parallel to $-5x + y = 2$ containing the point $(-6, -1)$ _____

4. a line containing the points $\left(\frac{1}{2}, \frac{-3}{4}\right)$ and $\left(\frac{2}{3}, \frac{-7}{4}\right)$ _____

5. a line perpendicular to $-3x - 4y = 10$ and having the same y-intercept as $x + 3y = -5$ _____

6. a line having the same x-intercept as $y = 2x - 2$ and y-intercept as $\frac{1}{2}x - \frac{2}{3}y = -4$ _____

7. If $(5, 2)$, $(-2, 5)$, and $(-5, -2)$ are three vertices of a rectangle, what is the missing vertex? _____

8. If $(0, -4)$, $(5, 2)$, and $(-1, 1)$ are three vertices of a parallelogram, what are the three possibilities for the missing vertex? _____

Determine whether each pair of lines is parallel, perpendicular or neither.

9. l_1 is parallel to $6x - 2y = 11$
   l_2 is perpendicular to $3x - y = 1$

10. l_1 contains $(4, -3)$ and $(-5, 9)$
    l_2 is parallel to $4x + 3y = 18$

11. l_1 contains $(a, 0)$ and $(0, b)$
    l_2 is parallel to $-ax + by = -b$

12. l_1 is perpendicular to $y = 4$
    l_2 contains $(-1, -2)$ and $(-6, -2)$

13. A lawn-care service charges a start-up rate of $50 for each customer plus a charge of $25 per visit.

   a. Write an equation that models this situation. _____

   b. The lawn-care service has an initial investment of $2000. How many visits will it have to make to one customer in order to start making a profit? _____

Algebra 2   Practice Masters Levels A, B, and C   9

NAME _____ CLASS _____ DATE _____

Practice Masters Level A
1.4 Direct Variation and Proportion

Determine whether each equation describes a direct variation.

1. $y = 3x$ _____
2. $y = \dfrac{1}{4}x$ _____
3. $y = \dfrac{4}{x}$ _____

4. $y = -0.27x$ _____
5. $y = -3x + 1$ _____
6. $y + x = 0$ _____

In Exercises 7–14, *y* varies directly as *x*. Find the constant of variation, and write an equation of direct variation that relates the two variables.

7. $y = 4$ when $x = 2$ _____
8. $y = -7$ when $x = 3$ _____

9. $y = -12$ when $x = -20$ _____
10. $y = 5$ when $x = -5$ _____

11. $y = \dfrac{2}{3}$ when $x = -4$ _____
12. $y = -16$ when $x = \dfrac{4}{5}$ _____

13. $y = 0.02$ when $x = 0.06$ _____
14. $y = -1.4$ when $x = 0.007$ _____

Solve each proportion for the variable. Check your answers.

15. $\dfrac{x}{5} = \dfrac{14}{35}$ _____
16. $\dfrac{-3}{8} = \dfrac{x}{32}$ _____

17. $\dfrac{3x}{10} = \dfrac{13}{20}$ _____
18. $\dfrac{20}{55} = \dfrac{10x}{11}$ _____

19. $\dfrac{2x + 10}{5} = \dfrac{28}{35}$ _____
20. $\dfrac{4x - 6}{15} = \dfrac{12x}{90}$ _____

21. $\dfrac{x + 4}{10} = \dfrac{10}{50}$ _____
22. $\dfrac{x}{12} = \dfrac{27}{36}$ _____

In Exercises 23–27, *y* varies directly as *x*.

23. If $y = 4$ when $x = 2$, find y when $x = 5$. _____

24. If $y = 4$ when $x = 2$, find x when $y = 16$. _____

25. If $y = 5$ when $x = -5$, find y when $x = -3$. _____

26. If $y = 5$ when $x = -5$, find y when $x = 2$. _____

27. If $y = 6$ when $x = 3$, find x when $y = 4$. _____

Practice Masters Level B
1.4 Direct Variation and Proportion

In Exercise 1–6, *y* varies directly as *x*. Find the constant of variation, and write the equation of direct variation that relates the two variables.

1. $y = 6$ when $x = -4$ _____
2. $y = 4.8$ when $x = 1.8$ _____
3. $y = -10$ when $x = -4.5$ _____
4. $y = \frac{4}{9}$ when $x = \frac{-1}{3}$ _____
5. $y = -6.5$ when $x = -2.5$ _____
6. $y = 1.26$ when $x = 0.003$ _____

Determine whether the values in each table represent a direct variation. If so, write the equation for variation.

7.

x	y
-2	56
0	0
2	56
4	112
6	168

8.

x	y
0	0
2	4
3	10
4	18
5	28

Solve each proportion for the variable. Check your answers.

9. $\dfrac{3x+5}{10} = \dfrac{7}{2}$ _____
10. $\dfrac{2x-3}{9} = \dfrac{3x-4}{6}$ _____
11. $\dfrac{4x}{20} = \dfrac{x-3}{2}$ _____
12. $\dfrac{8x+2}{4} = \dfrac{4x}{3}$ _____

In Exercises 13–16, *a* varies directly as *b*.

13. If $a = 4.2$ when $b = 6.3$, find a when $b = 12.6$. _____

14. If $a = -15$ when $b = 2.25$, find b when $a = 20$. _____

15. If $a = \frac{1}{2}$ when $b = \frac{-3}{4}$, find a when $b = 16$. _____

16. If b is $\frac{2}{3}$ when $a = \frac{1}{9}$, find b when $a = 15$. _____

Algebra 2

Practice Masters Level C
1.4 Direct Variation and Proportion

In Exercises 1 and 2, *a* varies directly as *b*.

1. If $a = \frac{1}{4}$ when $b = 0.02$, find a when $b = 0.05$. _____

2. If $a = \frac{-2}{3}$ when $b = -15$, find b when $a = \frac{1}{5}$. _____

Determine whether the values in each table represent a direct variation. If so, write the equation for variation. If not, explain.

3.

x	y
10	−25
5	−12.5
0	0
−5	12.5
−10	25

4.

x	y
0	0
−3	−2
−6	−3
−9	−4
−12	−6

5. A spring stretches a distance that is directly proportional to the weight attached to the spring. A weight of 2.5 pounds stretches the spring 1.5 feet.

   a. Find the constant of variation. _____

   b. How much weight would be required to stretch the spring 10 feet? _____

   c. How much distance would the spring stretch if the weight was 6.4 pounds? _____

   d. How much weight would be required to stretch the spring 3 inches? _____

   e. Explain why the equation of this model would be a line through the origin.

   _____

6. At a given time, the height of an object is directly proportional to the length of the shadow that is cast on the ground. A person 6 feet tall casts a shadow 1 foot long at noon.

   a. What is the length of the shadow cast by a flagpole known to be 20 feet tall? _____

   b. A tree cast a shadow of 3 feet. How tall is the tree? _____

Practice Masters Level A

1.5 Scatter Plots and Least-Squares Lines

Match each correlation coefficient with one of the data sets graphed.

A. $r = 1$ B. $r \approx 0.6$ C. $r \approx 0$ D. $r \approx -0.1$ E. $r \approx -0.7$ F. $r = -1$

1.
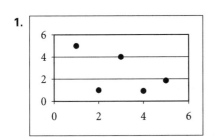

2.
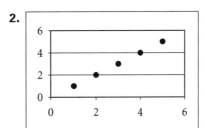

3.
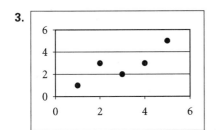

4.
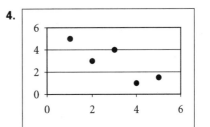

5.
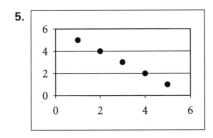

6.
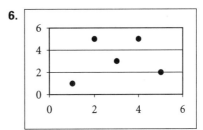

7. The equation of the least square line of the cost, C, of the number of machines, x, is given by the equation $C = 36x + 2500$. Based on the least-squares line, predict the cost of 24 machines. _____

Use a graphing calculator to create a scatter plot of the data in each table. Describe the correlation as positive, or negative, or no correlation. Then find an equation for the least-squares line.

8.
x	0	2	5	9
y	5	9	15	23

9.
x	0	2	4	10
y	−2.9	−11	−18.5	−44

10.
x	1	2	5	8
y	6	8.5	−10.2	−23.9

Practice Masters Level B

1.5 Scatter Plots and Least-Squares Lines

For each question, state whether the statement is true or false.

1. A correlation coefficient can be equal to −2. _____

2. If the slope of a least-squares line is negative, the correlation coefficient is negative. _____

3. A data set with a correlation coefficient of −0.72 has a stronger linear relationship than a data set with a correlation coefficient of 0.64. _____

Use a graphing calculator to create a scatter plot of the data in each table. Describe the correlation as positive, negative, or no correlation. Then find an equation for the least-squares line.

4.
x	4	3	5	6	8	8	5	2	1	7	9	7	2	1	3	4	7	5	2	1
y	6	5	9	10	13	14	10	3	2	12	17	11	4	1	5	7	11	9	2	1

5.
x	8	10	6	2	5	10	3	9	12	10	10	8	6	1	2	4	3	3	5	8
y	1	10	2	2	6	1	9	8	7	7	7	5	4	2	3	6	8	11	8	10

6. The average points per game for Larry Bird for a 10-year period is shown below.

1980	1981	1982	1983	1984	1985	1986	1987	1988	1989
21.3	21.2	22.9	23.6	24.2	28.7	25.8	28.1	29.9	19.3

a. Enter the data in a graphics calculator and find the equation of the least-squares line. _____

b. Find the correlation coefficient, r, to the nearest hundredth. _____

c. Use the least-squares line to predict Bird's scoring average in 1990. _____

d. Explain using the correlation coefficient whether your prediction for Bird's 1990 scoring average is accurate or inaccurate, given the fact that Bird's actual scoring averages was 24.3 points per game in 1990.

NAME _____ CLASS _____ DATE _____

Practice Masters Level C
1.5 Scatter Plots and Least-Squares Lines

The number of goals that Wayne Gretzky scored during a 10-year span of his hockey career is shown below.

1980	1981	1982	1983	1984	1985	1986	1987	1988	1989
51	55	92	71	87	73	52	62	40	54

1. Find the equation of the least-squares line. _____
2. Find the correlation coefficient, r, to the nearest hundredth. _____
3. Use the least-squares line to predict the number of goals Gretzky scored in 1990. _____
4. If Gretzky scored 40 goals in 1990, explain whether the prediction for Gretzky's goals scored in 1990 is accurate or not accurate. _____

The number of home runs hit by a player is shown below.

1990	1991	1992	1993	1994	1995	1996	1997	1998	1999
39	22	42	9	9	39	52	58	70	65

5. Find the equation of the least-squares line. _____
6. Find the correlation coefficient, r, to the nearest hundredth. _____
7. Use the least-squares line to predict the number of home runs the player hits in 2000. _____

The owner of a computer firm is studying the relationship between money spent on advertising and sales.

Month	Advertising Expense $, in thousands	Sales Revenue $, in millions
January	2	6
February	2	5.2
March	1	3
April	1.5	3.8
May	2.25	4.2
June	1.75	1.7

8. Find the equation of the least-squares line and the correlation coefficient, r to the nearest tenth. _____
9. Use the least-squares line to predict the sales revenue if advertising expenses were $3000 in July. _____

Practice Masters Level A
1.6 Introduction to Solving Equations

Solve each equation.

1. $3x - 5 = 15$ _____
2. $-2x + 7 = 29$ _____
3. $-5 - 4x = -11$ _____
4. $6 - 3x = 6$ _____
5. $5(x + 3) = -45$ _____
6. $\frac{1}{2}x - 4 = 2$ _____
7. $48 + 6x = -2x$ _____
8. $-6x - 7 = 20 + 3x$ _____
9. $\frac{3}{4}x + \frac{1}{2} = \frac{7}{2}$ _____
10. $-1.2 + 0.3x = 0.9x$ _____

Solve each equation by graphing. Give answers to the nearest hundredth.

11. $1.16x + 0.24 = 3.54 + 0.9x$ _____
12. $-2.4x - 2.6 = -3.8 - 1.4x$ _____

Solve each literal equation for the indicated variable.

13. $V = l \times w \times h$ for l _____
14. $A = \frac{1}{2}bh$ for b _____
15. $A = \frac{1}{2}(b_1 + b_2)h$ for h _____
16. $I = prt$ for t _____
17. $C = 2\pi r$ for r _____
18. $V = \frac{gt^2}{2}$ for g _____

Write and solve an appropriate equation for each situation.

19. Josie has a dog-walking service. She charges $10 per dog plus $2 per hour. If Josie walks Mr. Grove's 3 dogs for 2 hours how much money does Josie earn? _____

20. Harris earns $125 per week as a chef, plus $7 per hour for cleaning up. If Harris cleans 6 hours in a week in addition to his time as a chef, how much money does Harris earn? _____

NAME _____ CLASS _____ DATE _____

Practice Masters Level B
1.6 Introduction to Solving Equations

Solve each equation.

1. $7n + 10 = 3n + 2$ _____

2. $2(y + 1) = y - 8$ _____

3. $3(m - 2) - 5 = 6 - 2(m - 4)$ _____

4. $2x - \dfrac{2}{3} = -3x + \dfrac{7}{3}$ _____

5. $6(a + 2) - 8a = 10a + 9$ _____

6. $\dfrac{1}{3}x - \dfrac{7}{3} = \dfrac{1}{6}$ _____

7. $\dfrac{m}{6} + \dfrac{m}{4} + \dfrac{m}{3} = 1$ _____

8. $24 + 2(x + 4) = 2(x + 3)$ _____

Solve each equation by graphing. Give answers to the nearest hundredth.

9. $6.46 + 2.3x = 1.24x - 7$

10. $-2.05x + 1.8(x - 3) = 6.2$

11. $0.38 - 0.66x + 0.72x - 0.54 = 0$

12. $1.84 - 0.23x = 0.5(-0.46x + 3.68)$

Solve each literal equation for the variable indicated.

13. $A = \dfrac{1}{2}h(b_1 + b_2)$ for b_1

14. $P = 2l + 2w$ for l

15. $A = P(1 + RT)$ for T

16. $V = V_o + b(s + s_o)$ for s

17. $A = 2\pi r(r + h)$ for h

18. $A = \dfrac{1}{2}d_1 d_2$ for d_1

Write and solve an appropriate equation for each situation.

19. The measure of one complementary angle is 20° more than twice the measure of angle x. Find the measure of angle x. _____

20. Carson charges $18.75 per hour to fix computers, plus a flat service fee of $30. Carson billed Miss Belkis $255 for a job. How many hours did Carson spend fixing her computer? _____

Algebra 2 — Practice Masters Levels A, B, and C

NAME _____ CLASS _____ DATE _____

Practice Masters Level C
1.6 Introduction to Solving Equations

Solve each equation.

1. $-2 + 4.3x + 7(-5x - 2) = 12.6$

2. $\frac{2}{3}x - \frac{2}{5} = \frac{2}{5}x - \frac{2}{3}$

3. $-3(6 - 5x) + 6x = 10 - 4(2 - x) - 3x$

4. $\frac{1}{2}(3x + 6) - \frac{1}{4}(-8x - 32) = \frac{1}{9}(27x + 63)$

Solve each equation by graphing. Give answers to the nearest hundredth.

5. $12.1(3.8 + 7.2x) = 3.78x$

6. $-1.85 - 0.1x = 2.7x + 2.6$

7. $0.0005x - 0.001003 - 0.003x = 0.00007$

8. $-100 = 1.3(-2.5x + 5.8) - 80.2x$

9. The measure of one complementary angle is $\frac{4}{5}$ the measure of the other angle. Write an equation, and find the measure of each angle. _____

10. The purchase of a used car required a $1200 down payment and a monthly payment of $225.

   a. Write an equation representing the total amount of money paid. _____

   b. How much money would have been paid after the 16th month? _____

   c. If the buyer paid a total of $6600, how many months did it take to pay off? _____

11. Paul and Rene work in the computer department of a local appliance center. Paul earns a weekly base salary of $150 plus $5.75 per hour. Rene earns $175 as a base salary plus $5.25 per hour.

   a. Write an equation for the amount of money each person earns per week. _____

   b. If Paul and Rene work the same number of hours per week, how many hours must each work to earn the same amount? _____

   c. Who earns more if they both work 25 hours per week? _____

NAME _____ CLASS _____ DATE _____

Practice Masters Level A
1.7 Introduction to Solving Inequalities

Write an inequality that describes each graph.

1. _____
2. _____
3. _____
4. _____

Graph each inequality.

5. $x \geq -1$

6. $y \leq -2$

Solve each inequality, and graph the solution on the number line.

7. $5z - 6 > 14$

8. $3x + 10 \leq 1$

9. $4x - 15 > 73$

10. $5 - 2r \geq 11$

11. Graph $x > -6$ and $x < 2$ on the same number line.

12. Solve and graph $8x - 2 > 4$ or $3x + 6 < 12$.

Algebra 2   Practice Masters Levels A, B, and C   19

Practice Masters Level B

1.7 Introduction to Solving Inequalities

Write an inequality that describes each graph.

1. _____

2. _____

3. _____

4. _____

Solve each inequality, and graph the solution on the number line.

5. $2(x - 3) < 14$

   _____

6. $\frac{2}{3}(x - 12) \leq x + 5$

   _____

7. $-6(2y - 10) < 108$

   _____

8. $\frac{5}{3} - \frac{1}{4}x \geq \frac{-2}{3}x + \frac{1}{6}$

   _____

Graph the solution of each compound inequality on the number line.

9. $3x - 1 > 2$ or $2x + 6 < 2$

10. $4z + 6 > -10$ and $-3z - 5 > -17$

11. $10x - 4 < 6$ or $4x - 12 < 8$

12. $n + 1 < 4$ and $n - 1 > 4$

NAME _____ CLASS _____ DATE _____

Practice Masters Level C

1.7 Introduction to Solving Inequalities

Solve each inequality, and graph the solution on the number line.

1. $6(2 - t) - 3t < 4(1 - t) + 7$

2. $0.08m - 0.1(0.2m - 3) \geq 0.04m + 1.4$

3. $\dfrac{1}{3}\left(\dfrac{1}{4}t - \dfrac{5}{2}\right) - \dfrac{7}{12}t \leq 4 + \dfrac{2}{3}t$

4. $-5(-60 + 25m) - \dfrac{1}{3}m > 60\left(-4m + \dfrac{3}{2}\right) - 20m + 8$

5. $3p - 5 \leq 1 \text{ or } 6 > -2p + 4$

6. $-6(w + 2) > 12 \text{ and } 17 - 2w < 111$

7. $0.8 - 0.2z < 0.6z + 2.2 \text{ and } 4(z - 1.5) - 6 < 18$

8. $\dfrac{x}{4} + \dfrac{x}{2} \leq -3 \text{ or } \dfrac{x}{4} - \dfrac{x}{8} > 0$

Solve.

9. A season pass to an amusement park is $235. A daily ticket costs $38.50. How many visits to the park make the season pass a better buy? _____

10. The sum of the lengths of any two sides of a triangle must be greater than the length of the third side. If two sides of a triangle are 7.4 and 5.6, write an inequality to represent the length of the third side. _____

Practice Masters Level A
1.8 Solving Absolute-Value Equations and Inequalities

Match each statement on the left with a statement on the right.

1. $|x - 3| = 5$ _____   a. all real numbers

2. $|x - 3| < 5$ _____   b. $x < 8$ and $x > -2$

3. $|x - 3| > 5$ _____   c. no solution

4. $|x - 3| > -5$ _____   d. $x > 8$ or $x < -2$

5. $|x - 3| = -5$ _____   e. $x = 8$ or $x = -2$

Solve each equation. Graph the solution on the number line.

6. $|x - 4| = 2$ _____

7. $|2x + 3| = 7$ _____

8. $|3 - 2x| = 11$ _____

9. $|6x + 15| = 21$ _____

Solve each inequality. Graph the solution on the number line.

10. $|x + 6| > 1$ _____

11. $|3x - 5| \leq 10$ _____

12. $|5x| < -1$ _____

13. $|x - 10| \geq 15$ _____

For Exercises 14–15, write an absolute-value inequality.

14. $2x < 6$ and $-6 < 2x$

15. $10y - 12 < -10$ or $10y - 12 > 10$

22   Practice Masters Levels A, B, and C   Algebra 2

NAME _____ CLASS _____ DATE _____

Practice Masters Level B

1.8 Solving Absolute-Value Equations and Inequalities

Solve each equation, and graph the solution on the number line.

1. $|2x + 1| = 9$

2. $|5 - 2m| = 7$

3. $|-t| = 3$

4. $|-x| = 0$

5. $|4x - 7| = 1$

Solve each inequality, and graph the solution on the number line.

6. $|3n + 2| > -2$

7. $|5x - 4| \leq 6$

8. $|1 - 2b| < 5$

9. $\frac{1}{2}|x - 6| - 2 < 2$

10. $\left|\frac{4x + 3}{-5}\right| \geq -3$

11. A spindle is designed with a specification of a 102-centimeter diameter. The spindle will work if it is within 0.025 centimeters of the specified length. Write an absolute-value inequality to represent the measurement tolerance for the diameter, d. _____

Algebra 2   Practice Masters Levels A, B, and C   23

NAME _____ CLASS _____ DATE _____

Practice Masters Level C
1.8 Solving Absolute-Value Equations and Inequalities

Solve each equation or inequality, and graph the solution on the number line.

1. $1 < |2 - \frac{x}{3}|$

2. $|6 - 0.2t| \le 5$

3. $\frac{2}{3}|5x + 3| + 20 = 4$

4. $\frac{-1}{4}|8m - 4| \le -3$

5. $\frac{2}{3}|5(\frac{1}{5}r - 2)| = -12$

6. $|\frac{4}{3} - \frac{2}{3}x| > -2$

7. Solve $0.4|2.3x - 6.1| - 13.3 \le -11.7$ to the nearest tenth. Graph the solution on the number line.

8. Hannah weighs 155 pounds. To stay within her ideal weight, Hannah should not gain more than 6% or lose less than 4% of her current weight. What is the range of her ideal weight? _____

9. A woodworker charges $0.75 per minute to have a piece of wood milled. A milling machine can be purchased for $899. _____

   a. Write an inequality to find the number of minutes for which paying the woodworker is less than purchasing the machine. _____

   b. How many hours is this? _____

NAME _____ CLASS _____ DATE _____

Practice Masters Level A

2.1 Operations with Numbers

Classify each number in as many ways as possible.

1. $\dfrac{3}{5}$ _____

2. 2.14411444111... _____

3. -18.4 _____

4. $\sqrt{400}$ _____

5. $6.\overline{3}$ _____

6. $\sqrt{5}$ _____

State the property that is illustrated in each statement. Assume that all variables represent real numbers.

7. $4(x + 2) = 4x + 8$ _____

8. $14 - 31 + 16 = 14 + 16 - 31$ _____

9. $\dfrac{x}{7} \cdot \dfrac{7}{x} = 1$, for $x \neq 0$ _____

10. $-5 + (25 + 13) = (-5 + 25) + 13$ _____

11. $1(-47) = -47$ _____

12. $10 \times 7 \times 6.3 = 10 \times 6.3 \times 7$ _____

13. $14 + (-14) = 0$ _____

14. List the order in which algebraic operations must be performed.

   a. _____ b. _____

   c. _____ d. _____

Evaluate each expression using the order of operations.

15. $4^3 - 30 \div 3$ _____

16. $9 + 45 \div 5 \times 8$ _____

17. $2^2 + 6(4 + 5)$ _____

18. $6^2 - 14 + 6 \times 2$ _____

19. $(-4 + 6) \times -3 + 3$ _____

20. $\dfrac{4}{3} - \dfrac{2}{3} \cdot \dfrac{1}{9} \div \dfrac{1}{3}$ _____

Algebra 2   Practice Masters Levels A, B, and C

NAME _____ CLASS _____ DATE _____

Practice Masters Level B
2.1 Operations with Numbers

State whether each number is rational or irrational.
Explain your answer.

1. $\sqrt{324}$ _____ 2. $\sqrt{\dfrac{1}{9}}$ _____

3. $\sqrt{\dfrac{56}{8}}$ _____ 4. $\sqrt{0.0016}$ _____

5. π _____ 6. $100 \cdot 0.020022000222...$ _____

Write and justify each step in the simplification of each.

7. $(x + 6)3$ _____ 8. $(6 - x)(c + g)$ _____

9. $2 + x \div x - 1$ _____ 10. $\dfrac{2}{8} \cdot \dfrac{x}{2} \div 2$ _____

Evaluate each expression using the order of operations.

11. $(4^2 - 9) \cdot 8 - 6$ _____ 12. $12 \div 4(6 - 8)$ _____

13. $-3(10 - 18) \div 2$ _____ 14. $\dfrac{6^2 - 5}{3 - 5}$ _____

15. $6^2 - 6 \div 3$ _____ 16. $9(5 + (6 - 2) \times 5)$ _____

17. $0.75 \times 3 + 1 \div 2 - 0.4$ _____ 18. $\dfrac{2}{3} \times \dfrac{4}{5} \div \dfrac{1}{5} + \dfrac{4}{5}$ _____

Evaluate each expression using the order of operations.
Let $a = -3$, $b = 2$, and $c = -1$.

19. $a + b \div 2c$ _____ 20. $(b^2 - a)c + b$ _____

21. $|a| - 3a^2 \div c$ _____ 22. $a + (-bac - c)$ _____

23. $b^2 - a^2 \cdot c^2$ _____ 24. $\dfrac{b \cdot a}{c}(a + c)$ _____

25. $c + c \div b \cdot |c| - (a \cdot c)$ _____ 26. $a + bac - ab$ _____

27. $4c - 6cb + 3(2a)$ _____ 28. $12c - (3b) \div 2a + c^3$ _____

NAME _____ CLASS _____ DATE _____

Practice Masters Level C
2.1 Operations with Numbers

Evaluate each expression using the order of operations.

1. $\dfrac{9 \times 5 + 18 \div 3 - 3 \times 3 - 5}{5}$ _____

2. $6|4(-1) - 5| + 2|6 - 2(-1)|$ _____

3. $\dfrac{\frac{2}{3} + \frac{3}{4} - \frac{1}{6}}{\frac{10}{3} + 2}$ _____

4. $|6 - 9| + 3|2 - 6| - |-5|$ _____

5. $5^2 - 2[4 - 5(6^2 + 4)]$ _____

6. $\dfrac{\frac{5}{8} + \frac{3}{8} - \frac{1}{4}}{\frac{6}{7} - \frac{3}{7} + 2}$ _____

**Evaluate each expression using the order of operations.
Let $a = -3$, and $b = -5$.**

7. $\dfrac{a^2 - 5 \div b \times (-3)}{2 + a}$ _____

8. $a[b^2 + 2(b - a) \div 4]$ _____

9. $|a|^2 - |b|^2$ _____

10. $\dfrac{1}{2}\left|\dfrac{a + b^2}{a^2}\right|$ _____

Insert one or more pairs of parentheses to make a true statement.

11. $3 \times 6 + 2 \times 5 - 1 = 96$

12. $12 - 4 \times 5 + 3 = -20$

13. $1.2 + 6.2 \times 3.4 - 1.5 \times 6 = 84.36$

14. $\dfrac{1}{4} \times \dfrac{2}{3} + \dfrac{5}{6} - \dfrac{3}{8} = 0$

15. $\dfrac{1}{2} + \dfrac{1}{3} \cdot \dfrac{1}{4} - \dfrac{1}{8} = \dfrac{11}{24}$

16. $8^2 + 20 \div 5 + 5 \times 2 = 68$

Explain why the given property would be useful in simplifying each expression.

17. Commutative for multiplication $6 \times 14 \times 1.5$ _____

18. Associative for addition $(13 + 39) + -19$ _____

Algebra 2   Practice Masters Levels A, B, and C   27

Practice Masters Level A
2.2 Properties of Exponents

Evaluate each expression.

1. 5^{-2} _____

2. $(4 \cdot 3)^2$ _____

3. 14^0 _____

4. $\left(\dfrac{1}{3}\right)^{-2}$ _____

5. $\left(\dfrac{3}{4}\right)^3$ _____

6. $\left(\dfrac{1}{4}\right)^{-4}$ _____

7. $27^{\frac{1}{3}}$ _____

8. $64^{\frac{2}{3}}$ _____

9. $25^{\frac{5}{2}}$ _____

10. $81^{\frac{1}{2}}$ _____

11. $100^{\frac{-1}{2}}$ _____

12. $32^{\frac{-1}{5}}$ _____

13. $-2(2 \cdot 5^2)^2$ _____

14. $(3^2 \cdot 2^4)^0$ _____

Simplify each expression, assuming that no variable equals zero. Write your answer with positive exponents.

15. $m^5 m^{-4}$ _____

16. $(x^3)^5$ _____

17. $x^6 x^{-10}$ _____

18. $(x^{-2})^3$ _____

19. $(r^{-3})^{-1}$ _____

20. $p^1 p^{-5}$ _____

21. $\dfrac{w^{15}}{w^3}$ _____

22. $\dfrac{w^{-4}}{w^{-2}}$ _____

23. $\left(\dfrac{2w^2}{w^{-6}}\right)$ _____

24. $\left(\dfrac{4x^{-2}}{x^3}\right)^{-3}$ _____

25. $(xy^2)(xy^4)$ _____

26. $(-t^3)(-t^4)(-t^2)$ _____

27. $(4xy)^2(-x^2y)^5$ _____

28. $(-2a^2b^3)^2(-3a^3b^4)^3$ _____

29. $\dfrac{x^{-10}}{2x^{-5}}$ _____

30. $(-y^3)\left(-\dfrac{y^6}{y^{-2}}\right)$ _____

Practice Masters Level B
2.2 Properties of Exponents

Evaluate each expression.

1. $\left(\dfrac{4}{5}\right)^2$ _____ 2. $(5 \cdot 2)^2$ _____

3. 7^0 _____ 4. 7^{-2} _____

5. $\left(\dfrac{-2}{3}\right)^3$ _____ 6. $\left(\dfrac{1}{5}\right)^{-3}$ _____

7. $36^{\frac{3}{2}}$ _____ 8. $125^{\frac{4}{3}}$ _____

9. $-(4 \cdot 5)^2$ _____ 10. $(6^2)^{\frac{-3}{2}}$ _____

11. $64^{\frac{-2}{3}}$ _____ 12. $100^{\frac{-5}{2}}$ _____

13. $(-3 \cdot 2^2)^2$ _____ 14. $(3^0 \cdot 5^{-1})^{-3}$ _____

Simplify each expression, assuming that no variable equals zero. Write your answer with positive exponents.

15. $(-a)^3(-a)^5(-a^4)$ _____ 16. $\dfrac{x^{-3}y^{-1}}{y^2}$ _____

17. $(5r^3s^4)(-3rs^2)(-rs)$ _____ 18. $x^{\frac{2}{3}} \cdot x^{\frac{4}{9}}$ _____

19. $\left(x^{\frac{3}{7}}\right)^{14}$ _____ 20. $\left(x^{\frac{1}{3}}\right)^{\frac{-3}{5}}$ _____

21. $\dfrac{\left(x^{\frac{-4}{3}}\right)^4}{x^{\frac{-2}{3}}}$ _____ 22. $\left(\dfrac{3a^{-2}b^3}{2a^4b^5}\right)^{-5}$ _____

23. $\left((100x^5)^{\frac{-1}{2}}\right)^{-4}$ _____ 24. $\left(\dfrac{m^{-3}n^{-4}}{m^{-2}n^{-5}}\right)^{\frac{-2}{3}}$ _____

25. $(r^4s^2)(r^2s^3)^5$ _____ 26. $x^2(6x^3 + 3x^{-2} - x + 10)$ _____

27. $(x^{-6})^2(-x^2)^{-3}$ _____ 28. $a^{-2}y^{-2}(ay)^4$ _____

29. $\left(\dfrac{20x^5}{x^3}\right)^{-2}$ _____ 30. $\left[y^{\frac{10}{5}} \cdot y^{\frac{3}{5}}\right]^{-10}$ _____

Algebra 2

Practice Masters Level C
2.2 Properties of Exponents

Simplify each expression, assuming that no variable equals zero. Write your answers with positive exponents.

1. $y^4(y^{3m})^2$ _____

2. $x^{m+2} \cdot x^2$ _____

3. $z^{p+1} \cdot z^{p-3} \cdot z^{p+4}$ _____

4. $(x^y)^y$ _____

5. $\dfrac{w^x}{w^{x-3}}$ _____

6. $(r^{-x})^{-3} \cdot r^x$ _____

7. The frequency of a sound wave is given by the formula $f = V(w)^{-1}$, where f is the frequency, V is the velocity, and w is the length of the sound wave.

   a. Find the frequency of a sound wave with a length of 2.35 meters and a velocity of 235.23 meters per second. _____

   b. What is the velocity for a sound wave with a length of 2.56 meters and a frequency of 94.6 waves per second? _____

8. Coulomb's Law states the following: The magnitude of force that a particle exerts on another particle is directly proportional to the product of their charges and inversely proportional to the square of the distance between them.

   $$F = d^{-2}(9 \times 10^9)q_1 q_2,$$

   where: F is the force between the two particles, (in newtons);
   q_1 is the net charge on particle A, (in coulombs);
   q_2 is the net charge on particle B, (in coulombs);
   d is the distance between the particles, (in meters)

   a. Calculate the force between two particles, one with a charge of 3.0×10^{-5} coulombs and the other with a charge of 1.5×10^{-5} coulombs, that are 0.5 meters apart. _____

   b. What is the charge on one particle that is 0.25 meters from another particle with a charge of 2.6×10^{-6} coulombs? The force between the two particles is 13.07 newtons. _____

Practice Masters Level A

2.3 Introduction to Functions

State whether each graph represents a function.

1.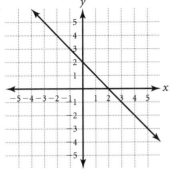

2.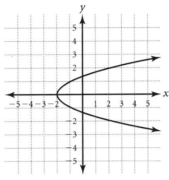

3.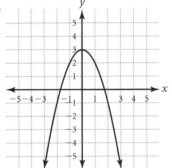

State whether the data in each table represents y as a function of x.

4.
x	0	1	2	3
y	4	5	6	7

5.
x	1	2	3	4
y	4	5	5	6

6.
x	−1	0	0	1
y	1	2	3	4

State whether each relation represents a function.

7. $\{(0, 3), (1, -2), (2, 3), (3, -2)\}$ _____

8. $\{(-5, 4), (-5, 3), (-4, 1), (0, 0)\}$ _____

State the domain and range of each function.

9. $\{(-1, 5), (-0.5, 0), (0, 1), (0.5, 3), (1, -4)\}$ _____

10. $\{(-3, 1), (-2, 0), (-1, 1), (0, 0)\}$ _____

11. $\{(5, -1), (4, -1), (3, -1), (2, -1)\}$ _____

12. The function graphed at the right. _____

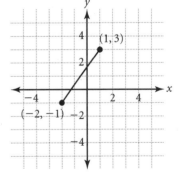

Evaluate each function for the given values of x.

13. $f(x) = 3x + 5$, for $x = -2$ and $x = 2$ _____

14. $f(x) = \frac{1}{2}x - 4$, for $x = 6$ and $x = 9$ _____

15. $f(x) = -2x - 5$, for $x = -3$ and $x = \frac{3}{2}$ _____

Practice Masters Level B
2.3 Introduction to Functions

State whether each graph represents a function.

1.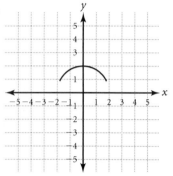

2.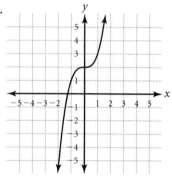

3.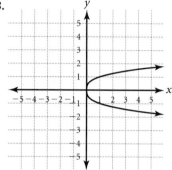

State whether each relation represents a function.

4. $\left\{\left(\dfrac{1}{6}, 3\right), \left(\dfrac{6}{15}, 6\right), \left(\dfrac{1}{6}, 1\right)\right\}$

5.
x	−6	−6	0	12
y	−30	−15	24	36

Evaluate each function for the given values of x.

6. $f(x) = \dfrac{1}{2}x - 7$, for $x = \{-2, -1, 0, 1, 2\}$ _____

7. $f(x) = -5 - 3x$, for $x = \{-6, -4, -2, 0, 2\}$ _____

8. $f(x) = 6 - 3x^2$, for $x = \left\{\dfrac{-3}{2}, \dfrac{-1}{2}, 0, \dfrac{3}{2}, \dfrac{1}{2}\right\}$ _____

9. $f(x) = \dfrac{2}{3}x^2 + 1$, for $x = \left\{\dfrac{-1}{3}, \dfrac{-1}{9}, 0, \dfrac{1}{9}, \dfrac{1}{3}\right\}$ _____

10. If the domain of $f(x) = -x^2$ is integer values of x such that $-3 \le x \le 0$, find the range. _____

11. Find $f(\sqrt{5})$ if $f(x) = -2x^2 + 6$. _____

Use a graphing calculator to graph each function, and state the domain and range.

12. $f(x) = -3x$

13. $f(x) = 2x^2$

14. $f(x) = -3x^2 + 5$

NAME _____ CLASS _____ DATE _____

Practice Masters Level C
2.3 Introduction to Functions

Evaluate each function for integral values of *x* in the given interval.

1. $f(x) = x^3 - 2x^2 + x - 3$ for $-2 \leq x < 2$ _____

2. $f(x) = -5x^2 + 2x - 3$ for $-8 < x < -3$ _____

3. $f(x) = \frac{1}{4}x^2 + 2x^3$ for $0 \leq x \leq 5$ _____

4. $f(x) = \frac{3x-2}{2^3}$ for $-4 \leq x \leq 0$ _____

Given $f(x) = \frac{1}{2}x^2 + 4$, find each indicated function.

5. $f(\sqrt{3})$ _____   6. $f(-\sqrt{2})$ _____

7. $f(\sqrt{2^3})$ _____   8. $f(\sqrt{a})$ _____

Use a graphing calculator to graph each function, and state the domain and range.

9. $f(x) = \frac{-1}{2}x^2 - 2x + 4$ _____

10. $f(x) = \frac{3x-2}{5}$ _____

11. $f(x) = \frac{1}{2}x^2 - 3$ _____

12. $f(x) = x^2 - 3x + 6$ _____

13. The selling price of items at a grocery store have a 35% markup from wholesale prices, *w*.

   a. Determine whether this relation is a function. If so, state the function. _____

   b. Find the selling price of an item with a wholesale cost of $1.50. _____

   c. What is the wholesale cost of an item with a selling price of $2.79? _____

   d. An item with a wholesale price of $3.99 is on sale for 20% off of the selling price. What is the sale price? _____

Algebra 2

NAME _____ CLASS _____ DATE _____

Practice Masters Level A
2.4 Operations with Functions

Find $f + g$ and $f - g$.

1. $f(x) = 9 - 3x;\ g(x) = 5x - 7$ _____

2. $f(x) = \frac{1}{2}x + 2;\ g(x) = x^2 + 2x$ _____

3. $f(x) = 3x^2 - 5x;\ g(x) = -4x + 7$ _____

Find $f \cdot g$ and $\frac{f}{g}$. State any domain restrictions.

4. $f(x) = x^2 + 9;\ g(x) = x - 9$ _____

5. $f(x) = 5x^2;\ g(x) = \frac{1}{2}x$ _____

Let $f(x) = 4x + 8$ and $g(x) = x + 2$. Find each new function, and state any domain restrictions.

6. $f + g$ _____   7. $f \cdot g$ _____

8. $f - g$ _____   9. $\frac{f}{g}$ _____

10. $g - f$ _____   11. $\frac{g}{f}$ _____

Use your answers to Exercises 6–11 to evaluate the indicated function for the given value.

12. $(f - g)(-3)$ _____   13. $(f + g)\left(\frac{-1}{5}\right)$ _____

14. $(g - f)(0)$ _____   15. $(f \cdot g)(-3)$ _____

16. $(f - g)\left(\frac{-3}{2}\right)$ _____   17. $\left(\frac{f}{g}\right)(10)$ _____

Find $f \circ g$ and $g \circ f$.

18. $f(x) = 6x;\ g(x) = 4x - 8$ _____

19. $f(x) = -3x + 3;\ g(x) = 6x$ _____

34   Practice Masters Levels A, B, and C   Algebra 2

Practice Masters Level B
2.4 Operations with Functions

Find $f + g$ and $f - g$.

1. $f(x) = 3x^2 + 2x;\ g(x) = -x^2 + 2x$ _____

2. $f(x) = 2x^2 + 4x - 1;\ g(x) = 3x^2$ _____

3. $f(x) = x^2 - x;\ g(x) = x + 1$ _____

Find $f \cdot g$ and $\dfrac{f}{g}$. State any domain restrictions.

4. $f(x) = -6x;\ g(x) = \dfrac{1}{2}x$ _____

5. $f(x) = \dfrac{1}{3}x^2 - 4;\ g(x) = -12x$ _____

6. $f(x) = x^3 - x^2;\ g(x) = 2x + 1$ _____

Let $f(x) = x^2 - 9$ and $g(x) = \dfrac{1}{2}x - 4$. Find each new function, and state any domain restrictions.

7. $f - g$ _____ 8. $f + g$ _____

9. $g - f$ _____ 10. $\dfrac{f}{g}$ _____

Find $f \circ g$ and $g \circ f$.

11. $f(x) = 2x - 3$ and $g(x) = 3x$ _____

12. $f(x) = x^2$ and $g(x) = 5x$ _____

13. $f(x) = \dfrac{1}{2}x - 3$ and $g(x) = \dfrac{1}{4}x$ _____

14. Use your answer to Exercise 11 to find $f \circ g(2)$.

_____

15. Use your answer to Exercise 12 to find $g \circ f(-3)$.

_____

16. Use your answer to Exercise 13 to find $f \circ g(-16)$.

_____

Algebra 2

NAME _____ CLASS _____ DATE _____

Practice Masters Level C
2.4 Operations with Functions

Let $f(x) = |10x - 4|$ and $g(x) = -8x^2$. Find each new function, and state any domain restrictions.

1. $f + g$ 2. $f \circ g$ 3. $f - g$ 4. $\dfrac{f}{g}$

_____ _____ _____ _____

Given the following functions, find each composite function value.

$h(x) = 2x + 5$ $g(x) = x^2 - 1$ $f(x) = \dfrac{-1}{2}x + \dfrac{3}{2}$

5. $(h \circ g)(-1)$ _____ 6. $(f \circ g)(-6)$ _____

7. $(g \circ g)(-3)$ _____ 8. $(g \circ f)(-6)$ _____

9. $(h \circ f)(5)$ _____ 10. $(h \circ h)\left(\dfrac{2}{3}\right)$ _____

11. If $h(x) = 4x^2 - 16$, find two functions f and g such that $f \circ g = h$. _____

_____

12. The cost function, in dollars, for producing microchips is given by $C(x) = 1.35x + 200$, where x is the number of chips produced. The number of chips produced is represented by $X(t) = 20t$, where t is the number of hours of production.

a. Give the cost, C, as a function of time, t. _____

b. Find the cost of the production run that lasts 8 hours. _____

c. How many microchips are produced in 8 hours? _____

13. A music store is offering a 10% discount on a CD. Assume a sales tax of 6.25%.

a. Write the function for the cost that represents the discount taken after the sales tax is applied. Then write the function before the sales tax is applied.

_____

b. Which function results in a lower final cost? Explain. _____

_____

36 Practice Masters Levels A, B, and C Algebra 2

NAME _____ CLASS _____ DATE _____

Practice Masters Level A
2.5 Inverses of Functions

State whether the relation is a function. Find the inverse, and state whether the inverse is a function.

1. $\{(0, 1), (1, 4), (2, 9), (3, 16)\}$ _____

2. $\left\{\left(\dfrac{1}{2}, 2\right), (1, 3), \left(\dfrac{3}{2}, 2\right), (2, 1)\right\}$ _____

3. $\{(4, 5), (5, 10), (4, 6), (3, 2)\}$ _____

4. $\{(2, 1), (3, 1), (4, 2), (2, 5)\}$ _____

5. $\{(-1, 6), (-2, 5), (-3, 4), (-4, 3)\}$ _____

Determine whether the inverse of each function graphed is also a function.

6. 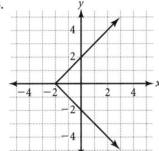 _____

7. 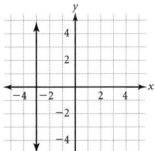 _____

For each function, find the equation of the inverse. Then use composition to verify that the equation you wrote is the inverse.

8. $y = 3x - 1$ _____

9. $y = 7x + 3$ _____

10. $y = -x$ _____

11. $y = \dfrac{2}{3}x - 3$ _____

12. $y = \dfrac{3}{4}x + 2$ _____

13. $y = -6.4x - 3.2$ _____

Use a graphing calculator to graph each function, and use the horizontal-line test to determine whether the inverse is a function.

14. $f(x) = x^2 + x$

15. $h(x) = -3 - x^2$

16. $g(x) = x^3$

_____   _____   _____

Algebra 2   Practice Masters Levels A, B, and C   37

NAME _____ CLASS _____ DATE _____

Practice Masters Level B
2.5 Inverses of Functions

State whether the relation is a function. Find the inverse, and state whether the inverse is a function.

1. $\{(1, 3), (2, 3), (4, 5), (9, 5)\}$ _____

2. $\{(0, 0), (1, 0), (2, 3), (1, 3)\}$ _____

3. $\{(2, -1), (3, 0), (4, -2), (5, -5)\}$ _____

4. $\left\{(-5, 3), \left(\frac{1}{2}, 1\right), \left(-\frac{3}{2}, 0\right), (2, 2)\right\}$ _____

Determine whether the inverse of each function graphed is also a function.

5. 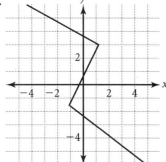 _____

6. 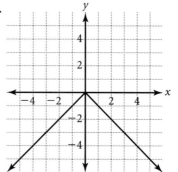 _____

For each function, find the equation of the inverse. Then use composition to verify that the equation you wrote is the inverse.

7. $y = \dfrac{x - 5}{3}$ _____

8. $y = \dfrac{2x + 7}{4}$ _____

9. $y = \dfrac{-3x - 2}{8}$ _____

10. $y = \dfrac{-3}{2}x + 6$ _____

11. $y = 4(2x + 3)$ _____

12. $y = \dfrac{-1}{4}(6x + 6)$ _____

Use a graphing calculator to graph each function, and use the horizontal-line test to determine whether the inverse is a function.

13. $f(x) = x^3 + 4$

14. $h(x) = -x^3 + x$

15. $g(x) = -x^2 - 5$

Practice Masters Level C
2.5 Inverses of Functions

Determine whether each pair of equations are inverses.

1. $y = 4x - 6$; $y = \dfrac{x}{4} + \dfrac{3}{2}$ _____

2. $y = -3x + 5$; $y = \dfrac{-1}{3}x + \dfrac{5}{3}$ _____

3. $y = -2x - 1$; $y = \dfrac{1}{2}x + \dfrac{1}{2}$ _____

4. $y = \dfrac{9}{5}x + 32$; $y = \dfrac{5}{9}(x - 32)$ _____

5. $y = \dfrac{-2}{3}x + 3$; $y = \dfrac{-3}{2}x - \dfrac{9}{2}$ _____

Find an equation for the inverse of each function.

6. $y = \dfrac{-x - 8}{4}$ _____

7. $y = 5\left(\dfrac{1}{2}\left(\dfrac{x-2}{3^3}\right)\right)$ _____

8. $y = 3\left(-x - \dfrac{1}{2}\right)$ _____

9. $y = 4.6 - 1.7x$ _____

10. $y = \sqrt{x}$ _____

11. $y = \dfrac{x + 2}{x}$ _____

Determine whether each function is a one-to-one function.

12. $y = x^2$ _____

13. $y = \dfrac{1}{4}x^3$ _____

14. $y = -5 - x - 2x^2$ _____

15. $y = -x^3 + 2x + 3$ _____

16. The price of a computer in 1980 was $6000. In the year 2000, a similar but updated version cost $1200. Assume the price follows a linear pattern.

   a. Write a function that expresses the cost of the computer, c, in terms of the year, y. _____

   b. Find the inverse of this function. Explain what the inverse function represents. _____

Algebra 2

Practice Masters Level A
2.6 Special Functions

Graph each function.

1. $f(x) = \begin{cases} 2x & \text{if } x < 0 \\ -x & \text{if } x \geq 0 \end{cases}$

2. $g(x) = 2|x| + 1$

3. $h(x) = 4[x]$

4. $f(x) = \begin{cases} x & \text{if } 0 \leq x \leq 2 \\ 2 & \text{if } x > 2 \end{cases}$

5. $g(x) = -|x| - 2$

6. $h(x) = -2[x]$

Write the piecewise function represented by each graph.

7.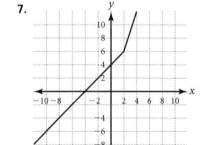

8.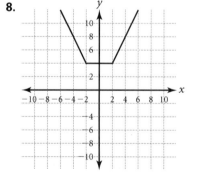

9.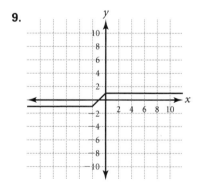

Practice Masters Level B
2.6 Special Functions

Graph each function.

1. $f(x) = \dfrac{1}{2}[x]$

2. $g(x) = \begin{cases} 2x \text{ if } x < 0 \\ 0 \text{ if } x = 0 \\ 2x \text{ if } x > 0 \end{cases}$

3. $h(x) = -2|x| + 1$

Write the special function represented by each graph.

4.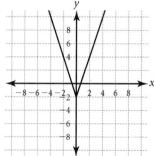

5.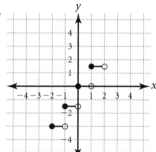

6.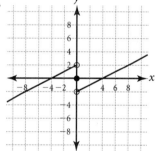

_____ _____ _____

Evaluate.

7. $|-4.7|$ _____

8. $-|-3.8|$ _____

9. $[16.93]$ _____

10. $[7.4]$ _____

11. $[0.7]$ _____

12. $[-8.6] + [3.1]$ _____

13. $-|-13.2| + [-0.1]$ _____

14. $-[4.7] + |4.7|$ _____

15. $[0.99] + [-0.99]$ _____

16. $[-|7.4|]^2$ _____

Algebra 2   Practice Masters Levels A, B, and C   41

NAME _____ CLASS _____ DATE _____

Practice Masters Level C
2.6 Special Functions

Graph each function.

1. $y = \left[\dfrac{1}{2}x\right]$

2. $y = |x - 2| + 1$

3. $y = \begin{cases} -3x & \text{if } x < 1 \\ 2 & \text{if } x = 1 \\ 2x + 1 & \text{if } x > 1 \end{cases}$

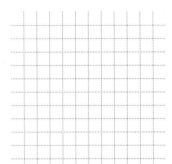
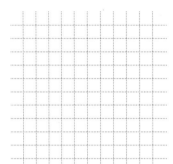
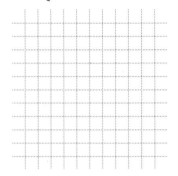

Evaluate.

4. $-[-1.7]^2$ _____

5. $-|-|5 - 6.2 + [-1.2]||$ _____

6. $\left[|-3.1|^2\right]^3$ _____

7. $\dfrac{|4.8|}{[-0.2]^2} - [1.4]^{-2}$ _____

8. A cross country coach only keeps track of the mileage, m, of her runners in whole numbers. In other words, a run of 6.4 miles, x, only counts as 6 miles. Express the mileage, m, the coach counts as a function of x, the mileage the runners actually run.

_____

9. A mason charges $100 per day plus $4.50 per block, b, for up to and including 200 blocks. The rate per block for over 200 blocks is $4.25 per block with no flat fee per day.

   a. Complete the table showing the total cost, c, for the given number of blocks.

   b. Write the function between cost, c, and the number of blocks, b.

b	c
100	
200	
201	
300	

_____

Practice Masters Level A
2.7 A Preview of Transformations

Identify each transformation from the parent function $f(x) = x$ to g.

1. $g(x) = x + 2$ _____

2. $g(x) = 3x$ _____

3. $g(x) = -x$ _____

4. $g(x) = -6x$ _____

5. $g(x) = (x - 5)$ _____

6. $g(x) = -3x + 2$ _____

Identify each transformation from the parent function $f(x) = x^2$ to g.

7. $g(x) = x^2 + 3$ _____

8. $g(x) = -x^2$ _____

9. $g(x) = (6x^2)$ _____

10. $g(x) = 2x^2 - 6$ _____

11. $g(x) = \frac{1}{2}x^2$ _____

12. $g(x) = (x - 2)^2$ _____

Write the function for each graph described below.

13. the graph of $f(x) = x^2$ reflected across the x-axis _____

14. the graph of $f(x) = x^2$ reflected across the y-axis _____

15. the graph of $f(x) = x^4$ translated 5 units up _____

16. the graph of $f(x) = |x|$ stretched horizontally by a factor of 3 _____

17. the graph of $f(x) = x^3$ translated 3 units down _____

18. the graph of $f(x) = x^2$ translated 4 units left _____

Algebra 2   Practice Masters Levels A, B, and C

Practice Masters Level B
2.7 A Preview of Transformations

Identify each transformation from the parent function $f(x) = x^3$ to g.

1. $g(x) = x^3 - 2$ _____

2. $g(x) = (x - 2)^3$ _____

3. $g(x) = (-4x^3)$ _____

4. $g(x) = -x^3 + 1$ _____

5. $g(x) = \frac{1}{3}x^3$ _____

6. $g(x) = 2\left(x - \frac{1}{2}\right)^3$ _____

Identify each transformation from the parent function $f(x) = \sqrt{x}$ to g.

7. $g(x) = \sqrt{x} + 4$ _____

8. $g(x) = \sqrt{(5x)}$ _____

9. $g(x) = \frac{1}{2}\sqrt{x}$ _____

10. $g(x) = -\sqrt{x} - 2$ _____

11. $g(x) = \sqrt{x - 3}$ _____

12. $g(x) = \sqrt{x - 2} + 4$ _____

Write the function for each graph described below.

13. the graph of $f(x) = x^3$ reflected across the y-axis _____

14. the graph of $f(x) = -x^2$ translated horizontally 5 units to the left _____

15. the graph of $f(x) = \sqrt{x}$ stretched vertically by a factor of 4 _____

16. the graph of $f(x) = x^3$ stretched horizontally by a factor of 5 _____

17. the graph of $f(x) = \frac{1}{3}x + 9$ reflected across the y-axis _____

18. the graph of $\sqrt{2x}$ stretched vertically by a factor of $\frac{1}{2}$ _____

Practice Masters Level C
2.7 A Preview of Transformations

Identify each transformation from the function $f(x) = \sqrt{x + 2}$ to g.

1. $g(x) = \sqrt{x + 4}$ _____

2. $g(x) = \sqrt{x}$ _____

3. $g(x) = -\sqrt{x + 3}$ _____

4. $g(x) = \sqrt{x - 1} + 2$ _____

5. $g(x) = 3\sqrt{x + 2}$ _____

6. $g(x) = \sqrt{3(x + 2)}$ _____

Identify each transformation from the function $f(x) = (x + 3)^3$ to g.

7. $g(x) = x^3$ _____

8. $g(x) = 2(x + 3)^3$ _____

9. $g(x) = (2(x + 3))^3$ _____

10. $g(x) = (x + 3)^3 + 2$ _____

11. $g(x) = (-x - 3)^3$ _____

12. $g(x) = -(-x - 3)^3$ _____

Write the function for each graph described below.

13. the graph of $f(x) = x^5$ across the y-axis _____

14. the graph of $f(x) = \sqrt{x + 4}$ stretched horizontally by a factor of 2 _____

15. the graph of $f(x) = \sqrt{x + 10}$ stretched vertically by a factor of 2 _____

16. the graph of $f(x) = \sqrt{x - 6}$ compressed horizontally by a factor of $\frac{1}{3}$ _____

NAME _____ CLASS _____ DATE _____

Practice Masters Level A

3.1 Solving Systems by Graphing or Substitution

Graph and classify each system. Then find the solution from the graph.

1. $\begin{cases} y = -x + 2 \\ y = 2x - 1 \end{cases}$

2. $\begin{cases} y = -x + 6 \\ y = \dfrac{1}{3}x + 2 \end{cases}$

3. $\begin{cases} x + y = 0 \\ 3x + y = 6 \end{cases}$

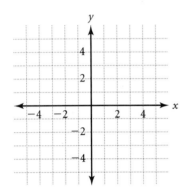
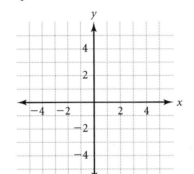
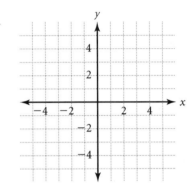

4. $\begin{cases} 2x + 4y = 12 \\ x + y = 2 \end{cases}$

5. $\begin{cases} y = x \\ y = 5x + 0 \end{cases}$

6. $\begin{cases} y = -2x + 1 \\ y = -2x + 5 \end{cases}$

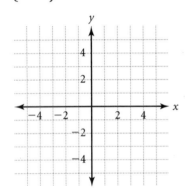
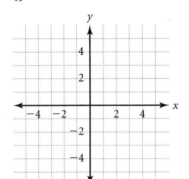
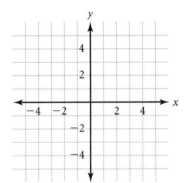

Use substitution to solve each system of equations. Check your solution.

7. $\begin{cases} y = x + 3 \\ 10x + 2y = 18 \end{cases}$

8. $\begin{cases} x = y - 8 \\ 4x - y = -2 \end{cases}$

9. $\begin{cases} 2x + y = 7 \\ 3x + 2y = 10 \end{cases}$

10. $\begin{cases} x - 3y = 0 \\ 3x - 2y = -7 \end{cases}$

11. $\begin{cases} 2x - y = -12 \\ -2x - 4y = 2 \end{cases}$

12. $\begin{cases} x - 6y = -18 \\ -3x + 2y = 6 \end{cases}$

46   Practice Masters Levels A, B, and C   Algebra 2

Practice Masters Level B

3.1 Solving Systems by Graphing or Substitution

Graph and classify each system. Then find the solution from the graph.

1. $\begin{cases} -x - y = 10 \\ -4x + 3y = -9 \end{cases}$

2. $\begin{cases} 2x + 3y = 12 \\ y = \dfrac{-1}{2}x + 3 \end{cases}$

3. $\begin{cases} 2x + y = 5 \\ -4x - 2y = -2 \end{cases}$

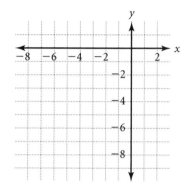
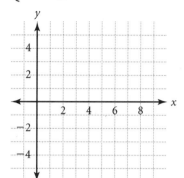
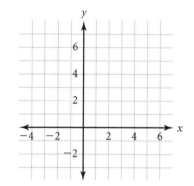

Use substitution to solve each system of equations. Check your solution.

4. $\begin{cases} 2x + 4y = 42 \\ 3x + y = 8 \end{cases}$

5. $\begin{cases} 3m + 6n = 12 \\ 4m + 5n = 28 \end{cases}$

6. $\begin{cases} -3x + 6y = 18 \\ y = \dfrac{1}{2}x + 3 \end{cases}$

7. $\begin{cases} 2x + 3y = 3 \\ -6x - 12y = -11 \end{cases}$

8. $\begin{cases} \dfrac{1}{2}x - \dfrac{1}{3}y = \dfrac{1}{6} \\ 6x - 4y = 1 \end{cases}$

9. $\begin{cases} 4.5x - 3y = -1.65 \\ -2x + 1.5y = 0.8 \end{cases}$

10. The perimeter of a pasture is 1250 feet. The pasture is four times longer than it is wide. Find the dimensions by setting up a system of equations.

11. A chemist mixes some 20% acid solution with pure 100% acid to increase the concentration. How much pure acid and acid solution should be mixed to form 100 milliliters of 40% acid solution.

12. A grocer mixes pecans priced at $4.50 per pound with almonds priced at $5.25 per pound. How many pounds of each type of nut should he use to have a total mixture of 14 pounds priced at $4.80 per pound?

Algebra 2

NAME _____ CLASS _____ DATE _____

Practice Masters Level C

3.1 Solving Systems by Graphing or Substitution

Use a graphing calculator to solve each system of equations.
Round final answers to the nearest hundredth, if necessary.

1. $\begin{cases} 0.007x - 2.3y = 4 \\ 2.8x + 1.06y = 3.02 \end{cases}$ _____

2. $\begin{cases} \dfrac{2}{5}x + \dfrac{1}{3}y = \dfrac{3}{4} \\ \dfrac{-1}{3}x - \dfrac{2}{5}y = -1 \end{cases}$ _____

Use substitution to solve each system of equations.
Check your solution.

3. $\begin{cases} 3x - 2y - 5z = 27 \\ -2x + 5y - 4z = 4 \\ x = 1 \end{cases}$ _____

4. $\begin{cases} -4p + 3q + 2r = -8 \\ p = 2q \\ 2q - 4r = 14 \end{cases}$ _____

5. $\begin{cases} 2x + 3y - z = 0 \\ -2x - 2y + 4z = 2 \\ 3x - y + 2z = 9 \end{cases}$ _____

6. $\begin{cases} 5x - 2y + 3z = -52.5 \\ -2x + y + 4z = 9 \\ 3x + 2y - z = -8.5 \end{cases}$ _____

Write a system of equations for Exercises 7–10, and solve.

7. With a head wind, an airplane travels 480 miles in 2 hours.
Returning with a tail wind takes 1 hour 30 minutes. Find the
plane's air speed and the speed of the wind.

_____

8. A chemist mixes a 30% chlorine solution with pure water to dilute
the mixture. How much chlorine solution and water must be mixed
to make 20 milliliters of a 10% chlorine solution?

_____

9. A person invests in a mutual fund, part of which earned 11%
interest and part of which earned 6.5%. If the total amount invested
was $5000 and the interest earned was $345, how much was
invested at each rate?

_____

10. A boat travels with the current a certain distance, d, in 1 hour 30
minutes. On the return trip, it takes the boat 2 hours 40 minutes.
In terms of d, what is the rate of the boat and the current?

_____

NAME _____ CLASS _____ DATE _____

Practice Masters Level A
3.2 Solving Systems by Elimination

Use elimination to solve each system of equations.
Check your solution.

1. $\begin{cases} 2x + y = 41 \\ 2x + 5y = 13 \end{cases}$ _____

2. $\begin{cases} 3x + 4y = 3 \\ -3x - 2y = -9 \end{cases}$ _____

3. $\begin{cases} x - 2y = 5 \\ 2x + y = 10 \end{cases}$ _____

4. $\begin{cases} 8x + 3y = 7 \\ -4x + 5y = -23 \end{cases}$ _____

5. $\begin{cases} x + 3y = 3 \\ 3x - y = -11 \end{cases}$ _____

6. $\begin{cases} 7c + 3d = 25 \\ 7c + 3d = 12 \end{cases}$ _____

7. $\begin{cases} 3x - 2y = 4 \\ -6x + 3y = -6 \end{cases}$ _____

8. $\begin{cases} -5x - 2y = 44 \\ -3x + 3y = 39 \end{cases}$ _____

9. $\begin{cases} 4x + 2y = -4 \\ 16x + 8y = -16 \end{cases}$ _____

10. $\begin{cases} -3x - y = -1 \\ 4x + 8y = -2 \end{cases}$ _____

11. $\begin{cases} 3x - 4y = 16 \\ 2x + 5y = 3 \end{cases}$ _____

12. $\begin{cases} 2x - 5y = \dfrac{8}{7} \\ 7x + \dfrac{1}{2}y = -2 \end{cases}$ _____

13. $\begin{cases} 6x - 5y = \dfrac{1}{2} \\ -3x + 2y = 0 \end{cases}$ _____

14. $\begin{cases} \dfrac{1}{3}x + 2y = 4 \\ -4x - y = 21 \end{cases}$ _____

15. $\begin{cases} 11x + 12y = 112 \\ 11x + 12y = 324 \end{cases}$ _____

16. $\begin{cases} 9x - 18y = 45 \\ 18x + 9y = 90 \end{cases}$ _____

Algebra 2

NAME _____ CLASS _____ DATE _____

Practice Masters Level B
3.2 Solving Systems by Elimination

Use elimination to solve each system of equations.
Check your solution.

1. $\begin{cases} 3x - 3y = 18 \\ -9x + 14y = -6 \end{cases}$ _____

2. $\begin{cases} x + 3y = -9 \\ 4y = x + 16 \end{cases}$ _____

3. $\begin{cases} -3(5 + x) = y \\ 5(x - 3) = 2y + 4 \end{cases}$ _____

4. $\begin{cases} 5x - \dfrac{28}{3}y = 2 \\ -3x = 4y \end{cases}$ _____

5. $\begin{cases} 2b = 2a + b - 4 \\ 3a = 3b - a + 2 \end{cases}$ _____

6. $\begin{cases} 8n = 6m + 3 \\ 9m = 12n - 4.5 \end{cases}$ _____

7. $\begin{cases} 0.06x - 0.04y = 0.06 \\ 0.18x - 0.12y = 0.16 \end{cases}$ _____

8. $\begin{cases} 0.01x + 0.02y = 1.8 \\ 0.3x + 0.4y = 0.2 \end{cases}$ _____

9. $\begin{cases} 2(2x + 3y) = 37 \\ 7x = 3(2y + 3) - 2 \end{cases}$ _____

10. $\begin{cases} 4x + 5y = 0 \\ \dfrac{3}{10}x + \dfrac{1}{4}y = 2 \end{cases}$ _____

11. $\begin{cases} 6x - 8y = 27 \\ -\dfrac{3}{4}x - \dfrac{1}{3}y = \dfrac{5}{8} \end{cases}$ _____

12. $\begin{cases} \dfrac{1}{5}x - \dfrac{2}{5}y = \dfrac{3}{2} \\ \dfrac{1}{8}x - \dfrac{8}{7}y = 4\dfrac{1}{16} \end{cases}$ _____

13. $\begin{cases} 2y + 3(4 - x) = 26 \\ x - 2(x + 5y) = 2x - 19y \end{cases}$ _____

14. $\begin{cases} \dfrac{1}{3}x + \dfrac{3}{4}y = 6 \\ \dfrac{2}{3}x + \dfrac{3}{2}y = 12 \end{cases}$ _____

50   Practice Masters Levels A, B, and C   Algebra 2

NAME _____ CLASS _____ DATE _____

Practice Masters Level C
3.2 Solving Systems by Elimination

Use elimination to solve each system of equations.

1. $\begin{cases} \dfrac{1}{2}x + \dfrac{3}{10}y = \dfrac{1}{2} \\ \dfrac{-5}{3}x - y = \dfrac{4}{3} \end{cases}$

2. $\begin{cases} \dfrac{\frac{1}{3}x - 4}{8} - \dfrac{y - 2}{6} = 2 \\ \dfrac{x + 2}{2} + \dfrac{y - 4}{8} = 1 \end{cases}$

_____

3. $\begin{cases} -0.4x + 1.2y = 1 \\ 3.6x - 10.8y = -9 \end{cases}$

4. $\begin{cases} 2x + y = a \\ -3x - 2y = b \end{cases}$

_____

5. $\begin{cases} -0.01x - 0.02y = 0.0011 \\ 0.24x + 0.12y = 0.0024 \end{cases}$

6. $\begin{cases} 3x + 2y = c \\ -4x - 5y = d \end{cases}$

_____

Write a system of equations for Exercises 7–9, and solve.

7. A factory employs skilled laborers at $25 per hour and unskilled laborers at $12 per hour. The hourly cost of the factory's 24 workers is $405. How many of each type of laborer works at the factory?

_____

8. A vendor at the county fair wishes to invest $300 in his milkshake booth and would like his total revenue to be $900. He figures he can make a small shake for $0.40 and a large shake for $0.68. He plans to sell a small shake for $1.25 and a large shake for $2.00. How many of each size shake will he need to make and sell to realize his revenue?

_____

9. A person wishing to invest $5000 and earn $800 in interest decides on a combination of a high-risk investment, which earns 19% interest, and a low risk investment which pays 7.5% interest. How much should be invested in each to ensure the $800 in interest?

_____

Algebra 2   Practice Masters Levels A, B, and C   51

NAME _____ CLASS _____ DATE _____

Practice Masters Level A
3.3 Linear Inequalities In Two Variables

Graph each linear inequality.

1. $y \geq x$

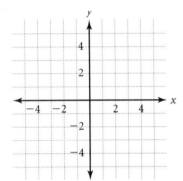

2. $y < 2x + 4$

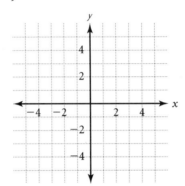

3. $y > -x$

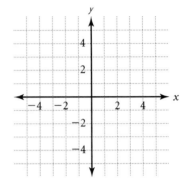

4. $y \leq \dfrac{-1}{2}x + 2$

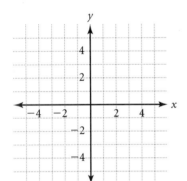

5. $3x + y - 5 > 0$

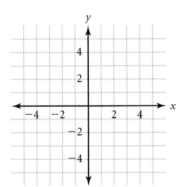

6. $y < -2$

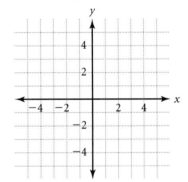

7. For a fund raiser, The Honor Society is selling two sizes of candy bars. They earn a profit of $1.00 on small candy bars, s, and $2.00 on large candy bars, l.

   a. Write an inequality that represents a profit of at least $300.

   _____

   b. Graph the inequality on the coordinate grid to the right.

   c. Find three ordered pairs that represent achieving the goal of making a $300 profit.

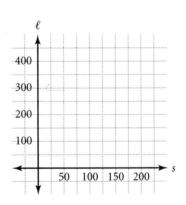

NAME _____ CLASS _____ DATE _____

Practice Masters Level B

3.3 Linear Inequalities In Two Variables

Graph each linear inequality.

1. $4x + 2y < 4$

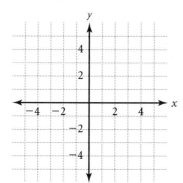

2. $-5x - 3y < 6$

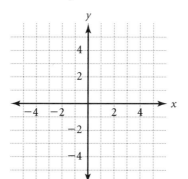

3. $x > 1$

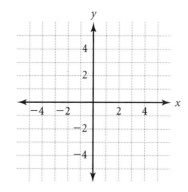

4. $y \geq 0$

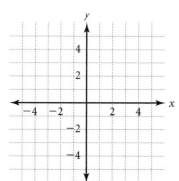

5. $-3x + y > 0$

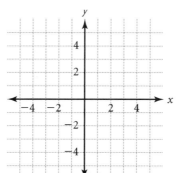

6. $4x - 3y > -5$

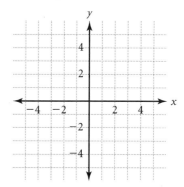

7. The dairy association is thinking of opening a milkshake stand at the fair. A goal of making a profit of over $800 makes it worthwhile to open.

 a. If they sell 400 small shakes, x, and 250 large shakes, y, write an inequality representing the profit they would have to make on each size in order to reach at least an $800 profit.

 _____

 b. Graph the inequality on the coordinate grid to the right.

 c. If they want to make the same profit on each size of shake, at least how much profit would they need to make selling either size shake?

 _____

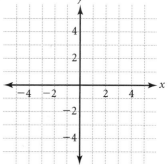

Algebra 2　　　Practice Masters Levels A, B, and C　　　53

NAME _____ CLASS _____ DATE _____

Practice Masters Level C

3.3 Linear Inequalities In Two Variables

Graph each linear inequality.

1. $\dfrac{4}{3}x + \dfrac{1}{2}y < \dfrac{-1}{4}$

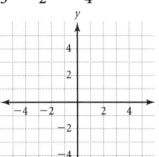

2. $-2x \leq 5$

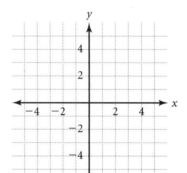

3. $\dfrac{-4}{5}y > 3$

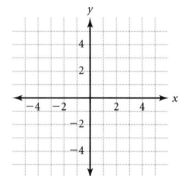

4. Ann and John plan to take $30 to the movies. Tickets cost $7.75 each, and the remainder of the money can be spent on popcorn and soft drinks. The price of a large popcorn, L, is $3.75, and the price of a small popcorn, P, is $2.75. The price of a large soft drink, D, is $2.50 and the price of a small soft drink, S, is $1.75. If the couple buys two tickets, write and graph the inequality that represents the number of popcorns and soft drinks that can be purchased for each given combination.

   a. large popcorns and large soft drinks

   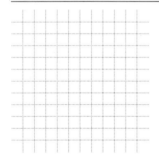

   b. large popcorns and small soft drinks

   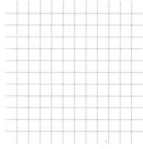

   c. small popcorns and large soft drinks

   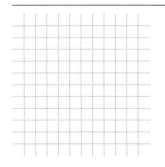

   d. small popcorns and small soft drinks

Practice Masters Level A

3.4 Systems of Linear Inequalities

Graph each system of linear inequalities.

1. $\begin{cases} y \leq 3x + 1 \\ y \geq -x \end{cases}$

2. $\begin{cases} y > 4x - 2 \\ y > 3x - 1 \end{cases}$

3. $\begin{cases} y \leq 2x \\ y \geq -x + 2 \end{cases}$

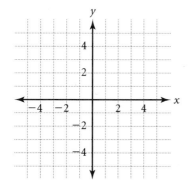 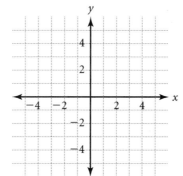 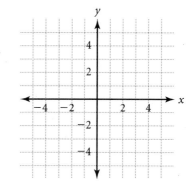

Write the system of inequalities whose solution is graphed.

4.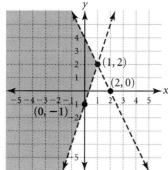

5.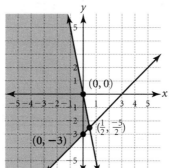

6.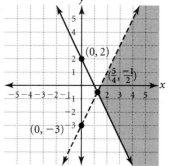

7. Graph the system of linear inequalities.

$\begin{cases} y \leq -4x + 6 \\ y < 3x - 1 \\ y \geq -2x - 3 \end{cases}$

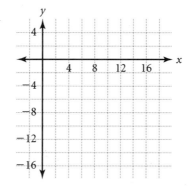

Algebra 2 Practice Masters Levels A, B, and C 55

Practice Masters Level B

3.4 Systems of Linear Inequalities

Graph each system of linear inequalities.

1. $\begin{cases} y \geq \frac{3}{2}x + 1 \\ y \geq -2x - 3 \end{cases}$

2. $\begin{cases} 2y < x - 4 \\ x - 3y < 6 \end{cases}$

3. $\begin{cases} 2x + 5y \geq 0 \\ -3x + 2y \leq 6 \end{cases}$

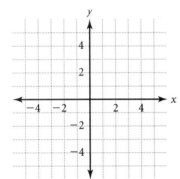
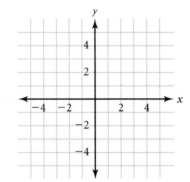
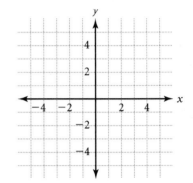

Write the system of inequalities whose solution is graphed.

4.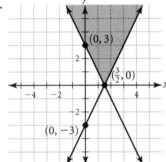

5.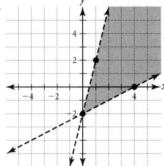

6.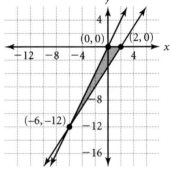

7. Graph the system of linear inequalities.

$\begin{cases} y \leq 3x - 6 \\ y > -4x + 1 \\ -2x - y > -1 \\ 2y \leq x - 4 \end{cases}$

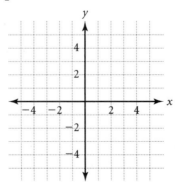

Practice Masters Level C

3.4 Systems of Linear Inequalities

Write the system of inequalities whose solution is graphed.

1.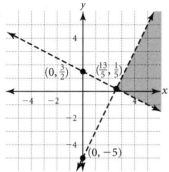

2.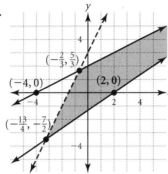

3.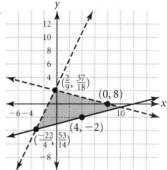

_____   _____   _____

Solve.

4. A teacher works for a textbook company during the summer. He makes $35 per page to write the material and $25 per page to type it. The teacher would like to earn at least $3500, but does not want more than 120 pages of work. Let x be the number of pages written and y be the number of pages typed. Write a system of inequalities to represent this situation, and graph the system.

_____

5. An investor wants to earn at least $1400 from investments. She plans to invest at most $10,000 in two funds. One fund pays 13% interest and the other fund pays 15% interest, but has a little more risk. Let x be the amount invested at 13% and y be the amount invested at 15%. Write and graph a system of inequalities to represent this situation.

_____

NAME _____ CLASS _____ DATE _____

Practice Masters Level A
3.5 Linear Programming

1. Graph the feasible region for the set of constraints.

$$\begin{cases} y \leq x - 2 \\ y \geq -x + 4 \\ y \geq 0 \\ x \leq 6 \end{cases}$$

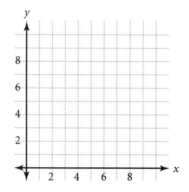

2. Refer to the information given on page 187 of the student textbook. Suppose Mr. Desmond wants to plant a number of acres of cotton, x, and a number of acres of beans, y, according to the following constraints:

   - at most 500 acres of cotton and soybeans
   - between 100 and 400 acres of cotton
   - at least 80 acres of soybeans

   Write a system of inequalities to represent the constraints. _____

The feasible region for a set of constraints has vertices (3, 0), (4, 3), (−1, 6) and (−4, 0). Given this feasible region, find the maximum and minimum values of each objective function.

3. $H = 3x + y$

   maximum: _____

   minimum: _____

4. $I = -2x + 3y$

   maximum: _____

   minimum: _____

5. $J = 2x - 4y$

   maximum: _____

   minimum: _____

Find the maximum and minimum values, if they exist, of each objective function for the given constraints.

6. $R = 2x - y$
   Constraints:
   $$\begin{cases} y \geq -x + 1 \\ y \leq 2x + 3 \\ y \geq 0 \\ x \leq 4 \end{cases}$$

   maximum: _____

   minimum: _____

7. $M = -3x + y$
   Constraints:
   $$\begin{cases} x + y \leq 5 \\ x - y \geq 6 \\ x \geq 3 \\ x \leq 5 \end{cases}$$

   maximum: _____

   minimum: _____

8. $N = -2x - 3y$
   Constraints:
   $$\begin{cases} 2x - y \geq -1 \\ y \geq x \\ y \leq 6 \\ y \geq 1 \end{cases}$$

   maximum: _____

   minimum: _____

NAME _____ CLASS _____ DATE _____

Practice Masters Level B
3.5 Linear Programming

1. Graph the feasible region for the set of constraints.

$$\begin{cases} y \leq 2x - 1 \\ 1 + y \geq 0 \\ x + y \geq 0 \\ x \geq 3 \end{cases}$$

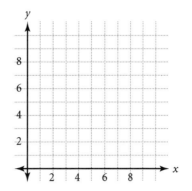

Find the maximum and minimum values, if they exist, of each objective function for the given constraints.

2. $P = -x + \dfrac{1}{2}y$

Constraints: $\begin{cases} y \geq x - 7 \\ x - y \leq 5 \\ x + y \geq 1 \\ y \leq 8 \end{cases}$

maximum: _____

minimum: _____

3. $R = -2x + \dfrac{3}{2}y$

Constraints: $\begin{cases} 2x + y \geq 4 \\ 2x - y \geq 4 \\ x - y \geq 4 \end{cases}$

maximum: _____

minimum: _____

4. Refer to the information given on page 187 of the student textbook. Suppose Mr. Desmond wants to plant a number of acres of cotton, x, and a number of acres of beans, y, according to the following constraints:

- at most 400 acres of cotton and soybeans
- between 50 and 250 acres of soybeans
- at least 100 acres of cotton

a. Write a system of inequalities to represent the constraints.

_____

b. Graph the feasible region.

c. Write a function for the revenue from Mr. Desmond's harvest.

_____

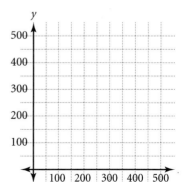

d. What is the maximum revenue? _____

Algebra 2 Practice Masters Levels A, B, and C 59

Practice Masters Level C

3.5 Linear Programming

1. Graph the feasible region for the set of constraints.

$$\begin{cases} 2x + y \leq 8 \\ x \geq 0 \\ x \leq 1 \\ y \geq 3 \end{cases}$$

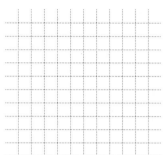

The feasible region for a set of constraints has vertices (−1, 2), (0, 3), (4, 2) and (3, −2). Given this feasible region, find the maximum and minimum values of each objective function.

2. $A = -2x + \frac{1}{2}y$

   maximum: _____

   minimum: _____

3. $B = \frac{-3}{2}x + 2y$

   maximum: _____

   minimum: _____

4. A chauffer must decide between driving his clients in the Rolls Royce or the Mercedes Benz. The Rolls Royce costs $1.75 per mile to operate and the Mercedes Benz costs $2.00 per mile. The chauffer can charge $4.00 per mile for the Rolls Royce and $6.50 per mile for the Mercedes Benz. The chauffer wants his expenses to be no more than $200 for the day and his total charges to be at least $600 for the day. The Rolls Royce must travel at most 90 miles and the Mercedes Benz must travel at least 30 miles.

   a. Write a system of inequalities to represent the constraints.

   _____

   b. Graph the feasible region.

   c. Write an objective function for the profit in a day.

   _____

   d. What is the maximum profit in a day? _____

Practice Masters Level A

3.6 Parametric Equations

Graph each pair of parametric equations for the interval $-3 \leq t \leq 3$.

1. $\begin{cases} x(t) = t + 3 \\ y(t) = 2t \end{cases}$

2. $\begin{cases} x(t) = -t + 2 \\ y(t) = -3t + 1 \end{cases}$

3. $\begin{cases} x(t) = \dfrac{1}{2}t + 2 \\ y(t) = \dfrac{-1}{2}t - 1 \end{cases}$

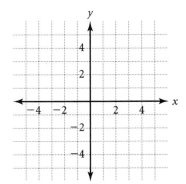
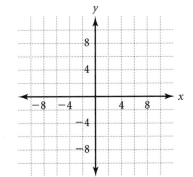
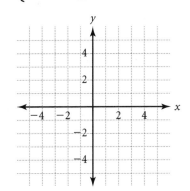

Write each pair of parametric equations as a single equation in *x* and *y*.

4. $\begin{cases} x(t) = t + 3 \\ y(t) = 2t \end{cases}$

5. $\begin{cases} x(t) = -t + 2 \\ y(t) = -3t + 1 \end{cases}$

6. $\begin{cases} x(t) = t \\ y(t) = 4 - t^2 \end{cases}$

7. $\begin{cases} x(t) = 2t - 2 \\ y(t) = -t \end{cases}$

8. $\begin{cases} x(t) = t - 1 \\ y(t) = t^2 \end{cases}$

9. $\begin{cases} x(t) = 2(t - 3) \\ y(t) = -3(t + 1) \end{cases}$

10. $\begin{cases} x(t) = \dfrac{2}{3}t + 2 \\ y(t) = \dfrac{-3}{4}t - 1 \end{cases}$

11. $\begin{cases} x(t) = \dfrac{1}{2}t + 2 \\ y(t) = \dfrac{-1}{2}t - 1 \end{cases}$

12. $\begin{cases} x(t) = \dfrac{1}{2}t \\ y(t) = \dfrac{1}{2}t^2 - 4t + 1 \end{cases}$

13. A catcher throws the baseball from home plate to second base. The following parametric equations describe the ball's path:

$$x(t) = 80t \text{ and } y(t) = 15t - 16t^2$$

The variables *x* and *y* are measured in feet per second, and *t* is measured in seconds. What horizontal distance will the baseball travel before it hits the ground? _____

Algebra 2

NAME _____ CLASS _____ DATE _____

Practice Masters Level B
3.6 Parametric Equations

Graph each pair of parametric equations for the interval $-3 \le t \le 3$.

1. $\begin{cases} x(t) = 4 - 2t \\ y(t) = \dfrac{1}{3}t \end{cases}$

2. $\begin{cases} x(t) = 5 + 3t \\ y(t) = \dfrac{-2}{3}t \end{cases}$

3. $\begin{cases} x(t) = \dfrac{1}{2}t + 3 \\ y(t) = \dfrac{3}{4}t + 1 \end{cases}$

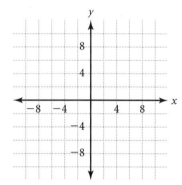
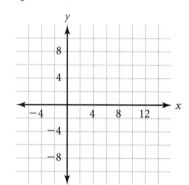
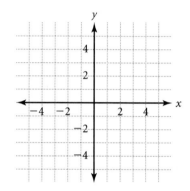

Write each pair of parametric equations as a single equation in x and y.

4. $\begin{cases} x(t) = t^2 \\ y(t) = t - 2 \end{cases}$

5. $\begin{cases} x(t) = -3t + 1 \\ y(t) = -2t - 2 \end{cases}$

6. $\begin{cases} x(t) = \dfrac{-1}{2}t \\ y(t) = \dfrac{1}{2}t^2 - 4t + 1 \end{cases}$

7. $\begin{cases} x(t) = -t^2 + 1 \\ y(t) = t - 3 \end{cases}$

Graph the function represented by the pair of parametric equations. Then graph the inverse on the same coordinate plane.

8. $\begin{cases} x(t) = -t \\ y(t) = -t - 4 \end{cases}$

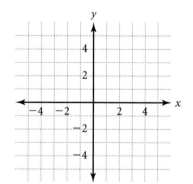

9. A golf ball is hit off the ground with a nine-iron. The following parametric equations describe the ball's path:

$$x(t) = 30t,\ y(t) = 120t - 16t^2.$$

The variables x and y are measured in feet per second, and t is measured in seconds. What distance will the golf ball travel before it hits the ground? _____

Practice Masters Level C

3.6 Parametric Equations

Write each pair of parametric equations as a single equation in x and y.

1. $\begin{cases} x(t) = -2t + 5 \\ y(t) = -3t + 3 \end{cases}$ _____

2. $\begin{cases} x(t) = t + 4 \\ y(t) = t^2 + t - 1 \end{cases}$ _____

3. $\begin{cases} x(t) = 0.1t - 0.4 \\ y(t) = 0.6t^2 + 0.3t \end{cases}$ _____

4. $\begin{cases} x(t) = 3t^2 \\ y(t) = 4t^2 + 2t - 1 \end{cases}$ _____

5. $\begin{cases} x(t) = 0.02t + 0.03 \\ y(t) = 0.1t^2 + 0.2 \end{cases}$ _____

6. $\begin{cases} x(t) = \frac{3}{4}t + \frac{2}{3} \\ y(t) = \frac{-2}{3}t + \frac{1}{5} \end{cases}$ _____

Graph the function represented by the pair of parametric equations. Then graph the inverse on the same coordinate plane.

7. $\begin{cases} x(t) = t^2 \\ y(t) = -t + 3 \end{cases}$ _____

8. $\begin{cases} x(t) = 2t - 1 \\ y(t) = t^2 \end{cases}$ _____

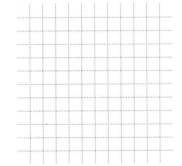
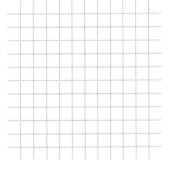

9. A tree has a height, x, of 30 inches and a trunk diameter, y, of 4 inches when planted. The height will increase at a rate of approximately 1 foot per year while the diameter of the trunk will increase by 0.75 inches per year.

   a. Write the parametric equations describing the height and trunk diameter for t years. _____

   b. How many years will it take for the height to reach 50 feet? _____

   c. What will the trunk diameter be when the tree is 50 feet tall? _____

NAME _____ CLASS _____ DATE _____

Practice Masters Level A
4.1 Using Matrices to Represent Data

For Exercises 1–12, let, $A = \begin{bmatrix} 6 & -4 \\ -2 & 3 \end{bmatrix}$, $B = \begin{bmatrix} 2 & 0 \\ 1 & -4 \end{bmatrix}$, and $C = \begin{bmatrix} 1 & -1 & 2 \\ 3 & 4 & -5 \end{bmatrix}$.

Give the dimensions of each matrix.

1. A _____   2. B _____   3. C _____

Find the indicated matrix.

4. $A + B$ 5. $-C$ 6. $3A$

7. $B - A$ 8. $\frac{1}{2}C$ 9. $-3B$

10. $3A - 2B$ 11. $0.01C$ 12. $\frac{3}{2}B + 2A$

Matrix F below represents the appliances sold at an appliance store during the first quarter of 2001. Use Matrix F for Exercises 13–17.

$$\begin{array}{c} \\ \text{Refrigerator} \\ \text{Stoves} \\ \text{TVs} \end{array} \begin{array}{ccc} \text{Jan.} & \text{Feb.} & \text{Mar.} \end{array} \\ \begin{bmatrix} 13 & 11 & 15 \\ 14 & 8 & 10 \\ 10 & 2 & 5 \end{bmatrix} = F$$

13. What are the dimensions of Matrix F? _____

14. Find the total number of stoves sold during the first quarter. _____

15. Find the number of all appliances sold during February. _____

16. What is in the location F_{32} and what does it represent? _____

17. In April of 2001, 2 refrigerators, 6 stoves and 9 TVs were sold. Write a new matrix that represents all four months. _____

64   Practice Masters Levels A, B, and C   Algebra 2

NAME _____ CLASS _____ DATE _____

Practice Masters Level B
4.1 Using Matrices to Represent Data

For Exercises 1–12, let $A = \begin{bmatrix} 3 & 0 & -4 \\ -1 & 6 & 2 \\ 2 & 4 & 2 \end{bmatrix}$, $B = \begin{bmatrix} 1 & \frac{2}{3} & \frac{3}{4} \\ \frac{2}{3} & -1 & -4 \\ \frac{3}{2} & 2 & 3 \end{bmatrix}$, and $C = \begin{bmatrix} 1 & 10 & 2 \\ -1 & 3 & -3 \\ 0 & -2 & 4 \end{bmatrix}$.

Give the dimensions of each matrix.

1. A _____ 2. B _____ 3. C _____

Find the indicated matrix.

4. $C - A$

5. $-2B$

6. $2A + 2C$

7. $12B$

8. $\frac{-1}{2}C$

9. $\frac{-1}{2}A + C$

10. $-2A + 3C$

11. $0.03A$

12. $\frac{1}{2}C + \frac{1}{2}A$

Matrix T shows different types of trees planted at a tree farm from 1996 to 1999. Use matrix T for Exercises 13–16.

$$\begin{array}{r} \\ \text{Black Walnut} \\ \text{Red Oak} \\ \text{Green Ash} \\ \text{Pecan} \end{array} \begin{array}{cccc} 1996 & 1997 & 1998 & 1999 \end{array} \\ \begin{bmatrix} 400 & 200 & 25 & 0 \\ 500 & 20 & 30 & 0 \\ 20 & 4 & 15 & 3 \\ 0 & 4 & 10 & 15 \end{bmatrix} = T$$

13. What are the dimensions of Matrix T? _____

14. In which year were the most trees planted, and how many were planted? _____

15. Which type of tree had the fewest number planted during this time? _____

16. In the Spring of 2000, the following trees were planted: 40 Black Walnut, 20 Red Oak, 600 Green Ash, 400 Pecan, and 25 Shagbark Hickory. Write a new matrix that includes the data for all five years.

Algebra 2   Practice Masters Levels A, B, and C

NAME _____ CLASS _____ DATE _____

Practice Masters Level C
4.1 Using Matrices to Represent Data

For Exercises 1–4, let, $A = \begin{bmatrix} 3 & -0.5 & 0.04 \\ -2.1 & -0.1 & 6.7 \end{bmatrix}$, $B = \begin{bmatrix} -4 & 1 & -3 & 4 \\ & 2 & 2 & 3 \end{bmatrix}$,

$C = \begin{bmatrix} -1.4 & 0.3 & -0.2 \\ 1.4 & 4.2 & -6.6 \end{bmatrix}$, and $D = \begin{bmatrix} 0 & \frac{2}{3} & \frac{-2}{3} & \frac{-2}{3} \end{bmatrix}$.

1. Explain why you cannot find $A + B$. _____

Find the indicated matrix.

2. $2B + 6D$ 3. $4(A + C)$ 4. $-2(A + 3C) - 5\left(C - \frac{1}{2}A\right)$

5. Book check-out information for the Fulton Branch public library from January to June, 2000, is given in Matrix F below. Similar data is given for the Heatherwood Branch in Matrix H, and the Darlington Branch D.

$$\begin{array}{r} \text{Jan. Feb. Mar. Apr. May June} \\ \begin{array}{r}\text{Fiction}\\ \text{Non-fiction}\\ \text{Reference}\end{array} \begin{bmatrix} 325 & 375 & 290 & 225 & 220 & 250 \\ 150 & 105 & 110 & 110 & 90 & 95 \\ 30 & 35 & 30 & 40 & 50 & 45 \end{bmatrix} = F \end{array}$$

$$\begin{array}{r} \text{Jan. Feb. Mar. Apr. May June} \\ \begin{array}{r}\text{Fiction}\\ \text{Non-fiction}\\ \text{Reference}\end{array} \begin{bmatrix} 500 & 250 & 200 & 100 & 100 & 150 \\ 75 & 75 & 25 & 50 & 100 & 125 \\ 10 & 30 & 50 & 40 & 50 & 55 \end{bmatrix} = H \end{array}$$

$$\begin{array}{r} \text{Jan. Feb. Mar. Apr. May June} \\ \begin{array}{r}\text{Fiction}\\ \text{Non-fiction}\\ \text{Reference}\end{array} \begin{bmatrix} 100 & 120 & 110 & 125 & 130 & 135 \\ 50 & 40 & 45 & 30 & 25 & 35 \\ 10 & 10 & 15 & 20 & 15 & 20 \end{bmatrix} = D \end{array}$$

a. Which month is the busiest for each branch? _____

b. What is the overall busiest month for all three branches together? _____

c. Write the matrix that represents the difference between the Fulton Branch and the other two branches combined. _____

66   Practice Masters Levels A, B, and C   Algebra 2

NAME _____ CLASS _____ DATE _____

Practice Masters Level A
4.2 Matrix Multiplication

Find each product, if it exists.

1. $\begin{bmatrix} 2 & -3 \\ 4 & 1 \end{bmatrix} \begin{bmatrix} 3 & -2 \\ 2 & 0 \end{bmatrix}$

2. $\begin{bmatrix} 2 & 1 & -5 \\ 3 & 6 & -1 \end{bmatrix} \begin{bmatrix} 1 & -5 \\ -4 & -6 \\ 4 & -2 \end{bmatrix}$

3. $\begin{bmatrix} 3 \\ 0 \\ 1 \end{bmatrix} \begin{bmatrix} -1 & -2 & 10 \end{bmatrix}$

4. $\begin{bmatrix} 6 & 1 & -4 \end{bmatrix} \begin{bmatrix} 2 \\ 2 \\ -2 \end{bmatrix}$

5. $\begin{bmatrix} 3 & 0 & 1 \\ -2 & 4 & 5 \\ -5 & 6 & -7 \end{bmatrix} \begin{bmatrix} 2 & 1 \\ 3 & -1 \end{bmatrix}$

6. $\begin{bmatrix} -1 & 5 \\ 2 & -3 \\ 4 & -2 \end{bmatrix} \begin{bmatrix} 3 \\ 1 \end{bmatrix}$

7. $\begin{bmatrix} 2 & 4 & 3 \\ -1 & 1 & 5 \\ 1 & -2 & -3 \end{bmatrix} \begin{bmatrix} 0 & 1 & 4 \\ -2 & -1 & -3 \\ 10 & 4 & 3 \end{bmatrix}$

Let $A = \begin{bmatrix} 2 & 0 & 5 \\ -1 & 3 & 2 \\ 4 & -2 & -3 \end{bmatrix}$ and $B = \begin{bmatrix} 6 \\ 10 \\ -2 \end{bmatrix}$. Find each indicated matrix, if it exists.

8. AB

9. BA

10. A^2

11. $(AB)^2$

12. Fred earns money by doing yard work for Mr. Capri. The number of times that Fred mowed and weeded Mr. Capri's lawn is given in Martrix A. The amount that Fred earns for each task is given in Matrix B. Determine the total amount of money that Fred earned in May and June.

$\begin{array}{c} \text{May} \\ \text{June} \end{array} \begin{bmatrix} \text{Mow} & \text{Weed} \\ 3 & 2 \\ 4 & 3 \end{bmatrix} = A \qquad \begin{array}{c} \text{Mow} \\ \text{Weed} \end{array} \begin{bmatrix} \text{Earned} \\ \$10 \\ \$8 \end{bmatrix} = B$

Algebra 2   Practice Masters Levels A, B, and C

NAME _____ CLASS _____ DATE _____

Practice Masters Level B
4.2 Matrix Multiplication

Find each product, if it exists.

1. $\begin{bmatrix} 2 & 4 & 0 \\ -2 & 1 & 3 \\ 0 & 1 & 2 \end{bmatrix} \begin{bmatrix} -2 & -3 \\ 5 & 1 \\ 0 & 2 \end{bmatrix}$ _____

2. $\begin{bmatrix} 2 & 1 & 3 \end{bmatrix} \begin{bmatrix} 1 & 10 \\ 4 & -2 \\ -6 & -1 \end{bmatrix}$ _____

3. $\begin{bmatrix} 2 & 1 \end{bmatrix} \begin{bmatrix} 3 \\ -1 \\ 2 \end{bmatrix}$ _____

4. $\begin{bmatrix} 2 & 4 \\ -1 & 3 \end{bmatrix} \begin{bmatrix} 2 & 4 \\ -1 & 3 \end{bmatrix}$ _____

5. $\dfrac{1}{2} \begin{bmatrix} 1 & 3 & -1 \\ -2 & 0 & 5 \\ 4 & 2 & -3 \end{bmatrix} \begin{bmatrix} 5 & 2 \\ -4 & 1 \\ -2 & 3 \end{bmatrix}$ _____

6. $-3 \begin{bmatrix} 6 & 4 & -4 \end{bmatrix} \begin{bmatrix} 1 \\ -1 \\ -2 \end{bmatrix}$ _____

Let $M = \begin{bmatrix} 1 & 3 & -1 \\ 0 & 2 & 4 \\ 2 & -1 & 1 \end{bmatrix}$, $N = \begin{bmatrix} 2 & 5 & -3 \\ 4 & 2 & 2 \\ 1 & 0 & 0 \end{bmatrix}$, $P = \begin{bmatrix} 2 & 6 \\ 3 & 2 \\ 1 & -4 \end{bmatrix}$, and $I = \begin{bmatrix} 1 & 0 & 0 \\ 0 & 1 & 0 \\ 0 & 0 & 1 \end{bmatrix}$.

Find each indicated matrix, if it exists.

7. $M^2 P$

8. PN

9. MI

10. IP

11. M^3

12. $N^2 M$

13. The directed network represents four towns and whom they may call without long distance charges. Everyone in a town may call anyone in their own town without long distance charges.

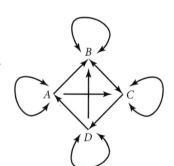

a. Create an adjacency matrix that represents the number of one-stage paths without long distance charges. _____

b. Create an adjacency matrix that represents the number of two-stage paths without long distance charges. _____

Practice Masters Level C

4.2 Matrix Multiplication

1. Matrix $Q = \begin{bmatrix} 3 & 2 & -4 & -5 \\ 0 & -3 & -1 & 4 \end{bmatrix}$ represents the vertices of quadrilateral ABCD. Let $J = \begin{bmatrix} -1 & 0 \\ 0 & -1 \end{bmatrix}$ and $K = \begin{bmatrix} -1 & 0 \\ 0 & 1 \end{bmatrix}$.

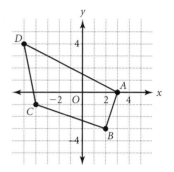

   a. Find JQ. _____   b. Find KQ. _____

   c. Graph $A'B'C'D'$, represented by JQ.   d. Graph $A'B'C'D'$, represented by KQ.

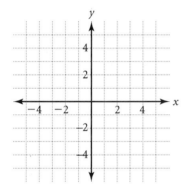
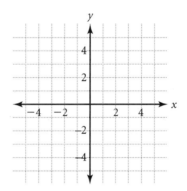

   e. What transformation did matrix J perform on ABCD? _____

   f. What transformation did matrix K perform on ABCD? _____

2. Matrix B shows the number of different types of baskets made by a college basketball player during the first four games of the season. Matrix P represents the points scored for each type of basket.

   2pt = 2 points
   3pt = 3 points
   FT = Free throws

$$\begin{array}{c} \\ \text{Game 1} \\ \text{Game 2} \\ \text{Game 3} \\ \text{Game 4} \end{array} \begin{array}{c} \text{2-pt 3-pt FT} \\ \begin{bmatrix} 10 & 4 & 6 \\ 8 & 3 & 10 \\ 9 & 3 & 8 \\ 12 & 1 & 2 \end{bmatrix} \end{array} = B \qquad \begin{array}{c} \text{Points} \\ \text{2pt} \\ \text{3pt} \\ \text{FT} \end{array} \begin{bmatrix} 2 \\ 3 \\ 1 \end{bmatrix} = P$$

   a. Find the product BP. _____

   b. What were the total points scored in all four games? _____

   c. What are the column and row labels of BP? _____

NAME _____ CLASS _____ DATE _____

Practice Masters Level A
4.3 The Inverse of a Matrix

Find the determinant, and tell whether the matrix has an inverse.

1. $\begin{bmatrix} 4 & 3 \\ 2 & 1 \end{bmatrix}$ _____

2. $\begin{bmatrix} -1 & 1 \\ 3 & 4 \end{bmatrix}$ _____

3. $\begin{bmatrix} 1 & -3 \\ -1 & 3 \end{bmatrix}$ _____

4. $\begin{bmatrix} -3 & 1 \\ -2 & 3 \end{bmatrix}$ _____

5. $\begin{bmatrix} \frac{1}{2} & 1 \\ 3 & 6 \end{bmatrix}$ _____

6. $\begin{bmatrix} 5 & \frac{-1}{3} \\ 30 & 2 \end{bmatrix}$ _____

Determine if each pair of matrices are inverses of each other.

7. $\begin{bmatrix} 1 & 2 \\ 1 & 3 \end{bmatrix} \begin{bmatrix} 3 & -2 \\ -1 & 1 \end{bmatrix}$ _____

8. $\begin{bmatrix} 2 & 3 \\ 1 & 1 \end{bmatrix} \begin{bmatrix} -1 & -3 \\ -1 & -2 \end{bmatrix}$ _____

Find the inverse matrix, if it exists. If the matrix does not exist, write *no inverse*.

9. $\begin{bmatrix} 2 & 3 \\ -2 & 1 \end{bmatrix}$ _____

10. $\begin{bmatrix} -1 & 4 \\ -2 & 5 \end{bmatrix}$ _____

11. $\begin{bmatrix} \frac{1}{3} & \frac{1}{2} \\ \frac{1}{2} & \frac{1}{4} \end{bmatrix}$ _____

12. $\begin{bmatrix} -5 & -6 \\ -3 & -1 \end{bmatrix}$ _____

13. $\begin{bmatrix} 1 & 1 \\ 0 & 1 \end{bmatrix}$ _____

14. $\begin{bmatrix} 0 & 0 \\ 1 & 0 \end{bmatrix}$ _____

15. $\begin{bmatrix} -2 & \frac{1}{3} \\ \frac{1}{2} & \frac{1}{12} \end{bmatrix}$ _____

16. $\begin{bmatrix} 0.02 & 0.05 \\ 0.1 & -0.7 \end{bmatrix}$ _____

Practice Masters Level B

4.3 The Inverse of a Matrix

Find the determinant, and tell whether the matrix has an inverse.

1. $\begin{bmatrix} -3 & -5 \\ \dfrac{4}{5} & -1 \end{bmatrix}$ _____

2. $\begin{bmatrix} \dfrac{1}{5} & \dfrac{1}{3} \\ \dfrac{1}{3} & \dfrac{1}{4} \end{bmatrix}$ _____

3. $\begin{bmatrix} 1 & 4 & -1 \\ -2 & 5 & 2 \\ -7 & 5 & -4 \end{bmatrix}$ _____

4. $\begin{bmatrix} 0.2 & 0.6 & -0.3 \\ 0.4 & 0.2 & 0.1 \\ 0.3 & -0.5 & 0.2 \end{bmatrix}$ _____

5. $\begin{bmatrix} 0 & 0 & 0 \\ -2 & 3 & 10 \\ -7 & 5 & 4 \end{bmatrix}$ _____

6. $\begin{bmatrix} 0 & -2 & 4 \\ 0 & 1 & -3 \\ 2 & 3 & -1 \end{bmatrix}$ _____

Find the inverse matrix, if it exists. If the matrix does not exist, write *no inverse*.

7. $\begin{bmatrix} \dfrac{1}{4} & 2 \\ \dfrac{1}{3} & -1 \end{bmatrix}$ _____

8. $\begin{bmatrix} -2 & \dfrac{4}{5} \\ 3 & & \\ \dfrac{3}{4} & \dfrac{1}{8} \end{bmatrix}$ _____

9. $\begin{bmatrix} \dfrac{1}{3} & \dfrac{1}{2} \\ \dfrac{1}{2} & \dfrac{1}{4} \end{bmatrix}$ _____

10. $\begin{bmatrix} 0 & \dfrac{-1}{4} \\ -3 & 0 \end{bmatrix}$ _____

11. $\begin{bmatrix} 1 & 0 & -1 \\ 2 & 3 & 5 \\ -4 & -3 & 2 \end{bmatrix}$ _____

12. $\begin{bmatrix} 3 & 0 & 0 \\ 0 & 5 & -2 \\ 1 & 0 & 2 \end{bmatrix}$ _____

13. $\begin{bmatrix} 2 & -3 & 5 \\ 6 & 4 & -2 \\ 1 & 7 & 9 \end{bmatrix}$ _____

14. $\begin{bmatrix} 14 & 60 & -71 \\ 100 & 12 & -50 \\ -24 & 37 & 48 \end{bmatrix}$ _____

15. $\begin{bmatrix} 2 & 12 & 4 \\ -4 & 6 & 10 \\ 14 & 24 & -8 \end{bmatrix}$ _____

16. $\begin{bmatrix} 0 & 0 & 0 \\ 7 & 1 & 0 \\ -2 & 0 & -2 \end{bmatrix}$ _____

Algebra 2   Practice Masters Levels A, B, and C

Practice Masters Level C

4.3 The Inverse of a Matrix

Find the determinant and the inverse of each matrix, if it exists.

1. $\begin{bmatrix} \frac{1}{4} & \frac{2}{3} \\ \frac{1}{3} & -1 \\ 6 \end{bmatrix}$ _____

2. $\begin{bmatrix} 0.002 & 0.04 \\ 0.05 & 0.003 \end{bmatrix}$ _____

3. $\begin{bmatrix} 12 & 10 & -11 \\ 9 & -8 & 10 \\ -10 & 5 & 10 \end{bmatrix}$ _____

4. $\begin{bmatrix} 10 & -10 & 10 \\ -10 & 10 & -10 \\ 10 & -10 & 10 \end{bmatrix}$ _____

5. $\begin{bmatrix} -1 & 0 & \frac{-3}{4} \\ \frac{1}{4} & \frac{-1}{2} & \frac{1}{3} \\ \frac{2}{3} & \frac{1}{2} & \frac{-1}{3} \end{bmatrix}$ _____

6. $\begin{bmatrix} 1.6 & -4.1 & 1.5 \\ 2.5 & 0 & -2.3 \\ -1.3 & -2.7 & 0 \end{bmatrix}$ _____

The determinant of a 4 × 4 matrix can be found using the method shown.

If $A = \begin{bmatrix} a & b & c & d \\ e & f & g & h \\ i & j & k & l \\ m & n & o & p \end{bmatrix}$, then $\det(A) = a\begin{vmatrix} f & g & h \\ j & k & l \\ n & o & p \end{vmatrix} - b\begin{vmatrix} e & g & h \\ i & k & l \\ m & o & p \end{vmatrix} + c\begin{vmatrix} e & f & h \\ i & j & l \\ m & n & p \end{vmatrix} - d\begin{vmatrix} e & f & g \\ i & j & k \\ m & n & o \end{vmatrix}$

Each 3 × 3 determinant in the above formula may then be expanded.

Use the formula above to find the determinant of each matrix.

7. $\begin{bmatrix} 0 & 2 & -3 & 0 \\ 1 & 0 & 2 & 4 \\ -2 & 1 & 0 & 3 \\ 2 & 2 & 1 & 1 \end{bmatrix}$ _____

8. $\begin{bmatrix} 1 & 3 & 0 & -2 \\ 4 & 5 & -1 & 2 \\ 3 & -3 & -4 & 1 \\ 2 & 2 & 5 & 0 \end{bmatrix}$ _____

Practice Masters Level A

4.4 Solving Systems with Matrix Equations

Write the matrix equation that represents each system.

1. $\begin{cases} 2x + y - 2z = 1 \\ 3x - 2y + 4z = -2 \\ x + 3y + 2z = 4 \end{cases}$

2. $\begin{cases} 6x + 2y - z = 1 \\ -2x + 3y + 2z = -2 \\ 3x - 5y = 2 \end{cases}$

Write the system of equations represented by each matrix equation.

3. $\begin{bmatrix} 2 & 1 & -3 \\ -1 & -1 & 2 \\ -2 & 4 & 5 \end{bmatrix} \begin{bmatrix} x \\ y \\ z \end{bmatrix} = \begin{bmatrix} 1 \\ -2 \\ 3 \end{bmatrix}$

4. $\begin{bmatrix} 3 & -1 & 1 \\ 2 & 3 & -2 \\ 4 & 2 & -3 \end{bmatrix} \begin{bmatrix} x \\ y \\ z \end{bmatrix} = \begin{bmatrix} -4 \\ 2 \\ 10 \end{bmatrix}$

Write the matrix equation that represents each system, and solve the system, if possible, by using the matrix equation.

5. $\begin{cases} 2x + y + 3z = -1 \\ x - 3y - 2z = 3 \\ 6x + 4y - 2z = -16 \end{cases}$ _____

6. $\begin{cases} x + y + 2z = 4 \\ 4x - y - z = 2 \\ -3x + y + z = 5 \end{cases}$ _____

7. $\begin{cases} x + y - z = 2 \\ 2x - 3y + z = 5 \\ 3x + 2y - 4z = 3 \end{cases}$ _____

8. $\begin{cases} 8x - 2y + 2z = 6 \\ -x - 2y - z = 0 \\ 3x + 7y - 3z = 6 \end{cases}$ _____

9. $\begin{cases} -x + \frac{1}{2}y - \frac{3}{2}z = \frac{-11}{2} \\ 3x - 4y - z = 0 \\ x - 2y - 2z = -5 \end{cases}$ _____

10. $\begin{cases} 3x + 10y - 4z = 3 \\ x + 2y - z = 1 \\ -3x + 4y + 2z = 4 \end{cases}$ _____

Algebra 2   Practice Masters Levels A, B, and C

Practice Masters Level B

4.4 Solving Systems with Matrix Equations

Write the matrix equation that represents each system.

1. $\begin{cases} 0.2x + 0.04y - 0.2z = -0.03 \\ 1.1x - 0.6y + 0.7z = 0.5 \\ 0.25x + 0.3y - 0.35z = 0.1 \end{cases}$

2. $\begin{cases} 2x - 3y = 1 \\ 4y + 2z = -3 \\ -x + z = 0 \end{cases}$

Write the system of equations represented by each matrix equation.

3. $\begin{bmatrix} 4 & 0 & -\frac{1}{2} \\ -3 & \frac{2}{3} & 0 \\ 0 & -\frac{1}{2} & -\frac{1}{2} \end{bmatrix} \begin{bmatrix} x \\ y \\ z \end{bmatrix} = \begin{bmatrix} 1 \\ \frac{2}{3} \\ -\frac{3}{4} \end{bmatrix}$

4. $\begin{bmatrix} -1 & 1 & -1 \\ -1 & -1 & -1 \\ 1 & -1 & 1 \end{bmatrix} \begin{bmatrix} x \\ y \\ z \end{bmatrix} = \begin{bmatrix} 0 \\ 0 \\ 0 \end{bmatrix}$

Write the matrix equation that represents each system, and solve the system, if possible, by using the matrix equation.

5. $\begin{cases} 2x + y - 2z = 1 \\ 3x - 2y + 4z = -2 \\ x + 3y + 2z = 4 \end{cases}$

6. $\begin{cases} 3x + y + z = 5 \\ 2x - y + z = 6 \\ y - 2z = 2 \end{cases}$

7. $\begin{cases} x + 2y + z = 4 \\ -x - 2y - z = \frac{-2}{3} \\ 3x - 3y + 4z = 9 \end{cases}$

8. $\begin{cases} 5x + 2y - 3z = -6 \\ x - 4y + 4z = -2 \\ x + y - 2z = 8 \end{cases}$

9. $\begin{cases} x - y = 6 \\ x + z = 4 \\ 2y - 3z = -4 \end{cases}$

10. $\begin{cases} 2x + y - 3z = 1 \\ -4x - 2y + 6z = -2 \\ x + \frac{1}{2}y - \frac{3}{2}z = \frac{1}{2} \end{cases}$

NAME _____ CLASS _____ DATE _____

Practice Masters Level C
4.4 Solving Systems with Matrix Equations

Solve each system of equations, if possible, by using a matrix equation. Give answers to the nearest hundredth, if necessary.

1. $\begin{cases} -0.03x - 0.007y = 0.5 \\ 1.02x + 2.5y = -0.04 \end{cases}$

2. $\begin{cases} -\dfrac{1}{4}x - \dfrac{2}{3}y = \dfrac{1}{5} \\ \dfrac{1}{12}x + \dfrac{4}{9}y = \dfrac{-1}{8} \end{cases}$

3. $\begin{cases} 4x - 2y + 4z = 5 \\ -3x + 3y - 5z = -7 \\ 6x + 2y - z = 14 \end{cases}$

4. $\begin{cases} 2w - 3x + 2y + z = 16 \\ 4w + 4x - 3y + 5z = 12 \\ -w - 2x + y - 3z = -3 \\ 3w + x - 4y + 2z = 17 \end{cases}$

5. An investment analyst invests money in three different mutual funds:

   - Mutual fund A contains 30% low-risk, 50% medium-risk and 20% high-risk stocks.

   - Mutual fund B contains 20% low-risk, 40% medium-risk and 40% high-risk stocks.

   - Mutual fund C contains 20% low-risk, 20% medium-risk and 60% high-risk stocks.

   A total of $6500 is invested in low-risk stocks, $10,000 in medium-risk stocks and $11,500 in high-risk stocks.

   a. Write a system of equations that represents the amount of money invested in each mutual fund.

   b. Write the matrix equation that represents the system.

   c. Solve the matrix equation.

Algebra 2   Practice Masters Levels A, B, and C   75

Practice Masters Level A
4.5 Using Matrix Row Operations

Write the augmented matrix for each system of equations.

1. $\begin{cases} 3x + 4y - 5z = 2 \\ -x + 2y + 2z = 10 \\ 2x - y + 3z = 1 \end{cases}$ _____

2. $\begin{cases} 5x - 4y = 8 \\ 2x - 3z = 1 \\ -3y + z = -2 \end{cases}$ _____

Find the reduced row-echelon form of each matrix.

3. $\begin{bmatrix} 1 & 0 & 3 & \vdots & -4 \\ 0 & 2 & 0 & \vdots & 6 \\ 1 & 0 & 5 & \vdots & 2 \end{bmatrix}$ _____

4. $\begin{bmatrix} 1 & 2 & 1 & \vdots & -4 \\ 0 & 1 & 2 & \vdots & 4 \\ -1 & 2 & 3 & \vdots & 5 \end{bmatrix}$ _____

5. $\begin{bmatrix} 3 & -1 & 2 & \vdots & 3 \\ 1 & 5 & 0 & \vdots & 27 \\ 0 & 1 & 1 & \vdots & 6 \end{bmatrix}$ _____

6. $\begin{bmatrix} 2 & 1 & 3 & \vdots & 9 \\ -3 & 5 & -2 & \vdots & -41 \\ 1 & 2 & -5 & \vdots & -16 \end{bmatrix}$ _____

Solve each system of equations by using the row-reduction method.

7. $\begin{cases} 3x - 5y = 3 \\ 2x + y = -11 \end{cases}$ _____

8. $\begin{cases} 2x + 3y = 8 \\ -3x - 4y = -11 \end{cases}$ _____

9. $\begin{cases} x + 4y - 2z = 6 \\ 3x - 5y + 2z = -17 \\ 5x + 3y + 3z = 5 \end{cases}$ _____

10. $\begin{cases} 4x + 2y + 3z = 3 \\ 3x + 2y + 2z = 2 \\ 2x + 3y + 3z = 5.5 \end{cases}$ _____

11. $\begin{cases} 2x - 3y + z = 10 \\ -x + 2y + z = -1 \\ 3x + y - 2z = -1 \end{cases}$ _____

12. $\begin{cases} -2x + 3y + 3z = 17 \\ x - 2y + 5z = 17 \\ 3x + 2y + 2z = 7 \end{cases}$ _____

NAME _____ CLASS _____ DATE _____

Practice Masters Level B
4.5 Using Matrix Row Operations

Find the reduced row-echelon form of each matrix.

1. $\begin{bmatrix} 2 & -2 & 3 & \vdots & -9 \\ 1 & -1 & 4 & \vdots & -12 \\ -4 & -3 & 5 & \vdots & -29 \end{bmatrix}$ _____

2. $\begin{bmatrix} 0.4 & -0.6 & -0.3 & \vdots & 3.8 \\ -0.2 & 0.1 & -0.1 & \vdots & -0.6 \\ 0.3 & -0.5 & 0.2 & \vdots & 2.2 \end{bmatrix}$ _____

3. $\begin{bmatrix} \frac{1}{2} & 2 & -4 & \vdots & 24 \\ 1 & \frac{-1}{2} & 3 & \vdots & \frac{-15}{2} \\ \frac{3}{2} & -1 & \frac{1}{2} & \vdots & \frac{-1}{2} \end{bmatrix}$ _____

Solve each system of equations by using the row-reduction method.

4. $\begin{cases} 2x - 2y = 6 \\ 3x - 4z = 3 \\ -x + 2y + 3z = -5 \end{cases}$ _____

5. $\begin{cases} 8x - 10y + 6z = 17 \\ -4x + 4y - 2z = -7 \\ 6x - 2y + 4z = 7 \end{cases}$ _____

6. $\begin{cases} \frac{1}{2}x - \frac{3}{4}y + \frac{1}{3}z = 6 \\ \frac{-1}{2}x + \frac{1}{4}y - \frac{1}{3}z = -2 \\ \frac{1}{3}x + \frac{5}{4}y + \frac{2}{3}z = -6 \end{cases}$ _____

7. $\begin{cases} 6.2x - 3.7z = 6.1 \\ -4y + 1.2z = 11.04 \\ -2.8x + 1.5y = -8.36 \end{cases}$ _____

8. $\begin{cases} 120x + 200y - 180z = 19{,}600 \\ -240x - 300y + 150z = -45{,}000 \\ 200x + 200y + 100z = 50{,}000 \end{cases}$

9. $\begin{cases} 0.02x - 0.03y + 0.05z = -0.019 \\ 0.01x + 0.04y - 0.02z = 0.014 \\ -0.3x + 0.02y + 0.04z = -0.074 \end{cases}$

Algebra 2   Practice Masters Levels A, B, and C

NAME _____ CLASS _____ DATE _____

Practice Masters Level C
4.5 Using Matrix Row Operations

Find the reduced row-echelon form of each matrix.

1. $\begin{bmatrix} 2 & 5 & -3 & \vdots & -5 \\ 3 & 1 & 8 & \vdots & -13 \\ -2 & 3 & 6 & \vdots & -25 \end{bmatrix}$

2. $\begin{bmatrix} \frac{1}{2} & -7 & 3 & \vdots & -32.2 \\ 5 & 4 & -1 & \vdots & -11.1 \\ 2 & -1 & 3 & \vdots & -8.2 \end{bmatrix}$

3. $\begin{bmatrix} 1 & 6 & 3 & -1 & \vdots & -54 \\ 3 & -2 & 5 & 2 & \vdots & 32 \\ -4 & 3 & 10 & 6 & \vdots & -128 \\ -5 & 2 & -2 & 5 & \vdots & -64 \end{bmatrix}$

4. $\begin{bmatrix} \frac{1}{2} & 3 & \frac{2}{3} & -4 & \vdots & -8\frac{5}{18} \\ \frac{1}{3} & 5 & -2 & \frac{1}{4} & \vdots & 6\frac{5}{6} \\ \frac{-2}{3} & 3 & 1 & -1 & \vdots & -10\frac{1}{6} \\ 6 & -2 & 5 & \frac{1}{2} & \vdots & 74\frac{2}{3} \end{bmatrix}$

Solve each system of equations by using the row-reduction method.

5. $\begin{cases} 6x + 3y - z = -1\frac{1}{2} \\ -14x + 10y - 5z = -27 \\ \frac{-1}{2}x - \frac{2}{3}y + 4z = 12\frac{1}{12} \end{cases}$

6. $\begin{cases} \frac{1}{2}x - \frac{2}{3}y + \frac{1}{4}z = -12\frac{7}{12} \\ \frac{3}{4}x - \frac{1}{4}y - \frac{1}{12}z = -11\frac{19}{36} \\ \frac{-1}{3}x + \frac{2}{3}y - \frac{5}{12}z = 10\frac{19}{36} \end{cases}$

7. $\begin{cases} w + 3x - 2y - 4z = -4 \\ 3w + 2x - 3y + 4z = 25 \\ -2w - 4x + 2y - 3z = -12 \end{cases}$

8. $\begin{cases} 0.1w + 0.2x - 0.1y + 0.3z = 0 \\ -0.2w + 0.4y - 0.2z = 0 \\ 0.4w + 0.3x + 0.2y - 0.5z = -0.07 \\ -0.5w + 0.7x - 0.1y + 0.1z = -0.31 \end{cases}$

NAME _____ CLASS _____ DATE _____

Practice Masters Level A

5.1 Introduction to Quadratic Functions

Show that each function is a quadratic function by writing it in the form $f(x) = ax^2 + bx + c$ and identifying *a*, *b*, and *c*.

1. $f(x) = (x + 2)(x - 3)$ _____

2. $f(x) = x(x + 3) - 2$ _____

3. $f(x) = (-x - 2)(x - 3)$ _____

4. $f(x) = (x + 3)^2$ _____

5. $f(x) = (2x - 3)(3x + 1)$ _____

Identify whether each function is quadratic. Use a graph to check your answers.

6. $f(x) = 4x^2 + 2x - 1$ _____   7. $f(x) = x^3 + x + 2$ _____

8. $f(x) = \dfrac{3x^3 + 4x^2}{2x}$ _____   9. $f(x) = -4x + 2$ _____

10. $f(x) = \dfrac{1}{x^2}$ _____   11. $f(x) = 4x + x^2 - 2$ _____

State whether the parabola opens up or down and whether the y-coordinate of the vertex is the minimum value or maximum value of the function.

12. $f(x) = 3x^2 + x - 2$ _____   13. $f(x) = 10 - 2x - x^2$ _____

14. $f(x) = (x + 3)(2 - x)$ _____   15. $f(x) = (3 - x)(1 - x)$ _____

Graph each function and give the approximate coordinates of the vertex.

16. $m(x) = x^2 + 1$ 17. $n(x) = -x^2 + x + 3$ 18. $p(x) = 3x^2 + 5x - 2$

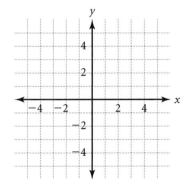   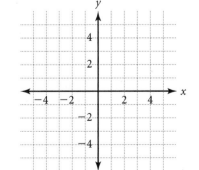   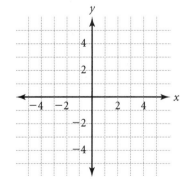

Algebra 2   Practice Masters Levels A, B, C   79

NAME _____ CLASS _____ DATE _____

Practice Masters Level B
5.1 Introduction to Quadratic Functions

Identify whether each function is a quadratic function. If the function is quadratic, identify *a*, *b* and *c* in the form $f(x) = ax^2 + bx + c$.

1. $f(x) = 3x^3 + 2x^2 + 5$ _____

2. $f(x) = x(5 - x) - 7$ _____

3. $f(x) = -(x + 2)(x - 3)$ _____

4. $f(x) = 3 - 2x$ _____

5. $f(x) = (x + 6)^3$ _____

6. $f(x) = (x - 3)^2 + 3x$ _____

State whether the parabola opens up or down. State whether the *y*-coordinate of the vertex is the minimum value or maximum value of the function. Then graph the function, and give the approximate coordinates of the vertex

7. $g(x) = 2x^2 - 5x + 3$

8. $h(x) = -3x^2 + 2x + 1$

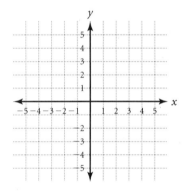
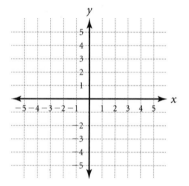

9. $g(x) = x(x - 3x) - 10$

10. $h(x) = -(x + 2)(x + 3)$

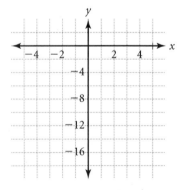
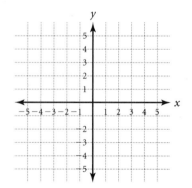

80   Practice Masters Levels A, B, C   Algebra 2

NAME _____ CLASS _____ DATE _____

Practice Masters Level C
5.1 Introduction to Quadratic Functions

Identify whether each function is a quadratic function. If the function is quadratic, identify *a*, *b* and *c* in the form $f(x) = ax^2 + bx + c$.

1. $f(x) = \dfrac{(x-3)^4}{(x-3)^2}$ _____

2. $f(x) = 4x + 2x(x+3) - 2$ _____

3. $f(x) = \dfrac{x^2}{5x+6}$ _____

4. $f(x) = \dfrac{3x^3 + 5x^2 - 6x}{2x}$ _____

5. $f(x) = (x-6)^3$ _____

6. $f(x) = (x-2)^2 + 5x$ _____

State whether the parabola opens up or down. State whether the *y*-coordinate of the vertex is the minimum value or maximum value of the function. Then graph the function, and give the approximate coordinates of the vertex.

7. $g(x) = \dfrac{1}{2}x^2 - \dfrac{3}{2}x + 2$

8. $h(x) = 4x^2 - \dfrac{1}{3}x - 1$

9. $g(x) = \dfrac{1}{3} + \dfrac{2}{3}x - \dfrac{2}{3}x^2$

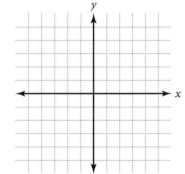
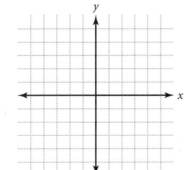
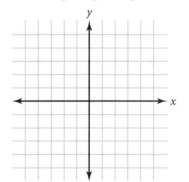

10. $g(x) = 0.01x^2 + 0.03x - 0.05$

11. $h(x) = \dfrac{-1}{3}(x+2)^2 - 4$

12. $m(x) = \dfrac{(x+2)^3}{x+2} - 3$

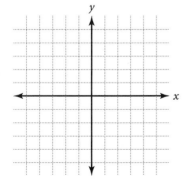
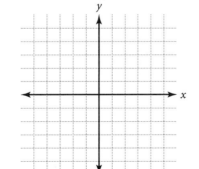
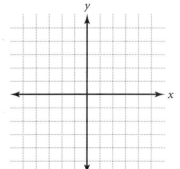

Algebra 2   Practice Masters Levels A, B, C   81

NAME _____ CLASS _____ DATE _____

Practice Masters Level A
5.2 Introduction to Solving Quadratic Equations

Solve each equation. Give exact solutions.

1. $x^2 - 49 = 0$

2. $4x^2 = 64$

3. $(x - 2)^2 = 100$

4. $2x^2 + 5 = 167$

5. $(x + 3)^2 = 25$

6. $3x^2 - 7 = 68$

Solve each equation. Give exact solutions. Then approximate each solution to the nearest hundredth.

7. $2x^2 - 5 = 12$ _____

8. $x^2 + 10 = 9$ _____

9. $4x^2 - 4 = 4$ _____

10. $3x^2 + 5 = 30$ _____

Find the unknown length in each right triangle, to the nearest tenth.

11.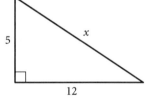

12.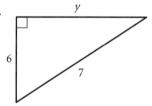

Find the missing side length in right triangle XYZ. Give answers to the nearest tenth, if necessary.

13. $x = 8$ and $y = 15$ _____

14. $y = 25$ and $z = 7$ _____

15. $x = 10$ and $z = 12$ _____

16. $x = 4$ and $y = 8$ _____

17. $x = 0.3$ and $z = 0.4$ _____

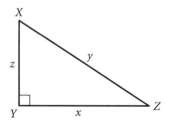

82   Practice Masters Levels A, B, C   Algebra 2

NAME _____ CLASS _____ DATE _____

Practice Masters Level B
5.2 Introduction to Solving Quadratic Equations

Solve each equation. Give exact solutions. Then approximate each solution to the nearest hundredth if necessary.

1. $\frac{1}{3}x^2 = \frac{4}{27}$ _____

2. $\frac{-3}{4}y^2 = \frac{-3}{64}$ _____

3. $\frac{-1}{2}x^2 - 5 = -5\frac{1}{8}$ _____

4. $4 - 3t^2 = -48.92$ _____

5. $-2(t^2 + 4) = -87.38$ _____

6. $9 + 4t^2 = 9$ _____

7. $4(x^2 - 2) + 3 = 11$ _____

8. $\frac{1}{2}(x^2 + 3) - 2 = 10$ _____

Find the missing side length in right triangle *XYZ*.
Give answers to the nearest tenth, if necessary.

9. $x = 2.6$ and $y = 6.4$ _____

10. $x = 10.2$ and $z = 0.8$ _____

11. $y = \frac{2}{3}$ and $z = \frac{1}{4}$ _____

12. $y = 5\frac{1}{2}$ and $z = 3\frac{1}{4}$ _____

13. $x = 1\frac{2}{5}$ and $z = 1\frac{2}{5}$ _____

14. $y = 4.5$ and $x = z$ _____

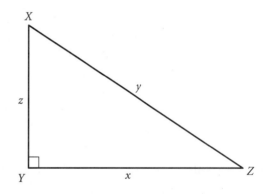

15. The bottom of a ladder is 6 feet from the base of a building and reaches a height of 12 feet on the side of the building. How long is the ladder? _____

16. A circle has an area of 25π square centimeters. What is the length of the radius of the circle? _____

17. A square has a diagonal of 13 inches. What are the dimensions of the square? _____

Algebra 2

NAME _____ CLASS _____ DATE _____

Practice Masters Level C

5.2 Introduction to Solving Quadratic Equations

Solve each equation. Give exact solutions. Then approximate each solution to the nearest hundredth, if necessary.

1. $(t - 7)^2 - 2 = 48.41$ _____

2. $-2(t - 6)^2 + 1 = 0$ _____

3. $-2(3x + 1)^2 + 2 = -16$ _____

4. $\dfrac{-1}{3}(5x - 3)^2 - 1 = -25$ _____

Write a quadratic equation for each pair of solutions.

5. 4 and -4

6. $\sqrt{15}$ and $-\sqrt{15}$

7. 6.48 and -6.48

_____   _____   _____

Solve.

8. A circle has an area of 100 square centimeters. What is the length of the diameter? _____

9. A 12-foot ladder is placed against a building such that the foot of the ladder is 3.5 feet from the base of the building. How far up the side of the building will the ladder reach? _____

10. Where will the base of the ladder in Exercise 9 be placed if a 6-foot tall man needs to reach a height of 15 feet by standing at the top of the 12-foot ladder? _____

11. In Fallow County, road systems are designed using a large grid system. Roads can run either due north and south or due east and west. Suppose a couple left their house and traveled 4 miles north and 10 miles west. How far from home are they? _____

12. The height of a falling object, in meters above ground, is given by $h(t) = -4.9t^2 + s$, where t is the time in seconds that the object has fallen, and s is the starting height in meters. If a person jumps off a 3-meter platform into a diving pool, how long will it take for the person to hit the water? _____

13. As part of their half-time show, the members of the band must walk diagonally from one corner of the football field to the opposite corner. If the field is 100 yards long and 35 yards wide, how far must the band members march? _____

84   Practice Masters Levels A, B, C   Algebra 2

Practice Masters Level A
5.3 Factoring Quadratic Expressions

Factor each expression.

1. $-12 - 3x$ _____
2. $-5x^2 - 10x$ _____
3. $6x^3 - 18x^2$ _____
4. $2x(3x + 4) - 3(3x + 4)$ _____
5. $-2x^2 - 5x$ _____
6. $-2x(x + 7) - (x + 7)$ _____

Factor each quadratic expression.

7. $x^2 + 5x + 4$ _____
8. $x^2 + x - 6$ _____
9. $x^2 + 3x - 18$ _____
10. $x^2 - 12x + 35$ _____
11. $x^2 - x - 2$ _____
12. $x^2 + 11x + 10$ _____
13. $x^2 - 25x + 100$ _____
14. $x^2 + 10x - 24$ _____
15. $x^2 - 12x + 32$ _____
16. $x^2 + 21x + 54$ _____
17. $2x^2 - x - 15$ _____
18. $3x^2 + 11x - 4$ _____
19. $8x^2 + 10x - 3$ _____
20. $25x^2 + 10x - 3$ _____
21. $9x^2 - 15x - 14$ _____
22. $12x^2 + 17x - 5$ _____

Solve each equation by factoring and applying the Zero-Product Property.

23. $x^2 - 225 = 0$ _____
24. $x^2 - 5x + 6 = 0$ _____
25. $x^2 + 7x = -10$ _____
26. $x^2 = 14x + 5$ _____
27. $-x^2 + 3x = -4$ _____
28. $x^2 - 8x = -16$ _____
29. $2x^2 - 32 = 0$ _____
30. $2x^2 + 14x + 20 = 0$ _____
31. $x^2 - 64 = 0$ _____
32. $4x^2 - 16 = 0$ _____

Algebra 2   Practice Masters Levels A, B, C

Practice Masters Level B
5.3 Factoring Quadratic Expressions

Factor each expression.

1. $3x^2 + 18x + 27$ _____
2. $x^2 + 8x - 20$ _____
3. $20x^2 + 7x - 6$ _____
4. $x^2 - 16x + 39$ _____
5. $x^2 + 48x + 47$ _____
6. $-8x^4 - 10x^3$ _____
7. $-1.5x^2 - 2.5x$ _____
8. $-7x^4 - 25$ _____
9. $6x^2 + 7x - 3$ _____
10. $x^2 + 0.3x + 0.2$ _____
11. $12x^2 - 47x + 40$ _____
12. $4x^2 - 4x - 120$ _____
13. $18x^2 + 33x - 216$ _____
14. $-5x^3 - 15x^2 + 140x$ _____
15. $x^4 - 1$ _____
16. $-3x(x + 2) - 8(x + 2)$ _____

Solve each equation by factoring and applying the Zero-Product Property.

17. $2x^2 + 8x = 0$ _____
18. $3x^2 - 1200 = 0$ _____
19. $4x^2 - 14x + 6 = 0$ _____
20. $12x^2 + x = 6$ _____
21. $18x^2 = -18x - 4$ _____
22. $x^2 - 625 = 0$ _____
23. $x^2 + 4x = 21$ _____
24. $120x^2 - 34x - 21 = 0$ _____
25. $\frac{1}{4}x^2 + 4x + 16 = 0$ _____
26. $\frac{1}{9}x^2 + \frac{1}{3}x + \frac{1}{4} = 0$ _____

86 Practice Masters Levels A, B, C Algebra 2

NAME _____ CLASS _____ DATE _____

Practice Masters Level C
5.3 Factoring Quadratic Expressions

Factor each expression.

1. $x^4 + 2x^2 + 1$ _____ 2. $x^4 - 16$ _____ 3. $x^8 - 256$ _____

Use factoring and the Zero-Product Property to solve each equation.

4. $-12x^2 - 8x + 4 = 0$ _____ 5. $10x - 3 = 8x^2$ _____

6. $x^2 - 0.04 = 0$ _____ 7. $2x^2 - 0.6x - 0.08 = 0$ _____

8. $6x^2 + 99 = 49x$ _____ 9. $5x^2 - \frac{5}{2}x + \frac{1}{5} = 0$ _____

10. $x^3 - 49x = 0$ _____ 11. $x^3 = -3x^2 - 2x$ _____

Use graphing to find the zeros of each function, to the nearest hundredth.

12. $2x^2 - 3x = 7$ _____ 13. $-4x^2 + 3x = -2$ _____

14. $9 - 2x = -3x^2$ _____ 15. $\frac{1}{4}x^2 - \frac{1}{3}x - \frac{1}{4} = 0$ _____

16. Joel and Wyatt toss a baseball. The height in feet, of the baseball, above the ground is given by $h(t) = -16t^2 + 55t + 6$, where t represents the time in seconds after the ball is thrown. How long is the ball in the air? _____

17. A rocket is launched into the air. The height in feet, of the rocket, above the ground is given by $h(t) = -4.9t^2 + 220t$, where t represents the time in seconds after the launch.

   a. How long is the rocket in the air? _____

   b. When is the rocket at its highest point? _____

   c. How high is the rocket at its highest point? _____

18. A 15-inch by 28-inch poster is to be framed using a border of uniform width. To make the framing visually appealing, the border should have an area equal to 60% of the area of the poster. How wide should the border be? _____

Algebra 2 Practice Masters Levels A, B, C

Practice Masters Level A

5.4 Completing the Square

Complete the square for each quadratic expression to form a perfect-square trinomial. Then write the new expression as a binomial squared.

1. $x^2 + 12x$

2. $x^2 - 14x$

3. $x^2 + 26x$

4. $x^2 + 5x$

5. $x^2 - 3x$

6. $x^2 + x$

Solve each equation by completing the square. Give exact solutions. Then give answers rounded to the nearest hundredth, if necessary.

7. $x^2 + 8x + 2 = 0$

8. $x^2 - 10x + 15 = 0$

9. $x^2 + 6x = 0$

10. $x^2 - 5x = 0$

11. $x^2 - 3x + 9 + 6 = 0$

12. $x^2 + 5x + 2 = 0$

Write each quadratic function in vertex form. Give the coordinates of the vertex and the equation of the axis of symmetry.

13. $f(x) = x^2 + 4x + 3$

14. $f(x) = x^2 - 6x - 16$

15. $f(x) = x^2 + 2x + 5$

16. $f(x) = x^2 + 10x + 1$

NAME _____ CLASS _____ DATE _____

Practice Masters Level B
5.4 Completing the Square

Solve each equation by completing the square. Give exact solutions.
Then round answers to the nearest hundredth, if necessary.

1. $x^2 + 6x - 10 = 0$

2. $x^2 = 5x + 5$

3. $2x^2 + 6x - 7 = 0$

4. $3x^2 - 9x + 5 = 0$

5. $2x^2 - 5x + 1 = 0$

6. $4x^2 + 7x - 10 = 0$

7. $x^2 - 7x - 20 = 0$

8. $2x^2 - 3x + 6 = 0$

9. $3x^2 = 4x + 15$

10. $-2x^2 + 5x + 21 = 0$

11. $5x^2 - x - 1 = 0$

12. $x^2 + 0.02x = 0.07$

Write each quadratic function in vertex form. Give the
coordinates of the vertex and the equation of the axis of
symmetry. Then describe the transformations from $f(x) = x^2$ to h.

13. $h(x) = x^2 - 3x - 15$

14. $h(x) = 2x^2 + 2x + 3$

15. $h(x) = -3x^2 - 6x + 10$

16. $h(x) = -5x^2 + x$

17. $h(x) = x^2 + 9x$

18. $h(x) = -10x - x^2$

Algebra 2   Practice Masters Levels A, B, C   89

NAME _____ CLASS _____ DATE _____

Practice Masters Level C
5.4 Completing the Square

Solve each equation by completing the square. Give exact solutions. Then round answers to the nearest hundredth, if necessary.

1. $-4x^2 + 2x + 5 = 0$

2. $6x^2 - 4x - 3 = 0$

3. $2x^2 + 5x - 1 = 0$

4. $-2x^2 - 9x + 13 = 0$

5. $\dfrac{1}{2}x^2 + 3x = -7$

6. $\dfrac{-1}{2}x^2 + 4x = -3$

7. $\dfrac{1}{3}x^2 + \dfrac{2}{3}x = 4$

8. $\dfrac{-1}{3}x^2 + 5x = 6$

9. $-2x^2 + 6x + 1$

10. $5x^2 = 6x - 10$

11. $\dfrac{1}{2}x^2 + 2x = 5$

12. $\dfrac{-1}{2}x^2 = 3x - 2$

Write each quadratic function in vertex form. Give the coordinates of the vertex. Then describe the transformation from $f(x) = x^2$ to h.

13. $h(x) = \dfrac{-1}{2}x^2 + \dfrac{1}{2}x - 1$

14. $h(x) = \dfrac{1}{4}x^2 - 2x + 1$

15. Write an equation for the quadratic function that has a vertex of $(3, 1)$ and contains the point $(2, 6)$.

16. Write an equation for the quadratic function that has a vertex of $(-2, 5)$ and contains the point $(3, -1)$.

Practice Masters Level A

5.5 The Quadratic Formula

Use the quadratic formula to solve each equation. Give exact solutions.

1. $x^2 + 7x = -10$

2. $-4x^2 + 5x - 1 = 0$

3. $8x^2 - 10x + 3 = 0$

4. $12x^2 + x = 6$

5. $8x = -2x^2$

6. $x^2 - 3x + 9 = 0$

7. $6x^2 + 7x - 3 = 0$

8. $x^2 = -5x - 2$

9. $2x^2 + 3x - 7 = 0$

10. $3x^2 + 5x + 1 = 0$

11. $x^2 - 9x + 15 = 0$

12. $5x^2 - 5x + 9 = 0$

For each quadratic function, write the equation for the axis of symmetry, and find the coordinates of the vertex.

13. $y = x^2 + 4x + 3$

14. $y = -3x^2 - 6x + 10$

15. $y = x^2 + 10x + 1$

16. $y = 2x^2 + 4x + 1$

17. $y = \frac{1}{2}x^2 - x + 4$

18. $y = -2x^2 + 5x + 2$

Algebra 2 — Practice Masters Levels A, B, C

Practice Masters Level B
5.5 The Quadratic Formula

Use the quadratic formula to solve each equation. Give exact solutions. Then round answers to the nearest hundredth, if necessary.

1. $6x^2 - 49x = -99$

2. $2x^2 = 0.6x + 0.08$

3. $9x + 6 = -x^2$

4. $(x - 3)^2 = 5$

5. $(x + 7)(x - 4) = 3$

6. $-5x^2 - 10x - 51 = 0$

7. $-6x^2 + 19 = 0$

8. $6x^2 + 11x = 13$

9. $2x(x - 3) - 7 = 10$

10. $-3(x + 2)(x - 4) = -4$

For each quadratic function, write the equation for the axis of symmetry, and find the coordinates of the vertex.

11. $y = x^2 + 8x$

12. $y = -2x^2 + 8x - 4$

13. $y = 5x^2 + 2x - 12$

14. $y = x^2 - 0.16x - 13.0336$

15. $y = 0.2x^2 - 0.1x + 0.3$

16. $y = x^2 - 5x + 6.29$

Practice Masters Level C
5.5 The Quadratic Formula

Use the quadratic formula to solve each equation. Give exact solutions. Then round answers to the nearest hundredth, if necessary.

1. $\dfrac{1}{2}x^2 + 4x = \dfrac{3}{4}$

2. $\dfrac{3}{4}x^2 = \dfrac{1}{2}x + \dfrac{1}{3}$

3. $(0.02x - 3)(0.1x + 2) = 0$

4. $\left(\dfrac{1}{2}x + \dfrac{3}{2}\right)\left(\dfrac{1}{3}x - \dfrac{1}{2}\right) = 2$

For each quadratic function, find the exact coordinates of the vertex.

5. $y = -x\left(\dfrac{1}{2}x - 12\right) - \dfrac{1}{10}$

6. $y = \dfrac{1}{3}x^2 + \dfrac{3}{4}x - \dfrac{1}{12}$

7. $y = \left(\dfrac{1}{4}x - 2\right)\left(x + \dfrac{1}{3}\right)$

8. $y = -3\left(x - \dfrac{1}{5}\right)\left(x + \dfrac{2}{3}\right)$

9. A ball is thrown in an upward direction off of a platform that is 45 feet high with an initial velocity of 60 feet per second. The height, in feet, of the ball at time t is given by $h(t) = -16t^2 + 60t + 45$. The time, t, is given in seconds.

   a. At what time after the start is the ball 45 feet high again?

   b. Find the time when the ball hits the ground.

10. A pasture is to be enclosed with 200 feet of fencing on three sides and a barn on the fourth side.

    a. Write a quadratic equation to model the area of the pasture.

    b. What is the maximum area that can be enclosed?

Algebra 2

Practice Masters Level A
5.6 Quadratic Equations and Complex Numbers

Identify the real and imaginary parts of each complex number.

1. $4 + 2i$ _____

2. $-6 - 3i$ _____

3. $4i$ _____

4. -5 _____

Simplify.

5. $\sqrt{-25}$ _____

6. $\sqrt{-17}$ _____

7. $(4i)^2$ _____

8. $(-5i)^2$ _____

Find the discriminant, and determine the number of real solutions. Then solve.

9. $4x^2 + 7x + 3 = 0$

10. $x^2 - 7x + 2 = 0$

11. $x^2 + 3x - 5 = 0$

12. $-2x^2 - 5x = 10$

Perform the indicated operations, and simplify.

13. $(-4 + 2i) + (6 - 3i)$

14. $(2 + 5i) - (5 + 3i)$

15. $(7 + 7i) - (-6 - 2i)$

Multiply, and graph the product in the complex plane.
Then find the conjugate and the absolute value of the product.

16. $(2 + i)(2 - i)$

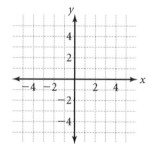

17. $(-2 + i)(3 + i)$

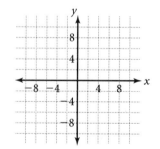

94   Practice Masters Levels A, B, C   Algebra 2

Practice Masters Level B
5.6 Quadratic Equations and Complex Numbers

Find the discriminant, and determine the number of real solutions. Then solve.

1. $3x^2 - 75 = 0$
2. $3x^2 + \frac{1}{3}x = \frac{1}{4}$
3. $x^2 + \frac{2}{3}x + \frac{1}{9} = 0$

4. $-2x = -3x^2 - 5$
5. $-2x^2 = 0.3x - 4$
6. $0.6x^2 - 2.8x = -0.5$

Perform the indicated operations, and simplify.

7. $(-6 - i) + \left(\frac{1}{2} + \frac{1}{2}i\right)$
8. $-9i\left(\frac{1}{3}i + 2\right)$

9. $(0.7 + 0.4i) + (2.3 - 0.6i)$
10. $(3 - 2i)(-7 - 3i)$

11. $\dfrac{3 + 2i}{i}$
12. $\dfrac{16}{3i}$

13. $\dfrac{4 + i}{6 + i}$
14. $\dfrac{2 - 3i}{4 - 5i}$

Graph each product and its conjugate in the complex plane. Then find the absolute value of the product.

15. $(2 + i)(2 - i)$
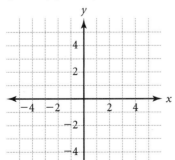

16. $(6 + 2i)(3 + 4i)$
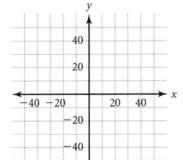

17. $i(6 - i)$
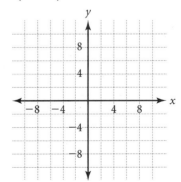

Algebra 2

NAME _____ CLASS _____ DATE _____

Practice Masters Level C
5.6 Quadratic Equations and Complex Numbers

Find the discriminant, and determine the number of real solutions. Then solve.

1. $(x - 3)(x + 2) = 12$

2. $-4x(x - 1) = -9$

3. $3x^2 = \dfrac{x + 3}{4}$

4. $x(x + 3) = -4$

Perform the indicated operations, and simplify. Then name the conjugate of your answer.

5. $\dfrac{1}{2}(3 - 2i) - \dfrac{1}{3}(6 + 5i)$

6. $(2 + i\sqrt{5})(-1 + i\sqrt{5})$

7. $\dfrac{15 + 5i}{25 - 5i}$

8. $i(-5i^2)^2$

9. $i^4(6 - 3i) + i^2(2 + i)$

10. $\dfrac{2 - 7i}{3 + 11i}$

11. $(i\sqrt{17} + 3)^2$

12. $(14 + i\sqrt{21})^2$

Evaluate. Then sketch a diagram that shows the absolute value.

13. $\left|\dfrac{1}{2} - 4i\right| =$ _____

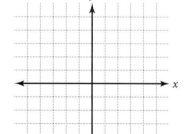

14. $|0.7 - 0.2i| =$ _____

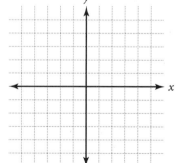

Practice Masters Level A
5.7 Curve Fitting with Quadratic Models

Solve a system of equations in order to find a quadratic function that fits each set of data points exactly.

1. $(2, 5), (3, 12), (0, -3)$

2. $(1, 3), (2, 2), (4, 6)$

3. $(-1, 4), (0, 4), (1, 6)$

4. $(-2, 1), (2, -7), (4, -23)$

5. $(-3, 27), (0, 6), (1, 11)$

6. $(-4, 75), (-2, 19), (-1, 6)$

7. Refer to the pattern of dots below, in which each set of dots is formed by making a square that is one row of dots wider and longer than the previous one.

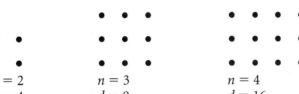

$n = 1$ $\quad$ $n = 2$ $\quad$ $n = 3$ $\quad$ $n = 4$
$d = 1$ $\quad$ $d = 4$ $\quad$ $d = 9$ $\quad$ $d = 16$

a. Find a quadratic function for the number of dots, d, in terms of the number of dots on each side, n.

b. How many dots are in a square with 9 dots on each side?

c. How many dots are on the side of a square with a total number of 441 dots?

8. A rocket is shot into the air. The table shows the height, y, of the rocket x seconds after it is thrown.

a. Use regression to find a quadratic model for the data.

Time (in seconds)	Height (in feet)
1	14
4	50
9	65
12	52

b. What was the maximum height reached by the ball?

c. How long did it take the ball to reach its maximum height?

Algebra 2

NAME _____ CLASS _____ DATE _____

Practice Masters Level B
5.7 Curve Fitting with Quadratic Models

Solve a system of equations in order to find a quadratic function that fits each set of data points exactly.

1. $(10, -660), (15, -1435), (20, -2510)$

2. $(-10, 552), (-20, 2092), (-30, 4632)$

3. $\left(\frac{1}{2}, -3\right), (2, 27), (4, 95)$

4. $(0, 150), \left(\frac{1}{2}, 99\frac{3}{4}\right), (1, 49)$

5. $\left(-2, \frac{3}{2}\right), \left(2, \frac{11}{2}\right), \left(4, \frac{27}{2}\right)$

6. $(-3, -1), (0, 4), (3, 3)$

7. $(-8, 5), (-4, -4), (4, 2)$

8. $(2, 3), (4, 6.4), (6, 11.4)$

9. $(-1, -1.1), (0, -0.6), (1, -0.9)$

10. $(-2, 11.1), (-1, 1.6), (1, -1.2)$

A baseball player throws a ball. The table shows the height, h, of the ball, t, seconds after it is thrown.

Time (in seconds)	Height (in feet)
1	21
3	25
5	10
6	1

11. Use regression to find a quadratic model for the data.

12. What was the maximum height reached by the ball?

13. How long did it take the ball to reach maximum height?

14. Determine the height of the ball 1.5 seconds after it was thrown.

15. Determine how many seconds it took for the ball to hit the ground.

NAME _____ CLASS _____ DATE _____

Practice Masters Level C
5.7 Curve Fitting with Quadratic Models

The following table lists the number of sport utility vehicles (SUVs) sold in the United States in certain years.

Year	Number (in millions)
1988	1.0
1991	0.8
1998	2.8

1. Find a quadratic function to model the data. _____

2. What was the minimum number of SUVs sold during the 1988-1998 time period? _____

3. What was the number of SUVs sold in 1997? _____

4. Predict the number of SUVs sold in 2010. _____

5. What year most likely had sales of about 1.5 million? _____

The population of Milwaukee is listed in the table. Use the data to answer Exercises 6–8.

Year	Population
1900	285,000
1960	637,400
1980	636,300
1990	630,000
1998	580,000

6. Use regression to find a quadratic model for population in terms of years from 1900. _____

7. Estimate the population in 2010. _____

8. What year was the population approximately 400,000? _____

The number of U.S. cellular telephone subscribers is listed in the table. Use the data to answer Exercises 11–13.

Year	Number of Subscribers
1987	1,230,855
1990	5,283,055
1992	11,032,753
1996	44,042,992
1998	69,209,321

9. Use regression to find a quadratic model for population in terms of years from 1985. _____

10. According to this model, what was the number of subscribers in 1988? _____

Algebra 2   Practice Masters Levels A, B, C   99

NAME _____ CLASS _____ DATE _____

Practice Masters Level A
5.8 Solving Quadratic Inequalities

Solve each inequality. Graph the solution on a number line.

1. $x^2 - x - 2 < 0$ _____

2. $x^2 - 7x + 10 \geq 0$ _____

3. $x^2 - 8x + 7 \leq 0$ _____

4. $x^2 + x - 6 \leq 0$ _____

5. $x^2 + 6x + 9 < 0$ _____

6. $x^2 + 20x + 100 > 0$ _____

Graph each inequality.

7. $y \geq (x - 2)^2 + 3$

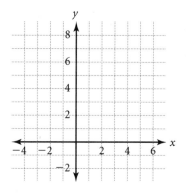

8. $y < x^2 - 7x + 10$

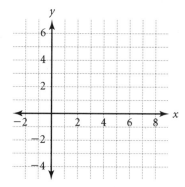

9. The cost, C, for producing a pair of gloves is based on the selling price, p. The cost function is given by $C(p) = 300p + 210$ and the revenue function is given by $R(p) = -10p^2 + 400p$.

   a. Use the cost function and the revenue function to determine the profit function, $P(p)$. _____

   b. At what price do the gloves need to be sold in order to make a profit? _____

Practice Masters Level B
5.8 Solving Quadratic Inequalities

Solve each inequality. Graph the solution on a number line.

1. $2x^2 + x - 3 > 0$ _____   2. $6x^2 - 19x + 10 < 0$ _____

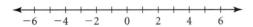   

3. $12x^2 + 5x - 3 \leq 0$ _____   4. $45x^2 + 16x - 16 \geq 0$ _____

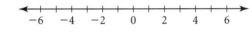   

5. $50x^2 + 25x + 2 > 0$ _____   6. $x^2 - x + \dfrac{2}{9} \leq 0$ _____

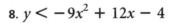

Graph each inequality.

7. $y \geq 6x^2 + 7x - 5$ 8. $y < -9x^2 + 12x - 4$

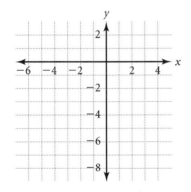   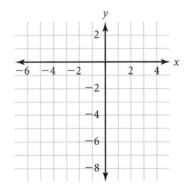

9. A golf club wholesaler uses the table to determine the revenue for an order of golf clubs from various retail outlets.

   a. Determine the revenue function.

   _____

   b. What is the maximum revenue?

   _____

   c. How many golf clubs would result in maximum revenue?

   _____

Number of clubs purchased	Price per club	Revenue per order
1	$112	$112
2	$110	$220
3	$108	$324
4	$106	$424
5	$104	$520

Algebra 2 — Practice Masters Levels A, B, C

NAME _____ CLASS _____ DATE _____

Practice Masters Level C

5.8 Solving Quadratic Inequalities

Solve each inequality. Graph the solution on a number line.

1. $3x^2 - x < 8 - 4x$ _____

2. $\frac{1}{2}x^2 + \frac{1}{2}x - \frac{3}{2} \geq 0$ _____

   ◄─┼─┼─┼─┼─┼─┼─┼─┼─┼─┼─┼─►
   $-6\ -4\ -2\ \ 0\ \ 2\ \ 4\ \ 6$

   ◄─┼─┼─┼─┼─┼─┼─┼─┼─┼─┼─┼─►
   $-6\ -4\ -2\ \ 0\ \ 2\ \ 4\ \ 6$

3. $0.7x^2 + 0.4x \leq 0.6$ _____

4. $3.24x - 2.32x^2 < 6.92 + 4.74x$ _____

   ◄─┼─┼─┼─┼─┼─┼─┼─┼─┼─┼─┼─►
   $-6\ -4\ -2\ \ 0\ \ 2\ \ 4\ \ 6$

   ◄─┼─┼─┼─┼─┼─┼─┼─┼─┼─┼─┼─►
   $-6\ -4\ -2\ \ 0\ \ 2\ \ 4\ \ 6$

Graph each inequality.

5. $y \leq \frac{-1}{3}x^2 - \frac{2}{3}x + \frac{4}{3}$

6. $y \leq \frac{3}{5}x^2 - \frac{1}{6}x - \frac{1}{9}$

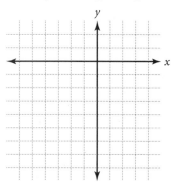
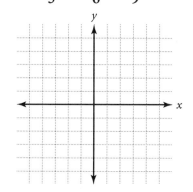

7. An athletic shoe wholesaler uses the table to determine the revenue for an order of pairs of shoes from various retail outlets.

Number of pairs of shoes	Price per pair of shoes	Revenue per order
1	$45	$40
5	$43	$215
100	$39	$3900
500	$35	$17,500

   a. What is the revenue function? _____

   b. What is the maximum revenue? _____

   c. How many pairs of shoes would result in maximum revenue? _____

   d. If it costs the wholesaler $12 to produce a pair of shoes and they have $150 in total costs per order, what is the cost function? _____

   e. What is the profit function? _____

   f. How many pairs of shoes must be sold in order to make a profit? _____

   g. What is the maximum profit per order? _____

NAME _____ CLASS _____ DATE _____

Practice Masters Level A
6.1 Exponential Growth and Decay

Find the multiplier for each rate of exponential growth or decay.

1. 7% decay _____
2. 3% growth _____
3. 5% decay _____
4. 10% growth _____
5. 13% growth _____
6. 15% decay _____
7. 5.2% growth _____
8. 4.1% decay _____

Given $x = 4$, $y = \frac{1}{3}$, $z = 4.2$, evaluate each expression.

9. 3^x _____
10. 4^y _____
11. 5^{2z} _____
12. $2(5^x)$ _____
13. $(2 \cdot 5)^x$ _____
14. 10^{z-2} _____
15. $4(10)^y$ _____
16. $3(0.2)^z$ _____
17. $5(0.5)^{4x}$ _____
18. 3^{2y} _____

Determine whether each table represents a linear, quadratic, or exponential relationship between x and y.

19.
x	1	2	3	4
y	2	4	8	16

20.
x	0	2	4	6
y	−1	3	7	11

21.
x	−2	0	2	3
y	4	0	4	9

_____   _____   _____

Write the exponential expression, and predict the population of bacteria for each time period.

22. 100 bacteria that double every hour _____

   a. after 1 hour   b. after 4 hours   c. 1 hour ago

23. 75 bacteria that triple every hour _____

   a. after 2 hours   b. after 5 hours   c. 2 hours ago

Algebra 2   Practice Masters Levels A, B, and C

NAME _____ CLASS _____ DATE _____

Practice Masters Level B
6.1 Exponential Growth and Decay

Find the multiplier for each rate of exponential growth or decay.

1. 7.1% growth _____   2. 7.1% decay _____

3. 6.3% growth _____   4. 6.3% decay _____

5. 0.14% decay _____   6. 0.07% growth _____

Given $x = 5$, $y = \frac{3}{4}$, $z = 2.7$, evaluate each expression.

7. -3^{x+1} _____   8. $(-3)^{x+1}$ _____

9. 2^z _____   10. 2^{3z} _____

11. 6^{2x+1} _____   12. 5^{xy} _____

Determine whether each table represents a linear, quadratic, or exponential relationship between x and y.

13.
x	2	4	6	8
y	12	48	108	192

14.
x	−1	0	1	4
y	0.33	1	3	81

15.
x	−2	0	2	4
y	−11	−5	1	7

_____   _____   _____

Write the exponential expression, and predict the population of bacteria for each time period.

16. 25 bacteria that triple every 2 hours _____

   a. after 1 hour   b. after 4 hours   c. 1 hour ago

   _____   _____   _____

17. 50 bacteria that double every 30 minutes _____

   a. after $1\frac{1}{2}$ hours   b. after 3 hours   c. 2 hours ago

   _____   _____   _____

18. 73 bacteria that triple every 30 minutes _____

   a. after 1 hour   b. after $2\frac{1}{2}$ hours   c. after 1 hour 20 min

   _____   _____   _____

19. $550 that doubles every 7 years _____

   a. after $3\frac{1}{2}$ years   b. after 6 years   c. after 42 years

   _____   _____   _____

Practice Masters Level C

6.1 Exponential Growth and Decay

Evaluate each expression to the nearest thousandth for the given value of x.

1. $2 \cdot 5^x$ for $x = 0.5$ _____

2. $-3 \cdot \left(\dfrac{1}{2}\right)^{3x}$ for $x = 2$ _____

3. $\left(\dfrac{1}{4}\right)^{2x+1}$ for $x = \dfrac{1}{4}$ _____

4. $(10)^{\frac{x}{2}+1}$ for $x = 5$ _____

5. $-4^{0.2x}$ for $x = 0.35$ _____

6. $-4 \cdot 5^{\left(\frac{x}{4}\right)}$ for $x = 0.1$ _____

7. 2^{3x} for $x = -2$ _____

8. 5^{4x-1} for $x = \dfrac{1}{2}$ _____

9. The present population of a country is 2,321,255. The population is growing at about 1% per year. Predict the population of the country for each given time.

   a. 2 years from now

   b. $3\dfrac{1}{2}$ years from now

   c. 3 years ago

   d. 2 months from now

10. A virus contains bacteria that grows at a rate of 2% per decade. Presently the virus contains 6000 bacteria. Find the number of bacteria for each given time.

    a. 30 years from now

    b. 5 years from now

    c. 10 years ago

    d. 1 year ago

11. If a virus becomes an epidemic when the count reaches 1,000,000, when will the virus in Exercise 10 be classified as an epidemic? _____

12. A population of bacteria contains 10,000 bacteria. It is decaying at a rate of 10% per hour, how many bacteria will be present after each given time.

    a. 2 hours _____

    b. 10 hours _____

Practice Masters Level A
6.2 Exponential Functions

Identify each function as linear, quadratic, or exponential.

1. $f(x) = 3^x + 2$ _____
2. $f(x) = \dfrac{x}{2} + 3^2$ _____
3. $f(x) = \dfrac{1}{2}x^2 - 3$ _____
4. $f(x) = (6^{3x})^2$ _____
5. $f(x) = (x + 2)(x - 3)$ _____
6. $f(x) = (x + 6(x - 4) - x)$ _____

Tell whether each function represents exponential growth or decay.

7. $y = 6^x$ _____
8. $y = 2 \cdot 0.1^x$ _____
9. $y = 2 \cdot 4^x$ _____
10. $y = 1.001^x + 6$ _____
11. $y = 100 \cdot 0.77^x$ _____
12. $y = 0.001 \cdot 2^x$ _____

Graph each function by plotting five points.

13. $y = 2^x$

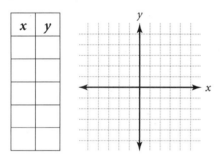

14. $y = -3^x$

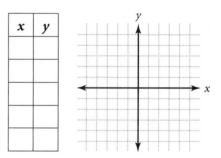

15. $y = 2(4)^x$

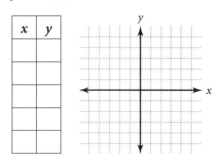

16. $y = -2(5)^x$

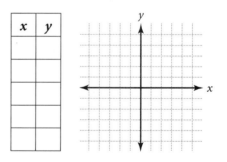

Find the final amount for each investment.

17. $100 earning 5% interest compounded annually for 3 years _____
18. $400 earning 6% interest compounded annually for 5 years _____

Practice Masters Level B
6.2 Exponential Functions

Tell whether each function represents exponential growth or decay. Then graph the function by plotting five points.

1. $y = \left(\dfrac{1}{3}\right)^x$

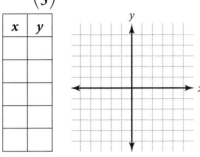

2. $y = \left(\dfrac{3}{2}\right)^x$

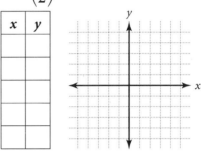

3. $y = 6(0.1)^x$

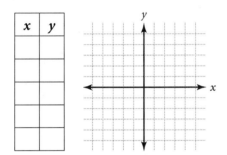

4. $y = 2(1.7)^x$

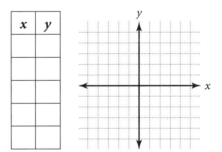

Find the final amount for each investment.

5. $1000 earning 5.25% interest compounded annually for 3 years _____

6. $1400 earning 3% interest compounded semi-annually for 5 years _____

7. $1400 earning 3.5% interest compounded semi-annually for 5 years _____

8. $2500 earning 6% interest compounded quarterly for 5 years _____

9. $2925 earning $5\dfrac{1}{4}$% interest compounded quarterly for 3 years _____

10. $700 earning 5% interest compounded annually for 25 years _____

11. $825 earning 4.7% interest compounded annually for 30 years _____

Practice Masters Level C

6.2 Exponential Functions

Tell whether each function represents exponential growth or decay. Then graph each function.

1. $g(x) = 4(2^x)$ _____

2. $f(x) = 4^x + 2$ _____

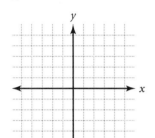
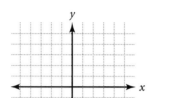

3. $g(x) = 3\left(\dfrac{1}{4}\right)^x$ _____

4. $f(x) = \dfrac{1}{2}\left(\dfrac{1}{3}\right)^x - 2$ _____

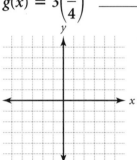
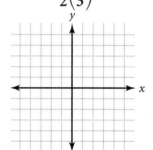

Use an exponential regression equation. Assume that interest is compounded only once per year.

5. An initial investment made in January of 1985 is worth $9000 in January of 2000, the rate of growth is 4%. What is the amount of the initial investment? _____

6. An initial investment made June 30, 1990 is worth $1472 on June 30, 1999, the rate of growth is 1.8%. What is the amount of the initial investment? _____

7. A car purchased August 3, 1989 for $14,592 is worth $6872 on August 3, 2000, what is the effective annual yield? _____

8. Consider the graphs of $y = 3^x$, $y = 4^x$, and $y = 6^x$.

   a. Which graph grows the fastest? _____

   b. Which graph grows the slowest? _____

   c. What is the y-intercept for each equation? _____

Practice Masters Level A
6.3 Logarithmic Functions

Write each equation in logarithmic form.

1. $3^4 = 81$

2. $2^6 = 64$

3. $5^{-2} = \dfrac{1}{25}$

4. $8^{\frac{1}{3}} = 2$

5. $\left(\dfrac{1}{5}\right)^4 = \dfrac{1}{625}$

6. $\left(\dfrac{1}{11}\right)^2 = \dfrac{1}{121}$

Write each equation in exponential form.

7. $\log_7 49 = 2$

8. $\log_{10} 10{,}000 = 4$

9. $\log_2 32 = 5$

10. $\log_{25} 5 = \dfrac{1}{2}$

11. $\log_{125} 5 = \dfrac{1}{3}$

12. $\log_9 9 = 1$

Solve each equation for *x*. Round your answers to the nearest hundredth.

13. $10^x = 100$

14. $10^x = 1$

15. $10^x = 0.1$

16. $10^x = 1040$

17. $10^x = 0.001$

18. $10^x = 2024$

Find the value of *n* in each equation.

19. $n = \log_{10} 10{,}000$

20. $n = \log_{13} 169$

21. $n = \log_{20} 400$

22. $2 = \log_5 n$

23. $4 = \log_3 n$

24. $10 = \log_2 n$

Algebra 2

Practice Masters Level B
6.3 Logarithmic Functions

Write each equation in logarithmic form.

1. $15^{-2} = \dfrac{1}{225}$ _____
2. $16^{\frac{-1}{4}} = \dfrac{1}{2}$ _____
3. $\left(\dfrac{1}{3}\right)^{-4} = 81$ _____
4. $\left(\dfrac{1}{4}\right)^{3} = \dfrac{1}{64}$ _____
5. $\left(\dfrac{1}{5}\right)^{-3} = 125$ _____
6. $1000^{\frac{-1}{3}} = \dfrac{1}{10}$ _____

Write each equation in exponential form.

7. $\log_{216} 6 = \dfrac{1}{3}$

8. $\log_{10,000} 10 = \dfrac{1}{4}$

9. $\log_4 \dfrac{1}{64} = -3$

10. $\log_4 \dfrac{1}{256} = -4$

11. $\log_{10} 0.01 = -2$

12. $\log_7 1 = 0$

Solve each equation for x. Round your answers to the nearest hundredth.

13. $10^x = 20$

14. $10^x = 51$

15. $10^x = 0.4$

16. $10^x = 1.5$

17. $10^x = 0.0075$

18. $10^x = 5075$

Find the value of n in each equation.

19. $-3 = \log_2 n$

20. $-4 = \log_5 n$

21. $6 = \log_n 64$

22. $-3 = \log_n \dfrac{1}{125}$

23. $-3 = \log_n 0.001$

24. $7 = \log_n 10,000,000$

Practice Masters Level C
6.3 Logarithmic Functions

For Exercises 1–4, write an equivalent exponential or logarithmic equation.

1. $\log_{81} 9 = \dfrac{1}{2}$ _____

2. $\left(\dfrac{1}{2}\right)^{-3} = 8$ _____

3. $-6 = \log_2 \dfrac{1}{64}$ _____

4. $\left(\dfrac{1}{4}\right)^{-4} = 256$ _____

Calculate $[H^+]$ for each of the following.

5. black tea with a pH of 5.5

6. baking soda with a pH of 9.3

7. apple juice with a pH of 3.5

8. pure water with a pH of 0

Graph each function and identify the transformations from f to g.

9. $f(x) = \log_{10} x; \quad g(x) = 3 \log_{10} x$

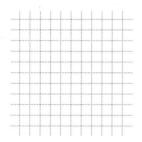

10. $f(x) = \log_{10} x; \quad g(x) = 5 + \log_{10} x$

11. $f(x) = \log_{10} x; \quad g(x) = -2 + \log_{10} x$

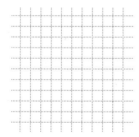

12. $f(x) = \log_{10} x; \quad g(x) = \dfrac{1}{2} \log_{10} x$

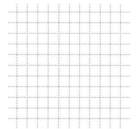

Identify the transformations from f to g.

13. $f(x) = \log_{10} x; \quad g(x) = 4 - \log_{10} x$

14. $f(x) = \log_{10} x; \quad g(x) = -5 - \dfrac{1}{4} \log_{10} x$

Practice Masters Level A
6.4 Properties of Logarithmic Functions

Write each expression as a sum or difference of logarithms. Then simplify, if possible.

1. $\log_3 (9 \cdot 25)$

2. $\log_4 \dfrac{16}{64}$

3. $\log_2 15m$

4. $3 \log_5 8$

5. $\log_4 \dfrac{4}{x}$

6. $\log_7 \dfrac{x}{49}$

Write each expression as a single logarithm. Then simplify, if possible.

7. $\log_2 3 + \log_2 6$

8. $\log_{10} 12 - \log_{10} 2$

9. $4 \log_5 2 + 3 \log_5 3$

10. $\log_6 36 - 2 \log_6 3$

11. $\log_9 8 - (\log_9 + \log_9 3)$

12. $(\log_3 x + \log_3 y) - \log_3 z$

Evaluate each expression.

13. $\log_6 6^3$

14. $11^{\log_{11} 20}$

15. $\log_2 2^{4.5}$

16. $12^{\log_{12} 100}$

Solve for x.

17. $\log_8 (2x + 5) = \log_8 11$

18. $\log_6 (12 - 2x) = \log_6 2$

19. $2 \log_4 x = \log_4 144$

20. $\log_{10} (3x + 4) = \log_{10} (6x + 7)$

21. $\log_2 x - \log_2 5 = 3$

22. $\log_3 8 + \log_3 2x = 96$

23. $3 \log_2 x = \log_2 2 + \log_2 32$

24. $\log_5 (x - 6) - \log_5 9 = \log_{10} 100$

NAME _____ CLASS _____ DATE _____

Practice Masters Level B
6.4 Properties of Logarithmic Functions

Use the values in the table to approximate the value of each logarithmic expression in Exercises 1–8.

$\log_2 5 = 2.322$	$\log_2 3 = 1.585$	$\log_2 10 = 3.322$
$\log_4 2 = 0.5$	$\log_4 10 = 1.661$	$\log_4 5 = 1.161$

1. $\log_2 15$ _____
2. $\log_4 50$ _____
3. $\log_4 2.5$ _____
4. $\log_2 \frac{10}{3}$ _____
5. $\log_4 25$ _____
6. $\log_2 9$ _____
7. $\log_2 150$ _____
8. $\log_4 100$ _____

Write each expression as a single logarithm. Then simplify, if possible.

9. $\log_5 4 + \log_5 x - \log_5 y$

10. $7 \log_3 m - \log_3 n$

11. $\log_3 5 + \frac{1}{2} \log_3 y$

12. $2 \log_6 (x + 8) - 4 \log_6 x$

13. $(\log_4 x - \log_4 y) + \frac{1}{2} \log_4 z$

14. $\frac{4 \log_5 n}{3}$

State whether each equation is true or false.

15. $y \log_3 x = \log_3 x^y$

16. $(\log_5 x - \log_5 y) - \log_5 z = \frac{xz}{y}$

Solve for x, and check your answers.

17. $\log_3 (x + x) + \log_3 5 = 5$

18. $\log_6 (x - 3) - \log_6 4 = \log_6 (x + 2)$

19. $2 - \log_{16} x^2 = 1$

20. $\log_5 (x^2 - 44) = 3$

21. $\log_2 (x^2 - 6x) = 4$

22. $\log_3 x + \log_3 (x + 24) = 4$

Algebra 2   Practice Masters Levels A, B, and C   113

Practice Masters Level C

6.4 Properties of Logarithmic Functions

1. The atmospheric pressure, P, in pounds per square inch and altitude, a, in feet, are related by the logarithmic equation $a = -55{,}555.56 \log_{10} \frac{P}{14.7}$.

   Find the ratio of air pressure at the top of each set of mountain ranges.

   a. Mount Kanchenjunga, India-Nepal, 28,208 feet
      Mount Rainer, Washington, 14,410 feet _____

   b. Pikes Peak, Colorado, 14,110 feet
      Grand Teton, Wyoming, 13,766 feet _____

   c. Mount Everest, Nepal-Tibet, 29,028 feet
      Black Mountain, Kentucky, 4145 feet _____

2. The relationship between decibels, B, and sound intensity, I, is given by the formula: $B = \log_{10}\left(\frac{I}{10^{-16}}\right)$ where I is measured in watts per square meter.

   a. Simplify the formula by using the laws of logarithms. _____

   b. Find the number of decibels if the sound intensity is measured at 10^{-12} watts per square meter. _____

   c. Find the sound intensity if the decibel level is 75. _____

Solve for x, and check your answers.

3. $\log_2 3 - (\log_2 x + \log_2 (x + 2)) = -4$

4. $\log_{10} (6 - x) - 2 \log_{10} x = \log_{10} 12$

5. $\log_7 x + \log_7 (x - 4) = 3$

6. $\log_3 (x + 3) + \log_3 (x - 2) = 5$

Practice Masters Level A
6.5 Applications of Common Logarithms

Solve each equation. Round your answers to the nearest hundredth.

1. $3^x = 10$

2. $4^x = 15$

3. $2^x = 100$

4. $5^x = 10$

5. $7^x = 35$

6. $9^x = 18$

7. $6^x = 0.35$

8. $4^x = 6.2$

9. $5^{-x} = 40$

10. $7^{-x} = 52$

11. $8^{-x} = 3.6$

12. $10^{-x} = 0.006$

Evaluate each logarithmic expression to the nearest hundredth.

13. $\log_3 100$

14. $\log_7 50$

15. $\log_3 82$

16. $\log_5 47$

17. $\log_3 16$

18. $\log_{10} 51$

19. $\log_4 1.5$

20. $\log_2 0.75$

21. $\log_6 0.005$

22. $\log_{10} 10^{-5}$

23. $\log_5 \dfrac{1}{5}$

24. $\log_6 \dfrac{1}{36}$

Algebra 2   Practice Masters Levels A, B, and C

Practice Masters Level B

6.5 Applications of Common Logarithms

Solve each equation. Round your answers to the nearest hundredth.

1. $3^{-x} = 14$

2. $12 - 6^x = 4$

3. $-10 + 5^x = -3$

4. $4^{x+2} = 20$

5. $6^{x-3} = 10$

6. $5^{2x} = 30$

7. $6^{-3x} = 55$

8. $7^{(-2x+2)} = 10$

9. $8^{-3x-5} = 5$

10. $\left(\dfrac{1}{2}\right)^{2x} = \dfrac{1}{4}$

11. $(3)^{\frac{x}{3}} = \dfrac{1}{3}$

12. $(4)^{\frac{-3}{4}x} = \dfrac{1}{2}$

Evaluate each logarithmic expression to the nearest hundredth.

13. $\log_5 \dfrac{1}{3}$

14. $\log_3 \dfrac{1}{4}$

15. $\log_{10} \dfrac{3}{4}$

16. $\log_4 \dfrac{1}{2}$

17. $\log_{\frac{1}{2}} 6.5$

18. $\log_{\frac{1}{4}} 8$

19. $\log_{\frac{1}{3}} 0.21$

20. $\log_{\frac{1}{5}} 14.5$

21. $1 + \log_{\frac{1}{4}} 81$

22. $\log_3 50 - 2$

23. $\dfrac{3}{2} - \log_{\frac{1}{2}} 20$

24. $\log_{\frac{1}{3}} \left(\dfrac{5}{3}\right)^{-1}$

Practice Masters Level C
6.5 Applications of Common Logarithms

1. Find the relative intensity, R, of the following sounds in decibels using the formula $R = 10 \log \frac{I}{I_0}$.

   a. soft music with intensity 1000 times the threshold of hearing, I_0 _____

   b. conversation with intensity 400,000 times the threshold of hearing, I_0 _____

   c. rock band with intensity 10^9 times the threshold of hearing, I_0 _____

   d. pain threshold with intensity 6×10^{10} times the threshold of hearing, I_0 _____

2. The formula for the pH of a solution is defined as $pH = -\log [H^+]$, where $[H^+]$ is the hydrogen ion concentration in moles per liter. Find the pH for solutions with the following $[H^+]$.

   a. an acidic solution with $[H^+]$ of 4×10^{-3} _____

   b. an acidic solution with $[H^+]$ of 3.2×10^{-2} _____

   c. an alkaline solution with $[H^+]$ of 5.2×10^{-9} _____

   d. an alkaline solution with $[H^+]$ of 1.7×10^{-10} _____

Write each expression as a single logarithm.

3. $\dfrac{4 \log_b m}{5} + \dfrac{2 \log_b n}{3} - 6 \log_b m$

4. $\dfrac{\log_a z}{\frac{1}{2}} - \dfrac{\log_a y}{2} - \dfrac{1 \log_a x}{3}$

Solve each equation.

5. $3 - \log_2 (x - 6) = \log_2 \dfrac{x}{4}$

6. $\log_5 |4x - 3| = 3$

Practice Masters Level A

6.6 The Natural Base, e

Evaluate each expression to the nearest thousandth. If the expression is undefined, write *undefined*.

1. e^2 _____ 2. e^4 _____

3. $e^{3.7}$ _____ 4. $e^{\frac{1}{4}}$ _____

5. $3e^6$ _____ 6. $5e^{0.01}$ _____

7. e^{-3} _____ 8. $e^{-0.04}$ _____

9. $-6e^{-2}$ _____ 10. $\ln 100$ _____

11. $\ln 2$ _____ 12. $\ln 0.02$ _____

13. $\ln 4.3$ _____ 14. $\ln \frac{1}{5}$ _____

15. $\ln(-5)$ _____ 16. $\ln \sqrt{2}$ _____

Simplify each expression.

17. $e^{\ln 4}$ _____ 18. $e^{6\ln 3}$ _____

19. $\ln e^3$ _____ 20. $e^{\ln|n|}$ _____

Solve each equation for *x* by using the natural logarithm function. Round your answers to the nearest hundredth.

21. $15^x = 30$ _____ 22. $5^x = 150$ _____

23. $4^x = 2.3$ _____ 24. $1.9^x = 5$ _____

25. $5^{2x} = 80$ _____ 26. $3^{\frac{1}{2}x} = 10$ _____

27. The amount of radioactive carbon-14 remaining after *t* years is given by the formula $N(t) = N_0 e^{-0.00012t}$. How much of a 20 milligram sample will remain after 50 years? _____

118   Practice Masters Levels A, B, and C   Algebra 2

Practice Masters Level B
6.6 The Natural Base, e

Evaluate each expression to the nearest thousandth. If the expression is undefined, write *undefined*.

1. e^{-6} _____
2. $e^{\frac{-1}{4}}$ _____
3. $6e^{\frac{-3}{2}}$ _____
4. $e^{\sqrt{5}}$ _____
5. $-4e^{\frac{\sqrt{3}}{2}}$ _____
6. $3e^{-\sqrt{2}}$ _____
7. $e^{-2} \cdot e^{\frac{\sqrt{6}}{3}}$ _____
8. $-e^{\frac{-2}{3}}$ _____
9. $\left(e^{\frac{1}{4}}\right)^{\frac{-1}{3}}$ _____
10. $\left(4e^{\frac{1}{2}}\right)^{\frac{1}{3}}$ _____
11. $\left(e^{\frac{-1}{3}}\right)^{\sqrt{3}}$ _____
12. $\left(5(2e)^{\frac{-1}{3}}\right)^{\frac{-1}{4}}$ _____

Simplify each expression.

13. $-4e^{\ln 4}$ _____
14. $\frac{2}{3}\ln e^{\frac{1}{2}}$ _____
15. $e^{\ln e^{\frac{1}{3}}}$ _____
16. $\ln e^{\ln e - 3}$ _____

Solve each equation for *x* by using the natural logarithm function. Round your answers to the nearest hundredth.

17. $0.6^{-x} = 4$ _____
18. $0.5^{-x} = 0.2$ _____
19. $7^{-x} = 10$ _____
20. $5^{\frac{-1}{2}x} = \frac{1}{4}$ _____
21. $3^{\frac{1}{3}x} = -4$ _____
22. $10^{\frac{-2}{3}x} = \frac{2}{3}$ _____

23. Suppose $2000 is invested at 5.75% annual interest compounded continuously for 20 months. How much will the investment be worth at the end of 20 months? Is this function an example of exponential growth or exponential decay?

_____

_____

Algebra 2   Practice Masters Levels A, B, and C   **119**

Practice Masters Level C

6.6 The Natural Base, e

Graph each function. Describe the transformations from $f(x) = e^x$ to g.

1. $g(x) = 4e^x$

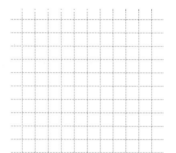

2. $g(x) = e^{-2x}$

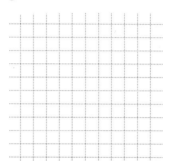

3. $g(x) = e^{\frac{1}{3}x}$

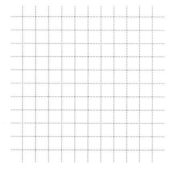

4. $g(x) = \frac{1}{3}e^{x+2}$

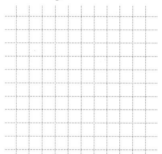

5. $g(x) = 2\ln x$

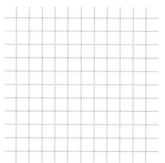

6. $g(x) = \ln(x - 3)$

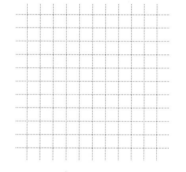

7. Evaluate: $\left[3(5e)^{\frac{1}{2}}\right]^{-\frac{5}{6}}$ _____

8. Evaluate: $(\ln\sqrt{31})^{\frac{2}{3}}$ _____

9. The population of Tokyo, Japan, the world's most populous city, in 1995 was 26,959,000. The annual growth rate from 1990 to 1995 was calculated to be 1.45%. Assume that this growth rate continues after 1995.

   a. Write the exponential growth model. _____

   b. Use the model to predict the population in 2015. _____

   c. Use the model to estimate the population in 1950. _____

Practice Masters Level A
6.7 Solving Equations and Modeling

Solve each equation for the missing variable. Write the exact solution and the approximate solution to the nearest hundredth, when appropriate.

1. $4^x = 256$

2. $3^{x-2} = 3^3$

3. $x = \log_2 \dfrac{1}{16}$

4. $\log_b \dfrac{1}{25} = -2$

5. $\log_3 y = -4$

6. $e^{2x} = 10$

7. $e^{-3x} = 12$

8. $\ln x = 9$

9. $e^{2x} - 7 = 5$

10. $4 \ln y = 100$

Use a graph to solve each equation.

11. $2 \ln y + 2 \ln 5 = 100$

12. $4 \log_{10} x - 5 = 13$

13. How much will $5000 be worth in 5 years if it is compounded continuously at 3% interest?

14. How much interest will be earned on a $6000 investment at 5.5% compounded continuously for 4 years?

15. Suppose a population grows according to the exponential function $P(t) = P_0 e^{rt}$, where P_0 is the initial population, r is the growth rate, t is the time in years, and $P(t)$ is the population at time t.

   a. What is the population after 10 years if the initial population is 900,000 and the growth rate is 2% per year?

   b. What is the population after $10\frac{1}{2}$ years if the initial population is 323,500 and the growth rate is $2\frac{1}{2}$%?

Algebra 2   Practice Masters Levels A, B, and C

Practice Masters Level B
6.7 Solving Equations and Modeling

Solve each equation for the missing variable. Write the exact solution and the approximate solution to the nearest hundredth, when appropriate.

1. $3^x = 9^2$

2. $4^{-x} = 64^{\frac{3}{2}}$

3. $5^{2x} = 625$

4. $2^{3x} = 128$

5. $-3 = \log_b \frac{1}{27}$

6. $\log_{\frac{1}{4}} y = \frac{-2}{3}$

Solve each equation for the missing variable. Give answers to the nearest thousandth.

7. $\ln(x - 4)\ln(x + 6) = \ln e^5$

8. $4e^{2x-1} = 3$

Use a graph to solve each equation.

9. $2e^{3x+4} = 24$

10. $\log x + \log(x + 15) = 2$

11. How much must be invested at $7\frac{1}{2}$% interest to have $3000 at the end of 3 years?

12. Suppose a population grows according to the exponential function $P(t) = P_0 e^{rt}$, where P_0 is the initial population, r is the growth rate, t is the time in years, and $P(t)$ is the population at time t. What was the population 5 years ago if the current population is 1,200,000 and the growth rate has been 1.7%?

For Exercises 13 and 14, use the radioactive decay formula $N(t) = N_0 e^{-0.00012t}$.

13. Tests on a fossil indicate that it contains 54% of its original carbon-14. How old is the fossil?

14. If an artifact contains 1% of its carbon-14, how old is the artifact?

Practice Masters Level C
6.7 Solving Equations and Modeling

In Exercises 1 and 2, use the earthquake magnitude formula
$M = \frac{2}{3} \log \frac{E}{10^{11.8}}$.

1. In 1989, an earthquake with a magnitude of 7.1 hit the San Francisco Bay area causing the delay of the World Series. The earthquake caused the death of 62 people. Find the energy released by the earthquake.

2. On August 17, 1999, Western Turkey suffered an earthquake which measured 7.4 on the Richter scale causing the death of about 17,000 people. Find the energy released by the earthquake.

In Exercises 3 and 4, use Newton's law of cooling,
$T(t) = T_s + (T_0 - T_s)e^{-kt}$.

3. A cup of hot coffee, 110°C, sits on a table at room temperature, 37°C. After 5 minutes, the coffee has a temperature of 90°C. Assume the room temperature remains constant at 37°C. What will the temperature be after 10 minutes?

4. An ice cube(s), 0°C, is dropped in a glass of lemonade, 36°C. Ten minutes later, the temperature of the lemonade is 34°C. Assume the temperature of the ice is still 0°C. When will the temperature of the lemonade be 30°C? (Assume the ice lasts that long.)

In Exercises 5 and 6, use the exponential growth/decay formula $P(t) = P_0 e^{kt}$.

5. The population of Detroit was 1,514,063 in 1970 and 1,027,974 in 1990. Predict the population in the year 2010.

6. A population of bacteria was found to be 1,000,000. Forty-eight hours later, the population was 870,000. What will the population be 6 days from the start?

7. The population of New Orleans was 465,538 in 1998 and 496,938 in 1990. Use an exponential model to estimate the population in the year 2000.

8. Use the algebraic method and the graphing method to solve $\log(3x - 50) + \log 2x = 2$ for x.

Algebra 2

Practice Masters Level A
7.1 An Introduction to Polynomials

Determine whether each expression is a polynomial. If so, classify the polynomial by degree and by number of terms.

1. $3x^2 + 2x + 1$ _____

2. $4x^3 + 5x^2 - 7x + 2$ _____

3. $\dfrac{6}{x^2} + \dfrac{2}{x} - 3$ _____

4. $\dfrac{1}{2}x^4 - 3x^2 + 5$ _____

Evaluate each polynomial expression for the indicated value of x.

5. $x^2 - 3x + 6$ for $x = 2$ _____

6. $x^3 - 3x^2 + 4x + 7$ for $x = 3$ _____

7. $2x^4 - 3x + 2$ for $x = 2$ _____

8. $x^2 + 7x - 10$ for $x = -2$ _____

Write each sum or difference as a polynomial in standard form.

9. $(6x^3 - 2x^2 + 7x + 6)$
   $+ (3x^3 + 2x^2 - 5x + 1)$

10. $(9x^3 + 7x^2 - 5x - 2)$
    $- (6x^3 + 4x^2 + x + 6)$

11. $(4x^2 + 2x + 1) - (6x^2 + 10x - 7)$

12. $(4x^4 + 3x - 5)$
    $+ (2x^4 + 6x^3 - 3x^2 + 9x - 5)$

13. $(3x - 5) - (2x^2 + 5x + 2)$

14. $(6.2x^3 + 3.1x - 2.1x^2)$
    $+ (3.1x^2 - 6.2x + 2.1x^3)$

Sketch the graph of each function. Describe the general shape of the graph.

15. $s(x) = x^3 + x^2 - 4x - 4$

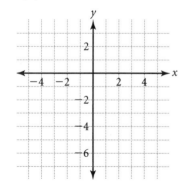

16. $r(x) = x^4 - 2x^3 - 4x^2 + x + 2$

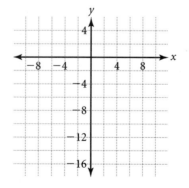

Practice Masters Level B
7.1 An Introduction to Polynomials

Determine whether each expression is a polynomial. If so, classify the polynomial by degree and by number of terms.

1. $\dfrac{x^2}{4} - \dfrac{x}{2} + 2$ _____

2. $4\sqrt{x} + 3x - 2$ _____

3. $6 - 2.1x^4$ _____

4. $3 + \dfrac{x^3}{2} - \dfrac{x^2}{7}$ _____

Evaluate each polynomial expression for the indicated value of x.

5. $-3x^3 + 4x^2 + 5x - 2$ for $x = 3$ _____

6. $3x^4 - 7x^3 - 2x^2 - 5x - 6$ for $x = -2$ _____

7. $3x^4 - 6.4x^3 + 2.7x^2 + 14.1x$ for $x = -3.1$ _____

Write each sum or difference as a polynomial in standard form.

8. $(4.7x^4 - 1) - (6.2x^3 - 5x^2 - 3x - 1)$ _____

9. $\left(\dfrac{1}{4}x^3 - \dfrac{1}{2}x^2 - \dfrac{1}{2}x + \dfrac{1}{3}\right) + \left(\dfrac{1}{2}x^3 - x^2 + \dfrac{3}{2}x + \dfrac{2}{9}\right)$ _____

10. $(5x - 2x^2 - 4x^4) - (3x^3 + 6x^4 - 2x^2 - 7)$ _____

11. $(3.2x^3 - 2.7\ldots\ldots + 2(-3.6x^3 + 5.4x^2 - 2.6x - 1.3)$ _____

Sketch the graph ... Describe the general shape of the graph.

12. $b(x) = -2x\ldots$

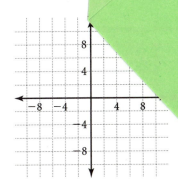

13. $f(x) = -x^3 + 5x^2 - 3x + 2$

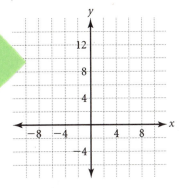

Algebra 2

NAME _____ CLASS _____ DATE _____

Practice Masters Level C
7.1 An Introduction to Polynomials

Evaluate each polynomial expression for the indicated value of x.

1. $\dfrac{-4}{11}x^3 - \dfrac{10}{11}x^2 + \dfrac{7}{11}x - \dfrac{1}{11}$ for $x = -11$ 2. $\dfrac{1}{2}x^4 - \dfrac{3}{2}x^3 + x - \dfrac{5}{2}$ for $x = \dfrac{1}{2}$

_____   _____

3. $-0.007x^4 + 0.006x^3 + 0.003x^2 - 0.005x$ for $x = 0.3$ _____

4. The sum of $7x^3 - 5x^2 - ax - 7$ and $5x^3 - bx^2 + cx + 5$ is
   $dx^3 + ax^2 + 4x - a$. Find a, b, c and d. _____

Sketch the graph of each function. Describe the general shape of the graph.

5. $m(x) = -4x^4 - 5x^3 + 4x^2 + 3x - 5$ 6. $c(x) = -\dfrac{1}{2}x^3 + \dfrac{1}{2}x^2 - 5$

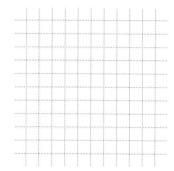   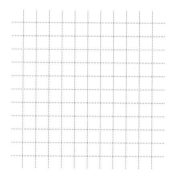

_____   _____

7. The cost of manufacturing a certain product is represented by the function
   $C(x) = 0.02x^3 - 0.01x + 100$, where x is the quantity of product.

   a. Is the function a polynomial? If so, classify the polynomial. _____

   b. What is the cost of manufacturing 15 items? _____

8. The profit from selling a certain product is represented by the function
   $M(p) = -0.002p^3 + 0.05p^2 - 0.01p$, where p is the selling price.

   a. What profit will be earned if the selling price is $10? _____

   b. What is the maximum profit? _____

   c. For what selling price will the profit be the largest? _____

   d. What selling price will result in $0 profit? _____

Practice Masters Level A
7.2 Polynomial Functions and Their Graphs

Graph each function and approximate any local maxima or minima to the nearest tenth.

1. $y = -3x^2 + 9x - 1$

2. $y = 2x^3 + 3x^2 - 4x + 1$

3. $y = x^2 - 2$

4. $y = x^3 + 3$

Graph each function. Find any local maxima or minima to the nearest tenth. Find the intervals over which the function is increasing or decreasing.

5. $y = -2x^2 - 3x + 4$

6. $y = 3x^3 - 2x + 1$

7. $y = -x^4 + 2$

8. $y = -x^3 + x^2$

Describe the end behavior of each function.

9. $y = 4x + 2x^2 - 6$

10. $y = 3x^4 - 2x^3 + 2x^2 + 4x - 5$

11. $y = 5x^3$

12. $y = -x + x^2 - 3x^3$

13. The percentage of the U.S. population that is foreign-born is shown in the table. Find a cubic regression equation for the data using $x = 0$ for 1900.

Year	1900	1910	1940	1970	1990
Percentage	13.6	14.7	8.8	4.7	8.0

Algebra 2   Practice Masters Levels A, B, and C

Practice Masters Level B
7.2 Polynomial Functions and Their Graphs

Graph each function and approximate any local maxima or minima to the nearest tenth.

1. $y = x^3 - 3x^2 + 6x + 3$

2. $y = -x^3 - 3x^2 + 4x + 5$

3. $y = x^4 - 2x^3 + 4x^2 + 3x + 6$

4. $y = 3x^3 - x^4$

Graph each function. Find any local maxima or minima to the nearest tenth. Find the intervals over which the function is increasing or decreasing.

5. $y = x^3 + 5x^2 - 2x + 1$ _____

6. $y = -x^3 + 3x^2 + 2x - 2$ _____

7. $y = x^4 + 3x^3 + 2x^2 - 4x - 1$ _____

Describe the end behavior of each function.

8. $y = x^2 - 3x + 2$

9. $y = 2x^3 - 3x^2 + 4x + 1$

10. $y = -x^4 + 5x^3 - 3x^2 + 6x + 8$

11. $y = 2 + 3x - 2x^2 - x^3$

12. The total number of high-school students taking a college entrance exam per year is shown in the table. Find a quartic regression equation for the data using $x = 0$ for 1990.

Year	1990	1991	1992	1993	1994	1995	1996	1997	1998	1999
Total number, in thousands	817	796	832	875	892	945	925	959	995	1019

128 Practice Masters Levels A, B, and C Algebra 2

NAME _____ CLASS _____ DATE _____

Practice Masters Level C
7.2 Polynomial Functions and Their Graphs

Graph each function.

a. Find any local maxima and minima.

b. Find the intervals over which the function is increasing or decreasing.

c. Describe the end behavior.

1. $y = -6x^4 + 5x^3 - 3x^2 + 7x + 1$

a. _____ b. _____

c. _____

2. $y = \frac{2}{3}x^3 - \frac{1}{6}x^2 + \frac{7}{9}x - \frac{1}{2}$

a. _____ b. _____

c. _____

3. $y = 0.2x^4 + 0.7x^2 - 0.5x + 0.1$

a. _____ b. _____

c. _____

4. $y = \frac{-1}{4}x^3 + \frac{1}{2}x^2 - \frac{1}{8}x + \frac{1}{2}$

a. _____ b. _____

c. _____

Find the appropriate regression model for each problem.

5. A function rises on both ends and contains the points
$(2, 5), (1, -3), (-1, -7), (-2, -15)$ and $(-3, 5)$. _____

6. A function falls on one end and rises on the other and contains the points $(-1, -15), (1, -5), (2, -42)$, and $(-2, -14)$. _____

7. The world motor vehicle production, in thousands, is given below.

Year	1960	1970	1980	1990	1991	1992	1993	1994	1995	1996	1997	1998
Vehicle	7.9	8.3	8.0	9.8	8.8	9.7	10.9	12.3	11.9	11.8	12.1	12.0

a. Using $x = 0$ for 1960, find a quartic regression model for the data. _____

b. Use the model to estimate the production of vehicles for the year 2001. _____

c. Using the model, when will the production of motor vehicles be 0? _____

Algebra 2   Practice Masters Levels A, B, and C

Practice Masters Level A

7.3 Products and Factors of Polynomials

Write each product as a polynomial in standard form.

1. $4x^3(2x^2 + 6x - 3)$

2. $(x + 4)(3x^2 + 2x + 2)$

3. $(x - 4)(x^3 + x - 2)$

4. $(2x + 5)(x^2 - 5x + 2)$

Use substitution to determine whether the given linear expression is a factor of the polynomial.

5. $x^2 + 7x - 18; x - 2$ _____

6. $x^2 + 15x + 44; x + 11$ _____

7. $x^2 + 12x + 32; x - 8$ _____

8. $x^3 - 3x^2 - 11x - 2; x + 2$ _____

Divide by using long division or synthetic division.

9. $(x^2 - 5x - 14) \div (x + 2)$

10. $(x^2 - 10x + 24) \div (x - 6)$

11. $(x^3 + 2x^2 - 3x + 20) \div (x + 4)$

12. $(x^3 - 7x^2 + 6x + 8) \div (x - 2)$

13. $(x^2 + 15x + 50) \div (x + 5)$

14. $(x^3 - 2x^2 - 5x - 12) \div (x - 4)$

15. $(x^3 - 9x^2 + 14x + 19) \div (x - 3)$

16. $(x^3 - 6x^2 - 48x - 43) \div (x + 4)$

For each function below, use synthetic division or substitution to find the indicated value.

17. $P(x) = x^2 + 2x + 3; P(4)$

18. $P(x) = x^2 - 6x + 3; P(5)$

19. $P(x) = x^3 - 2x^2 + 4x + 5; P(2)$

20. $P(x) = 3x^3 + 4x^2 - 5; P(-1)$

Practice Masters Level B
7.3 Products and Factors of Polynomials

Write each product as a polynomial in standard form.

1. $(3x + 4)(2x^3 - 4x^2 + 5x + 2)$

2. $(-2x + 5)(-4x^3 - 2x + 6)$

3. $(2x - 5)^3$

4. $(3x + 2)^2(2x - 3)$

Use substitution to determine whether the given linear expression is a factor of the polynomial.

5. $4x^3 - 3x^2 + 2x + 9;\ x + 1$ _____

6. $-5x^4 - 3x^3 - 6x;\ x - 3$ _____

7. $x^5 - 1;\ x - 1$ _____

8. $\frac{1}{2}x^3 + \frac{1}{3}3x^2 + \frac{1}{2}x + \frac{1}{3};\ x + 2$ _____

Divide by using long division or synthetic division.

9. $(3x - 7x^2 + 10) \div (x - 3)$

10. $(4x^3 - 2x^2 + 2x + 3) \div (-x - 2)$

11. $(-6x^4 + 2x^3 - 6x + 1) \div (x^2 + 3x - 1)$

12. $(x^5 + 2x) \div (5 - x)$

13. $(4 + 7x - 6x^2) \div (x + 3)$

14. $(-5x^3 + 3x^2 + 4x - 2) \div (x - 5)$

15. $(x^3 - 4) \div (x + 4)$

16. $\left(\frac{1}{2}x^3 - 2x^2 + \frac{1}{2}x - 1\right) \div \left(x - \frac{1}{2}\right)$

For each function below, use synthetic division or substitution to find the indicated value.

17. $P(x) = -x^4 + 6x^3 + 2x^2 - 10x - 8;\ P(2)$ _____

18. $P(x) = 2x^{47} + 6x^{15} + 3x^{12};\ P(-1)$ _____

19. $P(x) = 0.7x^4 - 0.2x^3 + 0.1x^2 + 0.3x - 4;\ P(2.1)$ _____

NAME _____ CLASS _____ DATE _____

Practice Masters Level C

7.3 Products and Factors of Polynomials

Write each product as a polynomial in standard form.

1. $(x + 3)^2(3x^2 - 2x + 2)$

2. $\left(x + \dfrac{2}{3}\right)\left(\dfrac{1}{2}x^2 - \dfrac{2}{3}x - \dfrac{1}{3}\right)$

3. $(x - 4)^4$

4. $(x - 3)^3(3x + 2)^2$

Factor each polynomial.

5. $x^3 - 12x^2 + 36x$

6. $x^3 + 64$

7. $x^3 - 125$

8. $x^3 - 2x^2 + x - 2$

Divide by using long division or synthetic division.

9. $\left(\dfrac{1}{4}x^2 - \dfrac{1}{2}x + 2\right) \div \left(x - \dfrac{1}{2}\right)$

10. $\left(\dfrac{1}{3}x^4\right) \div \left(x + \dfrac{1}{3}\right)$

11. $(0.05x^3 - 0.07x + 0.1) \div (x + 0.2)$

12. $(0.2x^4 - 0.5x) \div (x - 1.2)$

Use substitution to determine whether the given linear expression is a factor of the polynomial.

13. $x^3 - \dfrac{2}{3}x - \dfrac{1}{3}; \; x + \dfrac{1}{3}$ _____

14. $0.75x^4 + 0.5x^3 + 0.2x^2 - 0.35x + 0.65; \; x - 2$ _____

For each function below, use synthetic division or substitution to find the indicated value.

15. $P(x) = 6x^4 - 3x^3 + 4x^2 + 2x - 3; \; P(0.5)$

16. $P(x) = 1.5x^5 - 3.7x^3 + 2x^2 - 3; \; P(-1.2)$

NAME _____ CLASS _____ DATE _____

Practice Masters Level A
7.4 Solving Polynomial Equations

Use factoring to solve each equation.

1. $x^3 - 36x = 0$

2. $x^3 - 3x^2 - 28x = 0$

3. $x^3 - 2x^2 - 15x = 0$

4. $x^3 + 4x^2 + 4x = 0$

5. $x^3 - 9x^2 + 18x = 0$

6. $2x^3 + 2x^2 - 40x = 0$

Use a graph, synthetic division, and factoring to find all of the roots of each equation.

7. $x^3 - 2x^2 - 24x = 0$

8. $x^3 - 7x^2 + 14x - 8 = 0$

9. $x^3 + 2x^2 - 5x - 6 = 0$

10. $x^3 - 6x^2 - x + 30 = 0$

Use variable substitution and factoring to find all of the roots of each equation.

11. $x^4 - 13x^2 + 36 = 0$

12. $x^4 - 26x^2 + 25 = 0$

13. $x^4 - 21x^2 + 80 = 0$

14. $x^4 - 18x^2 + 77 = 0$

Use a graph and the Location Principle to find the real zeros of each function. Give approximate values to the nearest hundredth, if necessary.

15. $f(x) = x^3 - 7x - 6 = 0$

16. $f(x) = x^3 + 3x^2 - 10x$

17. $f(x) = x^4 + 2x^3 - 6x^2 + 10$

18. $f(x) = x^3 - 4x^2 + 7$

NAME _____ CLASS _____ DATE _____

Practice Masters Level B
7.4 Solving Polynomial Equations

Use a graph, synthetic division, and factoring to find all of the roots of each equation.

1. $x^4 - 3x^3 - 10x^2 + 24x = 0$

2. $15x^3 - 29x^2 + 17x - 3 = 0$

3. $60x^3 - 47x^2 + 12x - 1 = 0$

4. $18x^3 + 15x^2 - x - 2 = 0$

5. $6x^3 - 41x^2 + 58x - 15 = 0$

6. $6x^3 - 5x^2 - 151x - 210 = 0$

7. $8x^3 + 10x^2 + x - 1 = 0$

8. $12x^3 - 8x^2 - x + 1 = 0$

9. $x^3 + \dfrac{3}{2}x^2 + \dfrac{3}{4}x + \dfrac{1}{8} = 0$

10. $x^3 - 2x^2 - \dfrac{1}{4}x + \dfrac{1}{2} = 0$

Use variable substitution and factoring to find all of the roots of each equation.

11. $x^4 - 18x^2 + 65 = 0$

12. $36x^4 - 25x^2 + 4 = 0$

13. $16x^4 - 65x^2 + 4 = 0$

14. $2500x^4 - 125x^2 + 1 = 0$

Use a graph and the Location Principle to find the real zeros of each function. Give approximate values to the nearest hundredth, if necessary.

15. $f(x) = 8x^3 + 12x^2 - 2x - 3 = 0$

16. $f(x) = 4x^3 - 12x^2 - x + 15$

17. $f(x) = 9x^4 + 12x^3 + x^2 - 2x$

18. $f(x) = x^4 - 34x^2 + 225$

134   Practice Masters Levels A, B, and C   Algebra 2

NAME _____ CLASS _____ DATE _____

Practice Masters Level C
7.4 Solving Polynomial Equations

Use factoring to solve each equation.

1. $x^3 - 3x^2 - 9x + 27 = 0$

2. $x^3 + 2x^2 - 16x - 32 = 0$

3. $x^4 - 2x^3 + 25x^2 - 50x = 0$

4. $x^4 + 10x^3 - 4x^2 = 40x$

5. $x^4 - 19x^2 = -90$

6. $x^4 = 17x^2 - 30$

Use a graph, synthetic division, and factoring to find all of the roots of each equation.

7. $-56x^2 + 14x - 1 = -64x^3$

8. $125x^4 - 16x = -100x^3 + 20x^2$

9. $x^4 = 18x^2 - 77$

10. $12x^3 - 13x^2 + \dfrac{13}{3}x - \dfrac{5}{12} = 0$

Use variable substitution and factoring to find all of the roots of each equation.

11. $243x^3 - 243x^2 + 42x + 8 = 0$

12. $75 = 28x^2 - x^4$

13. $63x^3 + 184x^2 - 17x - 6 = 0$

14. $x^4 + 4x^3 - 10x^2 = 40x$

Use a graph and the Location Principle to find the real zeros of each function. Give approximate values to the nearest hundredth, if necessary.

15. $f(x) = \dfrac{1}{2}x^3 + \dfrac{1}{3}x^2 - \dfrac{1}{3}x + 2$

16. $f(x) = \dfrac{-1}{3}x^4 - \dfrac{1}{2}x^3 + x - \dfrac{1}{4}$

17. $f(x) = -0.1x^3 + 0.2x^2 - 0.6x - 0.05$

18. $f(x) = 0.6x^4 - 0.4x^2 + 0.2x - 0.3$

Algebra 2

NAME _____ CLASS _____ DATE _____

Practice Masters Level A
7.5 Zeros of Polynomial Functions

Find all of the rational roots of each polynomial equation.

1. $2x^3 - 3x^2 - 11x + 6 = 0$

2. $3x^3 - 10x^2 - 9x + 4 = 0$

3. $3x^3 + x^2 - 8x + 4 = 0$

4. $3x^3 + 5x^2 - 16x - 12 = 0$

5. $2x^3 - 3x^2 - 8x - 3 = 0$

6. $12x^3 - 31x^2 + 15x - 2 = 0$

Find all zeros of each polynomial function.

7. $P(x) = x^3 - 3x^2 - 7x + 21$

8. $P(x) = x^3 + 4x^2 - 4x - 16$

9. $P(x) = x^3 + 4x^2 - 5x - 20$

10. $P(x) = x^3 - 11x^2 - 11x + 121$

11. $P(x) = x^3 + 2x^2 - 6x + 3$

12. $P(x) = x^3 + 3x^2 - 7x - 18$

Find all real zeros of *x* for which the functions are equal. Give your answers to the nearest hundredth.

13. $P(x) = x^2;\ Q(x) = x^3 - 2x^2 - 12x + 10$ _____

14. $P(x) = -5x^2 + x;\ Q(x) = x^3 - x^2 - 6x - 30$ _____

15. $P(x) = 8x^2 - 19x + 24;\ Q(x) = x^3 + 10$ _____

Write a polynomial function, *P*, in factored form by using the given information.

16. *P* is of degree 3; zeros: $-2, 3, 1$;
    $P(0) = 6$

17. *P* is of degree 3; zeros: $2, 3i, -3i$;
    $P(0) = 36$

136   Practice Masters Levels A, B, and C   Algebra 2

Practice Masters Level B

7.5 Zeros of Polynomial Functions

Find all of the rational roots of each polynomial equation.

1. $6x^3 + x^2 - 4x + 1 = 0$

2. $12x^3 - 23x^2 - 3x + 2 = 0$

3. $9x^3 + 15x^2 - 32x + 12 = 0$

4. $9x^3 - 36x^2 - 4x + 16 = 0$

5. $25x^3 + 95x^2 + 64x + 12 = 0$

6. $24x^3 - 10x^2 - 3x + 1 = 0$

Find all zeros of each polynomial function.

7. $P(x) = x^4 - 9x^2 + 20$

8. $P(x) = 2x^3 + 5x^2 - 17x + 7$

9. $P(x) = 6x^3 - 10x^2 - 13x - 3$

10. $P(x) = 12x^3 + x^2 - 21x + 5$

11. $P(x) = 10x^3 - 3x^2 + 3x - 2$

12. $P(x) = 12x^3 + 14x^2 + 7x + 2$

Find all real zeros of x for which the functions are equal. Give your answers to the nearest hundredth.

13. $P(x) = 2x^3 - 3x$; $Q(x) = -2x^3 - 2x^2 + 2x - \dfrac{3}{2}$

14. $P(x) = x^4 - 7x^2 - 12x$; $Q(x) = -2x^3 + 8x^2 - 36$

15. $P(x) = 4x - 3x^2$; $Q(x) = 2x^3 - 4$

Write a polynomial function, P, in factored form by using the given information.

16. P is of degree 3; zeros: $\dfrac{1}{2}, \pm 4i$; $P(0) = 4$

17. P is of degree 4; zeros: 2(multiplicity of 2), -1(multiplicity of 2); $P(0) = -4$

Algebra 2   Practice Masters Levels A, B, and C   **137**

Practice Masters Level C

7.5 Zeros of Polynomial Functions

Find all of the rational roots of each polynomial equation.

1. $12x^3 - 4x^2 - 3x + 1 = 0$

2. $24x^3 + 6 = 22x^2 + 5x$

3. $36x^3 + 3x^2 + 2 = 11x$

4. $32x^3 = 14x + 3$

5. $81x^4 - 81x^3 + 9x^2 + 9x - 2 = 0$

6. $x^4 - \dfrac{17}{12}x^3 + \dfrac{7}{24}x^2 + \dfrac{1}{6}x - \dfrac{1}{24} = 0$

Find all zeros of each polynomial function.

7. $P(x) = 2x^3 - 13x^2 + 14x - 4$

8. $P(x) = 3x^3 + 7x^2 - 19x - 7$

9. $P(x) = 3x^4 - 5x^3 - 13x^2 + 16x - 4$

10. $P(x) = 4x^4 - 12x^3 - 17x^2 + 25x + 15$

11. The altitude of a hot-air balloon can be represented by the function $a(t) = 0.025t^2 + 2t$, where t is the time in seconds.

    a. How long will it take for the balloon to reach an altitude of 10 feet? _____

    b. How long will it take for the balloon to reach an altitude of 500 feet? _____

12. The profit made by selling an item is $P(x) = 0.005x^2 - 0.007x - 50$, where x is the number of items sold.

    a. How many items must be sold in order to start making a profit? _____

    b. How many items must be sold to make a profit of $600? _____

Write a polynomial function, P, in standard form by using the given information.

13. zeros: $\dfrac{1}{2}, \dfrac{1}{3}, \pm 5i$; $P(0) = 50$

14. zeros: $\dfrac{1}{2}$ (multiplicity of 2), $\pm 10i$; $P(0) = 50$

Practice Masters Level A

8.1 Inverse, Joint, and Combined Variation

Identify each equation as inverse, joint or combined variation.

1. $y = \dfrac{4}{x}$ _____

2. $y_1 = \dfrac{x_1 y_2}{x_2}$ _____

3. $y = \dfrac{a}{x}$ _____

4. $h = 6mn$ _____

For Exercises 5–8, y varies inversely as x. Write the appropriate inverse-variation equation.

5. $y = 45$ when $x = 3$

6. $y = -50$ when $x = 5$

7. $y = 108$ when $x = -4$

8. $y = 12$ when $x = 24$

For Exercises 9–12, y varies jointly as x and z. Write the appropriate joint-variation equation.

9. $y = 180$ when $x = 5$ and $z = 3$

10. $y = -126$ when $x = 2$ and $z = 7$

11. $y = -96$ when $x = -1$ and $z = 12$

12. $y = 30$ when $x = -15$ and $z = -6$

For Exercises 13–16, z varies jointly as x and y and inversely as w. Write the appropriate combined-variation equation.

13. $z = 5$ when $x = 2$, $y = 4$, and $w = 8$

14. $z = 2$ when $x = 3$, $y = -2$, and $w = 12$

15. $z = -3$ when $x = -5$, $y = 3$, and $w = 10$

16. $z = 1$ when $x = 20$, $y = 4$, and $w = 40$

17. A company finds that the lower the price of an item, the more the company can sell. For example, the company can sell 600 items if the price is $27.99. The company also discovers that the number of items sold varies inversely as the price. How many items can they sell if the price is $24.99?

Algebra 2   Practice Masters Levels A, B, and C

Practice Masters Level B
8.1 Inverse, Joint, and Combined Variation

Identify each equation as inverse, joint or combined variation.

1. $\dfrac{w}{rt} = 5$ _____

2. $xy = -6$ _____

3. $-3 = \dfrac{z}{xy}$ _____

4. $\dfrac{c_1 d_1}{d_2} = c_2$ _____

For Exercises 5–8, y varies inversely as x. Write the appropriate inverse-variation equation, and find y for the given value of x.

5. $y = 40$ when $x = 10$; $x = 13$

6. $y = -50$ when $x = -2$; $x = 6$

7. $y = 45$ when $x = -10$; $x = -4$

8. $y = 8$ when $x = 16$; $x = 8$

For Exercises 9–12, y varies jointly as x and z. Write the appropriate joint-variation equation, and find y for the given values of x and z.

9. $y = -126$ when $x = 3$ and $z = 7$; $x = 2$ and $z = 9$

10. $y = 120$ when $x = -5$ and $z = -2$; $x = 7$ and $z = -3$

11. $y = 150$ when $x = 10$ and $z = -6$; $x = 0.75$ and $z = 0.4$

12. $y = 12$ when $x = -6$ and $z = 8$; $x = \dfrac{1}{2}$ and $z = \dfrac{1}{4}$

For Exercises 13–16, z varies jointly as x and y and inversely as w. Write the appropriate combined-variation equation, and find z for the given values of x, y, and w.

13. $z = 3$ when $x = -2$, $y = 6$, and $w = 12$; $x = 5$, $y = -4$, and $w = \dfrac{1}{2}$

14. $z = 6$ when $x = -6$, $y = -9$, and $w = 3$; $x = -3$, $y = 6$, and $w = 5$

15. $z = -2.4$ when $x = -8$, $y = 15$, and $w = 5$; $x = 10$, $y = 20$, and $w = 4$

16. $z = -4.2$ when $x = 3$, $y = 4$, and $w = 2$; $x = 0.2$, $y = -0.3$, and $w = -0.2$

Practice Masters Level C
8.1 Inverse, Joint, and Combined Variation

For Exercises 1–4, y varies inversely as x. Write the appropriate inverse-variation equation. Find y for the given value of x.

1. $y = 52$ when $x = 7$; $x = 12$

2. $y = -6$ when $x = -8$; $x = \frac{1}{2}$

3. $y = 106$ when $x = 6$; $x = -3.5$

4. $y = 14.5$ when $x = 1.5$; $x = -2.25$

For Exercises 5–8, y varies jointly as x and z. Write the appropriate joint-variation equation. Find y for the given values of x and z.

5. $y = 100$ when $x = -3$ and $z = 14$; $x = 7$ and $z = 9$

6. $y = 95$ when $x = -15$ and $z = 4$; $x = -4$ and $z = 2$

7. $y = -16.5$ when $x = 1.5$ and $z = 2.5$; $x = 15$ and $z = -3$

8. $y = 0.75$ when $x = 0.25$ and $z = -0.25$; $x = 0.8$ and $z = 0.4$

For Exercises 9–10, write the appropriate combined-variation equation, and solve.

9. If x varies directly as t^2 and inversely as y, and $x = 192$ when $t = 8$ and $y = 3$, find y when $t = 9$ and $x = 486$.

10. If x varies directly as y and inversely as z, and $x = 120$ when $z = 5$ and $y = 4$, find x when $y = 8$ and $z = 2$.

If (x_1, y_1) and (x_2, y_2) satisfy $xy = k$, then $x_1 y_1 = x_2 y_2$. Find x or y as indicated. Round answers to the nearest hundredth, if necessary.

11. $(x, 7)$ and $(3, -5)$ _____

12. $(4, y)$ and $(5, -3)$ _____

13. $(6.2, -4)$ and $(x, 5)$ _____

14. $(-1.5, 2.5)$ and $(8, y)$ _____

15. A bicycle's pedal gear has 45 teeth and is rotating at 75 revolutions per minute. A chain links the pedal gear to a rear-wheel gear that has 24 teeth.

   a. How many revolutions per minute is the rear wheel gear moving? _____

   b. How fast, in miles per hour, is the bicycle moving if the wheel's diameter is 27 inches? _____

NAME _____ CLASS _____ DATE _____

Practice Masters Level A

8.2 Rational Functions and Their Graphs

Determine whether each function is a rational function. If so, find the domain. If the function is not rational, state why not.

1. $f(x) = \dfrac{x^2 + 3x + 2}{x - 2}$ _____

2. $f(x) = \dfrac{|x|}{x - 3}$ _____

3. $f(x) = \dfrac{x^2 - 2x - 5}{e^x}$ _____

4. $f(x) = \dfrac{x + 2}{x^2 - 1}$ _____

Identify all asymptotes and holes in the graph of each rational function.

5. $m(x) = \dfrac{3x + 1}{x + 5}$

6. $n(x) = \dfrac{x^2 - 2x}{(x - 2)^2}$

7. $p(x) = \dfrac{x + 5}{x^2 - 25}$

Find the domain of each rational function. Identify all asymptotes and holes in the graph of each rational function. Then graph.

8. $y = \dfrac{4x}{x + 2}$ _____

9. $y = \dfrac{x - 1}{x}$ _____

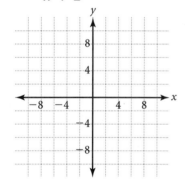
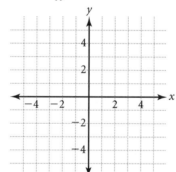

10. $y = \dfrac{x + 1}{x^2 + 3x + 2}$ _____

11. $y = \dfrac{x - 4}{x^2 - 16}$ _____

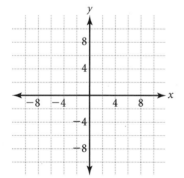
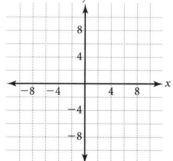

Practice Masters Level B

8.2 Rational Functions and Their Graphs

Determine whether each function is a rational function. If so, find the domain. If the function is not rational, state why not.

1. $f(x) = \dfrac{x^4 + 3x^3 - 2x^2 + 4x - 1}{x^2}$ _____

2. $f(x) = \dfrac{e^x(x-1)}{e^x}$ _____

3. $f(x) = \dfrac{x^2}{|x| - 2}$ _____

4. $f(x) = \dfrac{x^2 + 3x - 4}{x^2 + 5x - 6}$ _____

Identify all asymptotes and holes in the graph of each rational function.

5. $m(x) = \dfrac{3x + 2}{x^2 - 10}$

6. $n(x) = \dfrac{x + 7}{x^2 + 4x - 21}$

7. $p(x) = \dfrac{4x^4 - 1}{x^2 + 6x + 8}$

Find the domain of each rational function. Identify all asymptotes and holes in the graph of each rational function. Then graph.

8. $y = \dfrac{x^2 + 2x + 1}{x^2 - 3x - 4}$

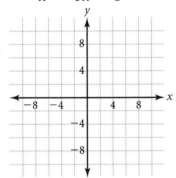

9. $y = \dfrac{x^2 + 4}{4x^2 - 1}$

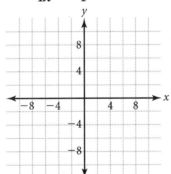

10. $y = \dfrac{x + 2}{x^2 - 3x - 10}$

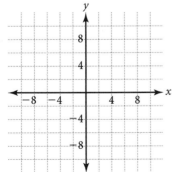

11. $y = \dfrac{x + 1}{x^3 + 1}$

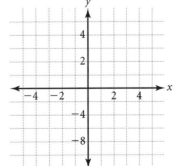

Algebra 2   Practice Masters Levels A, B, and C

Practice Masters Level C

8.2 Rational Functions and Their Graphs

Determine whether each function is a rational function. If so, find the domain. If the function is not rational, state why not.

1. $f(x) = \dfrac{x(x-1)+3}{x^3-1}$ _____

2. $f(x) = \dfrac{6x(3-x^2)}{x^4-16}$ _____

3. $f(x) = x^2 + 7x^3$ _____

4. $f(x) = e^x(x^5 - 3x^2 + 5)$ _____

Identify all asymptotes and holes in the graph of each rational function.

5. $m(x) = \dfrac{3x^2 - 2x + 1}{x^2 + 5x + 3}$

6. $n(x) = \dfrac{x^3 - 3}{x^3 + 5x^2 - 2x - 24}$

7. $p(x) = \dfrac{x^3 + 1}{x^4 - 1}$

Sketch the graph of each rational function. Identify all asymptotes and holes in the graph of the function.

8. $y = \dfrac{x^3}{x^2 - 4x - 3}$

9. $y = \dfrac{x^4}{x^2 - 1}$

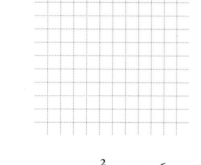

10. $y = \dfrac{x^2 - x - 6}{x^3 + 2x^2 - 9x - 18}$

11. $y = \dfrac{x^2 + 9x + 20}{x^4 - 41x^2 + 400}$

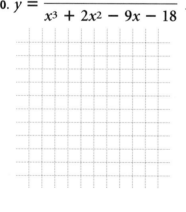
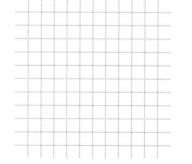

NAME _____ CLASS _____ DATE _____

Practice Masters Level A

8.3 Multiplying and Dividing Rational Expressions

Simplify each expression.

1. $\dfrac{x^2 + 6x + 8}{x^2 + x - 12}$

2. $\dfrac{x^2 + 11x + 30}{x^2 + 5x - 6}$

3. $\dfrac{x^2 - x - 12}{x^2 + 5x + 6}$

4. $\dfrac{x^2 - x - 42}{x^2 - 49}$

5. $\dfrac{x^{12}}{5} \cdot \dfrac{15}{x^4} \cdot \dfrac{x^3}{9}$

6. $\dfrac{42}{x} \cdot \dfrac{x^6}{14} \cdot \dfrac{x^2}{12}$

7. $\dfrac{6x^2 + 18x - 108}{x^2 + 4x - 12} \cdot \dfrac{x + 4}{10x - 30}$

8. $\dfrac{x + 5}{2x - 8} \cdot \dfrac{5x^2 - 5x - 30}{10x^2 + 70x + 100}$

9. $\dfrac{x^2 + 4x - 5}{18} \div \dfrac{x^2 - x}{6}$

10. $\dfrac{x^4}{x^2 + 15x + 54} \div \dfrac{x^2}{x + 9}$

11. $\dfrac{x^4 + 2x^3}{x^2 + 3x + 2} \cdot \dfrac{x^2 - 1}{5x - 5}$

12. $\dfrac{6x^2 - 24x + 24}{14x - 28} \div \dfrac{3x - 6}{x + 1}$

13. $\dfrac{\dfrac{x^2 + x - 6}{x + 5}}{\dfrac{3x^2 - 12}{3x + 15}}$

14. $\dfrac{\dfrac{3x + 6}{x^2 + 3x - 4}}{\dfrac{6x + 12}{x^2 - x - 20}}$

15. $\dfrac{\dfrac{5x - 35}{3x + 9}}{\dfrac{5x^2 + 15x}{x^2 + 6x + 9}}$

16. $\dfrac{\dfrac{x^2 + 5x + 6}{x^2 + 2x - 3}}{\dfrac{10x + 40}{10}}$

Algebra 2   Practice Masters Levels A, B, and C   145

Practice Masters Level B

8.3 Multiplying and Dividing Rational Expressions

Simplify each expression.

1. $\dfrac{x^2 + 10x + 24}{x^2 + 2x - 24}$

2. $\dfrac{x^3 + x^2 - 4x - 4}{x^2 + 4x + 3}$

3. $\dfrac{16x^4 + 112x^3 + 160x^2}{4x^2 + 8x}$

4. $\dfrac{x^3 + 4x^2 - 5x}{x^2 + 6x + 5}$

5. $\dfrac{14x^3}{27} \cdot \dfrac{-9}{2x^5} \cdot \dfrac{-6x^3}{3}$

6. $\dfrac{-3}{x^2} \cdot \dfrac{5}{x^3} \cdot \dfrac{3x^3}{30} \cdot \dfrac{1}{-6x^3}$

7. $\dfrac{x^2 - 9}{x^2 + 2x - 8} \cdot \dfrac{x^2 + 9x + 20}{x^2 - 3x}$

8. $\dfrac{x^2 + 8x + 16}{x^3 + 10x^2 + 32x + 32} \cdot x^2 + 8x + 16$

9. $\dfrac{6x^2 + 30x}{x^2 + 6x + 5} \div \dfrac{x^2 + 4x + 4}{x^2 - x - 6}$

10. $\dfrac{x^2 - 5x + 6}{x^2 + 7x + 12} \div \dfrac{x^2 - 3x + 2}{2x^4 + 6x^3}$

11. $\dfrac{x^2 - 4x - 12}{x^3 - 2x^2 - 4x + 8} \div \dfrac{1}{x^2 + 2x - 8}$

12. $\dfrac{1}{x^2 - 36} \cdot \dfrac{4x^2 + 32x + 48}{5x + 15}$

13. $\dfrac{\dfrac{x^2 + 6x + 9}{x^2 + 10x + 24}}{\dfrac{x + 3}{x^2 + 3x - 18}}$

14. $\dfrac{\dfrac{x^2 + 11x + 24}{x^2 - 12x + 35}}{\dfrac{5x + 40}{2x^2 - 14x}}$

15. $\dfrac{\dfrac{3x^3 + 24x^2}{x^2 + 18x + 80}}{\dfrac{x - 9}{x^2 + 20x + 100}}$

16. $\dfrac{\dfrac{4x^2 - 20x}{x^3 - 19x - 30}}{\dfrac{1}{x - 3}}$

146 Practice Masters Levels A, B, and C Algebra 2

Practice Masters Level C

8.3 Multiplying and Dividing Rational Expressions

Simplify each rational expression.

1. $\dfrac{x^3 - 3x^2 - 4x + 12}{x^3 + 3x^2 - 6x - 8}$

2. $\dfrac{6x^3 - 54x^2 + 120x}{x^3 - 15x^2 + 75x - 125}$

3. $\dfrac{x^2}{10} \cdot \left(\dfrac{x^3}{5}\right)^{-1} \cdot \dfrac{2x^{-2}}{x}$

4. $5xy \div \dfrac{10x^2}{y} \div \dfrac{4y^2}{x^3}$

5. $\dfrac{9x^2 + 3x - 2}{24x^2 - 2x - 1} \cdot \dfrac{6x + 1}{3x - 1}$

6. $\dfrac{4x^2 + 27x - 7}{9x^2 + 12x - 5} \cdot \dfrac{4x^2 - 7x + 3}{2x^2 + 13x - 7}$

7. $\dfrac{6x^2 + 13x - 5}{10x^2 + 29x + 10} \div \dfrac{6x^2 - 14x + 4}{8x^3 - 44x^2 + 48x}$

8. $(x^3 + 27) \cdot \dfrac{\dfrac{1}{x^3 - 37x - 84}}{\dfrac{1}{x^2 - 3x - 28}}$

9. $\dfrac{\dfrac{9x^2 + 15x}{28x^2 - x - 2}}{\dfrac{30x^2 + 35x - 25}{8x^2 - 2x - 1}}$

10. $\dfrac{5x^5 - 40x^2}{x^4 - 40x} \div \dfrac{15x^3 - 40x^2 - 300x}{3x^2 - 8x - 60}$

11. $\dfrac{(4x^2 + 2x - 30)^{-1}(2x^2 + x - 15)}{(2x - 5)^{-2}}$

12. $\dfrac{x^2 + 6x + 9}{x^2} \div \dfrac{4x^2 + 4x + 1}{x^2 - 3x} \cdot \dfrac{2x + 1}{x^2 - 9}$

13. What rational expression divided into $\dfrac{6x^2 + x - 1}{9x^2 + 12x - 5}$ equals $\dfrac{3x - 1}{4x + 7}$? _____

14. The quotient of $\dfrac{18x^2 - 3x - 1}{30x^2 + 29x + 4}$ and what rational expression equals $\dfrac{5x + 4}{x^2}$? _____

NAME _____ CLASS _____ DATE _____

Practice Masters Level A
8.4 Adding and Subtracting Rational Expressions

Write each expression as a single rational expression in simplest form.

1. $\dfrac{4}{x} + \dfrac{6}{x}$

2. $\dfrac{-3}{x+2} + \dfrac{6}{x+2}$

3. $\dfrac{7x}{12} - \dfrac{3x}{4}$

4. $\dfrac{2x+3}{9} - \dfrac{3x-5}{6}$

5. $\dfrac{4x-2}{10} - \dfrac{3x-2}{6}$

6. $\dfrac{-4}{x+2} + \dfrac{9}{x-3}$

7. $\dfrac{5x}{x+5} + \dfrac{-8x}{x-7}$

8. $\dfrac{x+1}{x+4} + \dfrac{x-2}{x-5}$

9. $\dfrac{x-2}{x^2+8x} + \dfrac{x+5}{x-3}$

10. $\dfrac{x}{x+7} - \dfrac{x+4}{x^3-3x^2}$

11. $\dfrac{4x}{x^2-16} + \dfrac{6}{x-4}$

12. $\dfrac{5}{x^2+7x+10} - \dfrac{4}{x^2-x-6}$

13. $\dfrac{x}{x^2-9} + \dfrac{x}{x^2+6x-27}$

14. $\dfrac{5}{x^2+7x+10} - \dfrac{4}{x^2-x-6}$

15. $\dfrac{x}{x^2+4x+4} + \dfrac{5}{x^2-4}$

16. $\dfrac{x+1}{x^2+11x+10} - \dfrac{x}{x^2+x}$

148 Practice Masters Levels A, B, and C   Algebra 2

Practice Masters Level B
8.4 Adding and Subtracting Rational Expressions

Write each expression as a single rational expression in simplest form.

1. $\dfrac{3}{x^2} + \dfrac{7}{x^4}$

2. $\dfrac{-4}{x^2 + 2x} + \dfrac{x}{x + 2}$

3. $\dfrac{2x}{x + 3} + \dfrac{-5x}{x - 2}$

4. $\dfrac{6}{x - 5} - \dfrac{10}{x + 7}$

5. $\dfrac{12}{x^2 + 6x} - \dfrac{x}{x - 2}$

6. $\dfrac{3x}{x^2 + 2x} + \dfrac{4}{x - 2}$

7. $\dfrac{x + 3}{x^2 + 9x + 14} + \dfrac{x - 3}{x^2 - 3x - 10}$

8. $\dfrac{x + 2}{x^2 - 6x - 7} - \dfrac{x - 2}{x^2 - x - 42}$

9. $\dfrac{x + 1}{x^2 - 16} + \dfrac{x + 2}{x^2 - 6x + 8}$

10. $\dfrac{x + 5}{x^2 - 5x - 50} - \dfrac{x + 3}{x^2 + 15x + 50}$

11. $\dfrac{x + 4}{x^3 + 5x^2 + 6x} - \dfrac{x + 2}{x^3 - 2x^2 - 8x}$

12. $\dfrac{x}{x^3 + 2x^2 + x} + \dfrac{x - 10}{x^3 + 12x^2 + 11x}$

13. $\dfrac{2x}{x^2 - x - 6} - \dfrac{x + 1}{x - 3} + \dfrac{x + 4}{x + 2}$

14. $\dfrac{x}{x^2 + 4x + 3} - \dfrac{4}{x^2 - 4x - 5} + \dfrac{2x}{x + 1}$

15. $\dfrac{\dfrac{5}{x - 1}}{\dfrac{x}{x - 1}} + \dfrac{2}{x^2}$

16. $\dfrac{\dfrac{x + 2}{3}}{\dfrac{x - 1}{4}} - \dfrac{1}{x^2 + x - 2}$

Algebra 2   Practice Masters Levels A, B, and C

NAME _____ CLASS _____ DATE _____

Practice Masters Level C
8.4 Adding and Subtracting Rational Expressions

Write each expression as a single rational expression in simplest form.

1. $\dfrac{3x}{2x+8} - \dfrac{4}{x-4} - \dfrac{2x}{x^3+x^2-16x-16}$

2. $\dfrac{x}{2x+6} + \dfrac{4}{x-3} - \dfrac{x+1}{x^2+x-12}$

3. $4x^{-1} + xy^{-1}$

4. $\dfrac{2x + x^{-1}y}{(xy)^{-2}}$

5. $\dfrac{\dfrac{4x+20}{5x}}{\dfrac{x^2-25}{10x^2}} + \dfrac{4}{x+5}$

6. $\dfrac{(x+y)^{-1}}{(x-y)^{-1}}$

7. $x^{-1} + y^{-1}$

8. $(x+1)^{-1} - (x-2)^{-1}$

9. $(x-3)^{-2} + (x-4)^{-2}$

10. $x^{-3} + (x+1)^{-3}$

Find expressions for A and B as indicated, to make the resulting equation true.

11. $\dfrac{x}{x+2} + \dfrac{A}{x^2-x-2} - \dfrac{2}{x^2-4} = \dfrac{x^3+2x+6}{B}$

12. $\dfrac{6a}{a-3b} - \dfrac{-a}{3b-a} + \dfrac{A}{a^2-6b+9b^2} = \dfrac{7a^2-19ab}{B}$

13. $\dfrac{A}{x^2-x-2} + \dfrac{x-4}{x^2+5x+4} + \dfrac{x+5}{x^2+2x-8} = \dfrac{3x^2+7x+25}{B}$

14. $\dfrac{x+5}{4x+1} - \dfrac{x}{6x^2-7x+2} - \dfrac{x-2}{8x^2-2x-1} = \dfrac{A}{(2x-1)(3x-2)(4x+1)}$

150   Practice Masters Levels A, B, and C   Algebra 2

Practice Masters Level A

8.5 Solving Rational Equations and Inequalities

Solve each equation. Check your solution. Give exact answers.

1. $\dfrac{x-2}{3x} = \dfrac{1}{4}$

2. $\dfrac{3x+2}{2x} = 2$

3. $\dfrac{3x}{x-2} = \dfrac{1}{3}$

4. $\dfrac{-6}{x+2} = \dfrac{5}{x-3}$

5. $\dfrac{2}{x} + \dfrac{1}{x+2} = \dfrac{1}{4}$

6. $\dfrac{4}{x-3} + \dfrac{3}{x+1} = 2$

7. $\dfrac{4}{x^2-16} + \dfrac{5}{x+4} = \dfrac{x+4}{x^2-16}$

8. $\dfrac{3x-6}{x} + 1 = \dfrac{10}{x} + \dfrac{4}{x}$

9. $\dfrac{7x+3}{x^2-8x+15} + \dfrac{3x}{x-5} = \dfrac{-1}{x-3}$

10. $\dfrac{3}{x-5} - \dfrac{2}{x+2} = 2x$

Solve each inequality. Check your solution. Give exact answers.

11. $\dfrac{x}{3} < 2x - 2$

12. $\dfrac{x+1}{x} > -3$

13. $-4 > \dfrac{x}{x-2}$

14. $\dfrac{x}{x+5} > x$

15. $\dfrac{x}{2} \leq \dfrac{x^2}{3}$

16. $\dfrac{x+4}{x-2} < x^2$

17. $\dfrac{x+5}{x} < -x^2$

18. $3x^2 > \dfrac{-(5+x)}{x^2}$

19. $\dfrac{1}{4x+1} > -x^3$

20. $-x^2 > \dfrac{1}{3x-6}$

Algebra 2   Practice Masters Levels A, B, and C   151

Practice Masters Level B
8.5 Solving Rational Equations and Inequalities

Solve each equation. Check your solution. Give exact answers.

1. $x^{-1} = \dfrac{1}{x^{-2}}$

2. $(x+1)^{-1} = \dfrac{1}{x^{-2}}$

3. $\dfrac{x+4}{x^2} = \dfrac{x+1}{x-5}$

4. $\dfrac{x}{x-3} = 4x^{-2}$

5. $x^{-3} = \left(\dfrac{1}{x^{-3}}\right)^{-1}$

6. $x^2 + 6x - 5 = -x^2 - 5x + 7$

7. $-x^2 + 2x - 3 = x^2 + 2x - 3$

8. $x^2 - 4 = (x+2)x^{-1}$

9. $\dfrac{1}{x+1} + \dfrac{2}{x+2} = \dfrac{x+3}{4}$

10. $\dfrac{2}{x+3} - \dfrac{3}{4-x} = \dfrac{2x-2}{x^2-x-12}$

Solve each inequality. Check your solution.

11. $\dfrac{x+2}{x} < \dfrac{x}{x-5}$

12. $x^2 + 3x - 2 > x^2 - 9$

13. $x^3 - 1 < x^2 + 4x + 8$

14. $x^3 + 2x > x^3 - x^2 - 2$

15. $x^3 + 2x^2 - 5x - 6 > x^4 + 1$

16. $x^4 - 3 < x^2$

17. $2x^3 - 3x < x^3 + x^2 - 5$

18. $x^3 + 2x^2 - 3x > 2x^3 + 3x - 4$

19. $x^4 - x^3 + 2x^2 > x^3 + 3x + 6$

20. $x^3 + 2x^2 - 3x + 4 > x^4 + 2x^3 - 6x^2 + x + 1$

Practice Masters Level C

8.5 Solving Rational Equations and Inequalities

State whether each equation or inequality is always true, sometimes true, or never true.

1. $\dfrac{x^2}{x+2} > 0.25x - 3$

2. $\dfrac{x+2}{x^2+3x+1} \geq -x^2 + 2x - 4$

3. $6x^2 - 3x + 7 = 2x^2 - 2x + 4$

4. $-x^2 + 2x + 1 \leq 3x^2 + 4$

Use a graphic calculator to solve each rational inequality. Round answers to the nearest tenth.

5. $\dfrac{x+3}{x-4} > x^3$

6. $\dfrac{x^2 + 2x - 3}{x} > 2x - 4$

7. $x^2 + x^3 < x^2 - 0.5$

8. An object weighing w_0 kilograms on Earth is h kilometers above Earth. The function that represents the object's weight at that altitude is $w(h) = w_0 \left(\dfrac{6400}{6400 + h}\right)^2$.

   a. Find the approximate altitude of a satellite that weighs 4000 kilograms on Earth and 1500 kilograms in space.

   b. Find the approximate altitude of a satellite that weighs 10,000 kilograms on Earth and 3000 kilograms in space.

9. The table shows the distances and speeds for the three segments of a triathlon.

   a. Write a rational equation to represent the total time, in hours, for the triathlon in terms of bicycling speed, s, in miles per hour.

   b. Find the speeds at which the athlete needs to complete in each event in order to complete the triathlon in $2\frac{1}{3}$ hours.

	Distance, in miles	Speed, in miles per hour
Swimming	$d_s = 1.5$	$\dfrac{s}{10}$
Bicycling	$d_b = 40$	s
Running	$d_r = 6.2$	$s - 25$

Practice Masters Level A
8.6 Radical Expressions and Radical Functions

Find the domain of each radical function.

1. $f(x) = \sqrt{3x - 1}$ _____ 2. $f(x) = \sqrt{4(x - 2)}$ _____

3. $f(x) = \sqrt{x^2 - 9}$ _____ 4. $f(x) = \sqrt{x^2 + 4}$ _____

Find the inverse of each quadratic function. Then graph the function and its inverse on the same coordinate plane.

5. $y = x^2 - 4$ 6. $y = x^2 + 1$ 7. $y = x^2 + 2x$

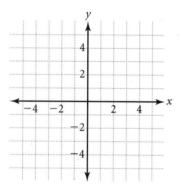
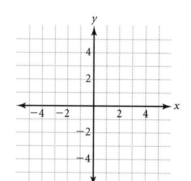
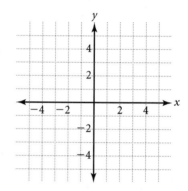

Evaluate each expression. Give exact answers.

8. $\sqrt[3]{\dfrac{125}{27}}$ _____ 9. $\sqrt[3]{\dfrac{-8}{64}}$ _____ 10. $3\sqrt[4]{81}$ _____

11. $-2\sqrt[5]{243}$ _____ 12. $\sqrt[3]{\dfrac{216}{1000}}$ _____ 13. $\dfrac{\sqrt[3]{1}}{\sqrt[4]{625}}$ _____

For the function, describe the transformation applied to $f(x) = \sqrt{x}$. Then graph both functions.

14. $g(x) = \sqrt{x + 3} - 2$

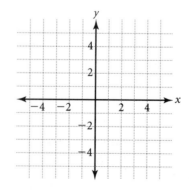

NAME _____ CLASS _____ DATE _____

Practice Masters Level B
8.6 Radical Expressions and Radical Functions

Find the domain of each radical function.

1. $f(x) = \sqrt{x^2 + 10x + 16}$ _____

2. $f(x) = \sqrt{x^2 - 2x - 3}$ _____

3. $f(x) = \sqrt{x^2 + 4x - 5}$ _____

4. $f(x) = \sqrt{x^2 + 6x + 9}$ _____

Find the inverse of each quadratic function. Then graph the function and its inverse on the same coordinate plane.

5. $y = x^2 + 5x + 6$

6. $y = x^2 - 3x - 40$

7. $y = x^2 + 3$

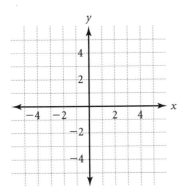
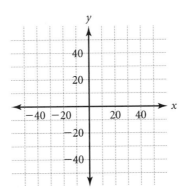
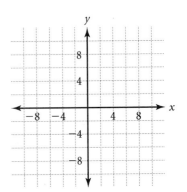

Evaluate each expression. Give exact answers.

8. $\sqrt[3]{\dfrac{-1000}{8}}$ _____

9. $-\sqrt[4]{\dfrac{256}{160,000}}$ _____

10. $\sqrt[3]{\dfrac{-27}{6}}$ _____

11. $2\sqrt[3]{729} + 3\sqrt[3]{125}$ _____

12. $-\sqrt[4]{16} - 3\sqrt[4]{10,000}$ _____

13. $(2\sqrt[3]{216})^2$ _____

For each function, describe the transformation applied to $f(x) = \sqrt{x}$. Then graph each transformed function.

14. $g(x) = \sqrt{3x - 6}$ _____

15. $g(x) = \sqrt{4x + 10}$ _____

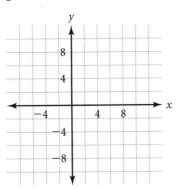
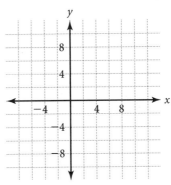

Algebra 2 — Practice Masters Levels A, B, and C

NAME _____ CLASS _____ DATE _____

Practice Masters Level C

8.6 Radical Expressions and Radical Functions

For each function, a) find the domain, b) find the inverse, and c) graph the function and its inverse.

1. $y = \sqrt{x^2 - 4x + 3}$

   a. _____
   b. _____
   c.

2. $y = \sqrt{x^2 + 10x + 25}$

   a. _____
   b. _____
   c.

3. $y = \sqrt{x^2 - 7x - 30}$

   a. _____
   b. _____
   c.

4. The volume, V, of a sphere with a radius of r is given by $V = \frac{4}{3}\pi r^3$.

   a. Find the radius of a sphere with a volume of $\frac{500\pi}{3}$ cubic units. _____

   b. Find the radius, to the nearest hundredth, of a sphere with a volume of 8000 cubic units. _____

Evaluate each expression. Give exact answers.

5. $\sqrt[3]{-120} + \sqrt[3]{400}$

6. $\left(\sqrt[4]{6}\right)\left(\sqrt[4]{24}\right)$

7. $\left(4\left(\sqrt[3]{-90}\right)^2\right)^3$

8. $\left(\sqrt[3]{45} + 2\right)^2$

For each function, describe the transformation applied to $f(x) = \sqrt{x}$.

9. $y = \frac{1}{4}\sqrt{2x + 5} - 3$

10. $y = -6\sqrt{-2x + 4} + 5$

Practice Masters Level A
8.7 Simplifying Radical Expressions

Simplify each radical expression by using the Properties of nth Roots.

1. $\sqrt{40}$

2. $\sqrt{32}$

3. $\sqrt[3]{56}$

4. $\sqrt[3]{54x^5}$

5. $\sqrt[4]{32x^5y^6}$

6. $(-64x^9y^6)^{\frac{1}{3}}$

Simplify each product or quotient. Assume that the value of each variable is positive.

7. $\sqrt{3x^5} \cdot \sqrt{4x^7}$

8. $(\sqrt{5x^3y})^2$

9. $\dfrac{\sqrt[3]{81x^{10}y^8}}{\sqrt[3]{3xy}}$

10. $\sqrt[3]{25x^4y} \cdot \sqrt[3]{10x^3y^5}$

Find each sum, difference, or product. Give your answer in simplest radical form.

11. $(\sqrt{3} + 2)(\sqrt{6} - 3)$

12. $(4\sqrt{2} + 5)(3 - 6\sqrt{2})$

13. $(25 + 5\sqrt{3}) + (13 - 8\sqrt{3})$

14. $(6 - 3\sqrt{2}) - (4 + 7\sqrt{2})$

15. $5\sqrt{3}(\sqrt{6} - 4\sqrt{8})$

16. $(4 + 6\sqrt{5}) - (8 - 5\sqrt{5})$

Write each expression with a rational denominator and in simplest form.

17. $\dfrac{5}{\sqrt{3}}$

18. $\dfrac{18}{\sqrt{8}}$

19. $\dfrac{2}{4 + \sqrt{2}}$

20. $\dfrac{6}{\sqrt{5} - 3}$

Algebra 2   Practice Masters Levels A, B, and C

Practice Masters Level B
8.7 Simplifying Radical Expressions

Simplify each radical expression by using the Properties of nth Roots.

1. $\sqrt{48}$

2. $\sqrt{72}$

3. $\sqrt[3]{-48}$

4. $\sqrt[4]{162x^7}$

5. $\sqrt[3]{128x^5y^4}$

6. $\left(64x^6y^{10}\right)^{\frac{1}{5}}$

Simplify each product or quotient. Assume that the value of each variable is positive.

7. $\sqrt[3]{2x^4} \cdot \sqrt[3]{12x^5}$

8. $\left(-7\sqrt[3]{x}\right)^3$

9. $\dfrac{\sqrt[3]{625x^{12}y^{16}}}{\sqrt[3]{5x^2y^4}}$

10. $\sqrt[4]{4x^2y^3} \cdot \sqrt[4]{8x^3y^6}$

Find each sum, difference, or product. Give your answer in simplest radical form.

11. $\left(4 - 6\sqrt{5}\right) - \left(-3 - 7\sqrt{5}\right)$

12. $\left(6 + 8\sqrt{3}\right) + \left(-5 - 16\sqrt{3}\right)$

13. $\left(3 + 2\sqrt{5}\right)\left(4 - 6\sqrt{5}\right)$

14. $\left(\sqrt{3} - 10\right)\left(5 + 8\sqrt{3}\right)$

15. $\left(2\sqrt{5} + 3\sqrt{7}\right)\left(-3\sqrt{5} - 6\sqrt{7}\right)$

16. $6\sqrt{3}\left(\sqrt{8} + 4\sqrt{6} - 9\sqrt{12}\right)$

Write each expression with a rational denominator and in simplest form.

17. $\dfrac{6}{3\sqrt{5}}$

18. $\dfrac{5}{\sqrt{5} - 6}$

19. $\dfrac{9}{\sqrt{11} + \sqrt{5}}$

20. $\dfrac{8}{\sqrt{6} - \sqrt{13}}$

NAME _____ CLASS _____ DATE _____

Practice Masters Level C
8.7 Simplifying Radical Expressions

Simplify each radical expression by using the Properties of *n*th Roots.

1. $\sqrt[5]{64}$

2. $\sqrt[6]{1{,}000{,}000x^5y^8}$

3. $\sqrt[5]{400{,}000x^{12}y^{15}}$

4. $\sqrt[5]{-x^5y^{13}z^8}$

5. $\sqrt[4]{500{,}000x^{120}}$

6. $\sqrt[5]{-128x^{32}y^{64}}$

Simplify each product or quotient. Assume that the value of each variable is positive.

7. $\sqrt[4]{5xy^9} \cdot \sqrt[4]{125x^7y^3}$

8. $\left(\sqrt[3]{-81x^2y^5}\right)^4$

9. $\dfrac{\left(500x^7y^{15}\right)^{\frac{1}{3}}}{\left(2xy^2\right)^{\frac{1}{3}}}$

10. $\left(\dfrac{1}{8}x^{10}y^3\right)^{\frac{1}{4}} \cdot \left(16x^2y\right)^{\frac{1}{4}}$

Find each sum, difference, or product. Give your answer in simplest radical form.

11. $(5^{\frac{1}{2}} - 4) \cdot (17 - 8^{\frac{1}{2}})$

12. $(4 - 2\sqrt{5}) - (6 + 3\sqrt{5}) + 9\sqrt{5}$

13. $\sqrt[3]{9x^2y^4} \cdot \left(x^2y^3\right)^{\frac{2}{3}}$

14. $4\sqrt{6}(3\sqrt{48} - 2\sqrt{12})$

15. $\sqrt{24}(3\sqrt{8} + 6\sqrt{10})$

16. $5^{\frac{2}{3}}\left(100^{\frac{2}{3}} - 20^{\frac{1}{3}}\right)$

Write each expression with a rational denominator and in simplest form.

17. $\dfrac{1}{3 - 2\sqrt{5}} + (3 - 6\sqrt{5})$

18. $\dfrac{1}{-4 + 2\sqrt{6}} - (4 - 5\sqrt{6})$

19. $\dfrac{2}{5 - 2\sqrt{3}} + \dfrac{3}{6 + 3\sqrt{3}}$

20. $\dfrac{3}{6 + \sqrt{2}} - \dfrac{5}{4 - 2\sqrt{2}}$

Algebra 2   Practice Masters Levels A, B, and C

NAME _____ CLASS _____ DATE _____

Practice Masters Level A
8.8 Solving Radical Equations and Inequalities

Solve each radical equation by using algebra. If the equation has no solution, write *no solution*.

1. $\sqrt{x-3} = 4$

2. $\sqrt{x+5} = 12$

3. $\sqrt{2x-6} = 2$

4. $2\sqrt{x+4} = 10$

5. $\sqrt{3x+4} - 2 = 8$

6. $\sqrt{4x-1} + 3 = 1$

Solve each radical inequality by using algebra. If the inequality has no solution, write *no solution*.

7. $\sqrt{x+2} \geq 5$

8. $\sqrt{x-3} \leq 8$

9. $\sqrt{2x+5} < 17$

10. $\sqrt{3x-2} > 13$

11. $\sqrt{x-2} > \sqrt{x+3}$

12. $2\sqrt{x-6} + 1 < 9$

Solve each radical equation or inequality by using a graph. Round solutions to the nearest tenth.

13. $\sqrt{x+3} = x$

14. $\sqrt{2x+5} = x$

15. $\sqrt{x^2+2} = \sqrt{x^4}$

16. $x - 4 > \sqrt{x^2+2x+1}$

17. $2\sqrt[3]{x} \geq x$

18. $x^2 - 4 \leq \sqrt{x^2+2}$

160   Practice Masters Levels A, B, and C   Algebra 2

NAME _____ CLASS _____ DATE _____

Practice Masters Level B
8.8 Solving Radical Equations and Inequalities

Solve each radical equation by using algebra. If the equation has no solution, write *no solution*. Check your solution.

1. $\sqrt{x+2} = x$

2. $3\sqrt{x+6} = 4\sqrt{x}$

3. $3x = \sqrt{4x+5}$

4. $\sqrt[3]{2x^2 + 3x} = -2$

5. $\sqrt{x+2} = x - 6$

6. $\sqrt{5x-3} = x - 4$

Solve each radical inequality by using algebra. If the inequality has no solution, write *no solution*. Check your solution.

7. $x > \sqrt{x+2}$

8. $\sqrt{3x+1} \geq 2x$

9. $\sqrt{x^2 - 9} \leq \sqrt{x^2 + 6x + 9}$

10. $\sqrt{x^2} \leq 5$

11. $4\sqrt{2x-1} \geq -3$

12. $3\sqrt{6x-2} \leq 1$

Solve each radical equation or inequality by using a graph. Round solutions to the nearest tenth. Check your solution using any method.

13. $4\sqrt{2x+3} = 3x$

14. $\frac{1}{4}\sqrt{3x - \frac{1}{2}} = \frac{x}{5}$

15. $\sqrt[3]{\frac{1}{2}x - 3} = \sqrt{x-2}$

16. $\sqrt{3x+1} > \sqrt{4x-2}$

17. $x - 3 < \sqrt{\frac{x}{2}}$

18. $\sqrt{2x+11} \leq x^2 + 2$

Algebra 2 Practice Masters Levels A, B, and C 161

NAME _____ CLASS _____ DATE _____

Practice Masters Level C
8.8 Solving Radical Equations and Inequalities

Solve each radical equation by using algebra. If the equation has no solution, write *no solution*. Check your solution.

1. $\sqrt{2x + 4} = 3\sqrt{x^2}$

2. $\sqrt{x^2 + 6x + 3} = 4 - x$

3. $3\sqrt{x^2 - 9} = x + 2$

4. $4\sqrt[3]{x + 3} = -2\sqrt[3]{x}$

5. $-\sqrt[3]{x - 2} = \sqrt[3]{-3x^2}$

6. $\sqrt{(x + 3)^2} = \sqrt{(x - 2)^2}$

Solve each radical inequality by using algebra. If the inequality has no solution, write *no solution*. Check your solution.

7. $\sqrt{x^2 - 5x + 4} \leq 2$

8. $5 < \sqrt{x^2 - 6x + 10}$

9. $\sqrt{x^2 + 3x - 5} \geq 9$

10. $4 > \sqrt{x^2 + 4x - 6}$

11. $\sqrt{4x - 2} > \sqrt{6x}$

12. $\sqrt{(x - 2)^2} \geq \sqrt{x + 4}$

Solve each radical equation or inequality by using a graph. Round solutions to the nearest tenth. Check your solution using any method.

13. $\sqrt[3]{x - 2} > \sqrt[3]{x^2 + 2x - 5}$

14. $\dfrac{x}{2} + 5 \geq \sqrt{x^2 - 20x + 10}$

15. $\sqrt[4]{x^2 + 6x + 3} \leq \sqrt[4]{x - 2}$

16. $\sqrt{x - 5} > \sqrt[4]{x + 3}$

17. $\sqrt[3]{x^2 + 2} < \sqrt{x^4 - x^2 + 6}$

18. $\sqrt[3]{x^3 - 2x^2 + 3x - 1} \geq -\sqrt[4]{x^2 - 6x + 5}$

162   Practice Masters Levels A, B, and C   Algebra 2

Practice Masters Level A

9.1 Introduction to Conic Sections

Solve each equation for y, graph the resulting equation, and identify the conic section.

1. $x^2 + y^2 = 16$

2. $y - 4x^2 = 0$

3. $3x^2 + 6y^2 = 12$

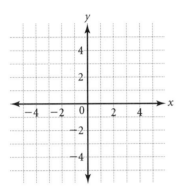
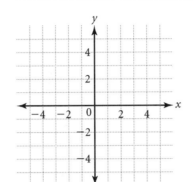
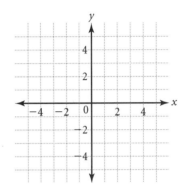

Find the distance between P and Q, and find the coordinates of M, the midpoint of $\overline{PQ}$. Give exact answers and approximate answers to the nearest hundredth when appropriate.

4. $P(2, 4)$ and $Q(5, 8)$

5. $P(5, 3)$ and $Q(10, 15)$

6. $P(0, 0)$ and $Q(-5, -5)$

7. $P(1, 4)$ and $Q(-2, -3)$

8. $P(-5, -4)$ and $Q(-6, -10)$

9. $P(-3, -5)$ and $Q(4, -7)$

Find the center, circumference, and area of the circle whose diameter has the given endpoints.

10. $P(4, 5)$ and $Q(10, 13)$

11. $P(0, 3)$ and $Q(9, 15)$

12. $P(-4, -6)$ and $Q(1, 6)$

13. $P(5, 2)$ and $Q(13, 17)$

Algebra 2   Practice Masters Levels A, B, and C   163

Practice Masters Level B
9.1 Introduction to Conic Sections

Solve each equation for y, graph the resulting equation, and identify the conic section.

1. $x^2 - 4y = 0$

2. $4x^2 + 5y^2 = 10$

3. $3x^2 - 2y^2 = 6$

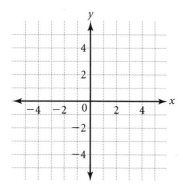
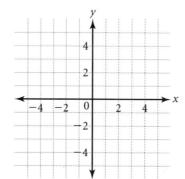
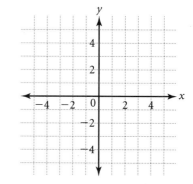

Find the distance between P and Q, and find the coordinates of M, the midpoint of $\overline{PQ}$. Give exact answers and approximate answers to the nearest hundredth when appropriate.

4. $P(-4, -12)$ and $Q(-12, -27)$

5. $P(-2, 4)$ and $Q(1, 7)$

6. $P\left(\dfrac{1}{2}, 6\right)$ and $Q\left(\dfrac{-1}{2}, 5\right)$

7. $P\left(3\dfrac{1}{2}, -2\right)$ and $Q(5, 5)$

8. $P\left(-1, \dfrac{3}{2}\right)$ and $Q\left(\dfrac{5}{2}, -3\right)$

9. $P\left(\dfrac{1}{2}, \dfrac{7}{2}\right)$ and $Q\left(\dfrac{5}{2}, \dfrac{-3}{2}\right)$

Find the center, circumference, and area of the circle whose diameter has the given endpoints.

10. $P(-3, 4)$ and $Q(-2, -5)$

11. $P(2, -5)$ and $Q(-3, 6)$

12. $P\left(\dfrac{1}{2}, \dfrac{-1}{2}\right)$ and $Q\left(\dfrac{5}{2}, \dfrac{-7}{2}\right)$

13. $P\left(\dfrac{1}{2}, 2\right)$ and $Q\left(-3, \dfrac{3}{2}\right)$

Practice Masters Level C

9.1 Introduction to Conic Sections

Solve each equation for y, graph the resulting equation, and identify the conic section.

1. $x^2 + 6y^2 = 8$

2. $2x^2 - 4y^2 = 10$

3. $4x^2 + 3y = 0$

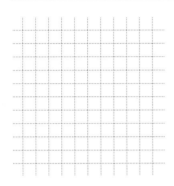
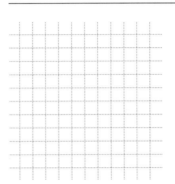
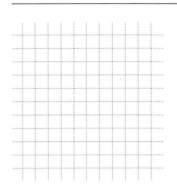

Determine whether the points form the vertices of a right triangle. If so, determine the midpoint of the hypotenuse.

4. $A(5, 1)$, $B(2, -3)$ and $C(2, 1)$

5. $A(5, 1)$, $B(1, 7)$ and $C(-5, -2)$

6. $A(6, 3)$, $B(-2, 4)$ and $C(4, 0)$

7. $A(-3, -3)$, $B(-8, 12)$ and $C(3, -1)$

For Exercises 8–11, use the formula $AB + BC = AC$, to determine whether the set of points are collinear.

8. $A(2, 3)$, $B(3, 8)$ and $C(1, -2)$

9. $A(4, -1)$, $B(-2, 3)$ and $C\left(1, 1\frac{1}{2}\right)$

10. $A\left(-3, -3\frac{1}{2}\right)$, $B(-1, -6)$ and $C(-5, -1)$

11. $A\left(-4\frac{1}{2}, 3\frac{1}{2}\right)$, $B\left(3\frac{1}{2}, 2\frac{1}{2}\right)$ and $C\left(\frac{-1}{2}, 3\right)$

12. An isosceles triangle has coordinates $X(1, 2)$, $Y(-4, 3)$, and $Z\left(-3\frac{1}{2}, -7\frac{1}{2}\right)$. Find the area of $\triangle XYZ$.

Algebra 2   Practice Masters Levels A, B, and C

Practice Masters Level A

9.2 Parabolas

Write the standard equation for each parabola graphed below.

1.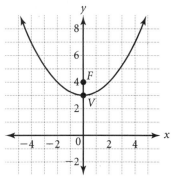

2.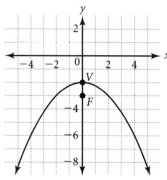

3.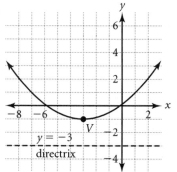

Graph each equation. Label the vertex, focus and directrix.

4. $y = \dfrac{1}{8}x^2$

5. $x = \dfrac{1}{12}y^2$

6. $y - 2 = \dfrac{1}{4}(x + 2)^2$

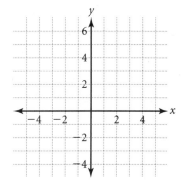

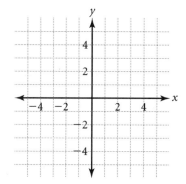

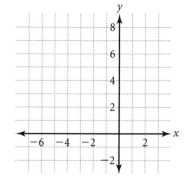

Write the standard equation for the parabola with the given characteristics.

7. vertex $(0, 0)$; focus $(0, 5)$ _____

8. vertex $(0, 0)$; focus $(-4, 0)$ _____

9. vertex $(-3, 0)$; directrix: $x = -6$ _____

10. vertex $(0, 4)$; directrix: $y = 9$ _____

11. focus $(2, 7)$; directrix: $y = -3$ _____

12. focus $(5, 1)$; directrix: $x = 0$ _____

166   Practice Masters Levels A, B, and C   Algebra 2

Practice Masters Level B
9.2 Parabolas

Write the standard equation for each parabola graphed below.

1.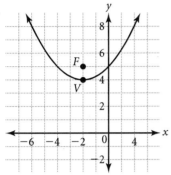

2.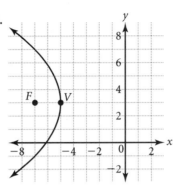

3.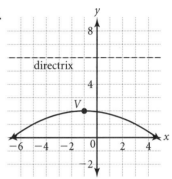

Graph each equation. Label the vertex, focus and directrix.

4. $y = \dfrac{-1}{8}(x - 3)^2$

5. $x + 2 = (y - 1)^2$

6. $y + 2 = -(x - 3)^2$

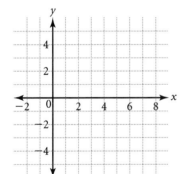
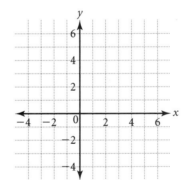
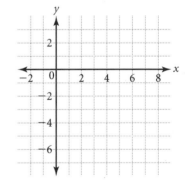

Write the standard equation for the parabola with the given characteristics.

7. vertex $(-3, 2)$; focus $(-3, -3)$ _____

8. vertex $(5, 1)$; directrix: $x = -1$ _____

9. focus $(4, 2)$; directrix: $y = -4$ _____

10. vertex $(-5, -1)$; directrix: $y = -3$ _____

11. focus $(3, 5)$; directrix: $x = 5$ _____

12. focus $(1, 6)$; vertex $(1, -2)$ _____

Practice Masters Level C
9.2 Parabolas

Write the standard equation for each parabola graphed below.

1.
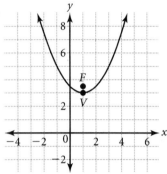

2.
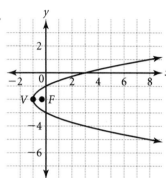

3.
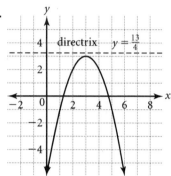

Graph each equation. Label the vertex, focus and directrix.

4. $y - 3 = -(x + 2)^2$

5. $y + \dfrac{1}{2} = 2(x - 3)^2$

6. $x + \dfrac{1}{2} = -2\left(y - \dfrac{3}{2}\right)^2$

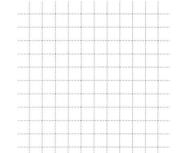

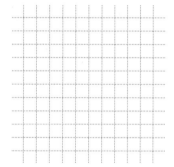

Write the standard equation for the parabola with the given characteristics.

7. vertex $\left(\dfrac{5}{2}, -1\right)$; directrix: $y = \dfrac{-3}{2}$ _____

8. directrix: $y = \dfrac{-1}{2}$; vertex $\left(\dfrac{5}{2}, \dfrac{3}{2}\right)$ _____

9. directrix: $x = \dfrac{3}{2}$; focus $\left(3, \dfrac{1}{2}\right)$ _____

10. vertex $\left(\dfrac{1}{2}, \dfrac{7}{2}\right)$; focus $\left(\dfrac{-3}{2}, \dfrac{7}{2}\right)$ _____

Practice Masters Level A

9.3 Circles

Write the standard equation for each circle graphed below.

1.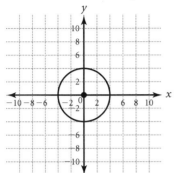

2.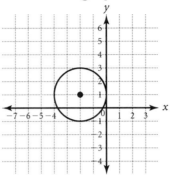

3.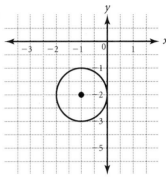

Write the standard equation of a circle with the given radius and center.

4. $r = 3$; $C(0, 0)$

5. $r = 5$; $C(2, 1)$

6. $r = 4$; $C(-3, 4)$

Graph each equation. Label the center.

7. $x^2 + y^2 = 144$

8. $(x - 3)^2 + (y - 1)^2 = 16$

9. $(x + 2)^2 + (y + 2)^2 = 64$

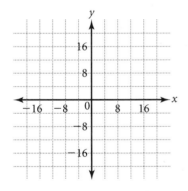
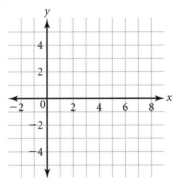
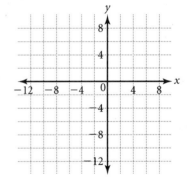

Write the standard equation for each circle. Then state the coordinates of its center and give its radius.

10. $x^2 + y^2 + 6x = 25$

11. $x^2 + y^2 + 4y = 12$

Algebra 2   Practice Masters Levels A, B, and C

Practice Masters Level B
9.3 Circles

Write the standard equation for each circle graphed below.

1.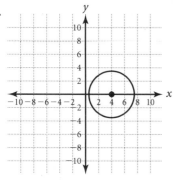

2.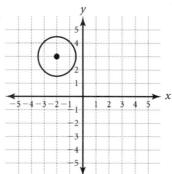

3.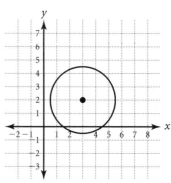

Write the standard equation of a circle with the given radius and center.

4. $r = 2; C\left(\dfrac{1}{2}, 4\right)$

5. $r = \dfrac{3}{4}; C\left(-2, \dfrac{3}{2}\right)$

6. $r = 0.7; C(0.6, -0.3)$

Graph each equation. Label the center.

7. $(x - 2)^2 + (y + 1)^2 = 12$

8. $\left(x + \dfrac{1}{2}\right)^2 + \left(y - \dfrac{1}{2}\right)^2 = \dfrac{9}{4}$

9. $\left(x - \dfrac{5}{2}\right)^2 + \left(y - \dfrac{7}{2}\right)^2 = \dfrac{25}{4}$

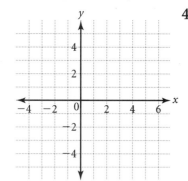
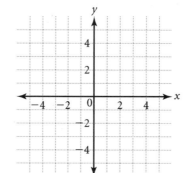
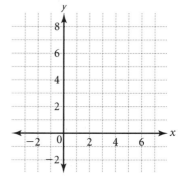

Write the standard equation for each circle. Then state the coordinates of its center and give its radius.

10. $x^2 + y^2 + 4x - 8y - 16 = 0$

11. $x^2 + y^2 - 3x - 5y - \dfrac{1}{2} = 0$

NAME _____ CLASS _____ DATE _____

Practice Masters Level C
9.3 Circles

Write the standard equation for each circle graphed below.

1.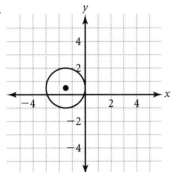

2.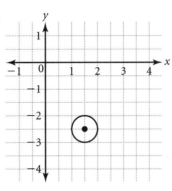

3.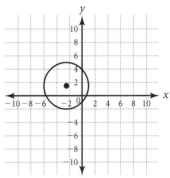

State whether the graph of each equation is a parabola or a circle.

4. $x^2 = 14 - y^2 - 3y$

5. $(x - 4)^2 + y = 6$

6. $-y^2 = x(x - 4)$

State whether the given point is inside, outside, or on the circle whose equation is given.

7. $P(-3, -5);\ (x + 3)^2 + (y - 2)^2 = 49$ _____

8. $P(3, 2);\ (x - 2)^2 + (y - 1)^2 = 16$ _____

9. $P(-5, 3);\ x^2 + (y + 6)^2 = 10$ _____

Graph each equation. Label the center and radius.

10. $x^2 + y^2 - 2x - 4y - 4 = 0$

11. $x^2 + y^2 + 5x - 3y - \dfrac{33}{2} = 0$

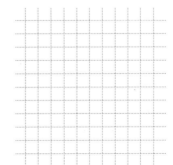

Algebra 2 — Practice Masters Levels A, B, and C

NAME _____ CLASS _____ DATE _____

Practice Masters Level A
9.4 Ellipses

Write the standard equation for each ellipse.

1.
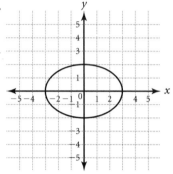

2.
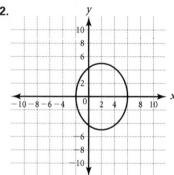

3.
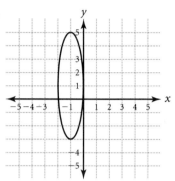

Sketch the graph of each ellipse. Label the center, foci, vertices, and co-vertices.

4. $\dfrac{x^2}{16} + \dfrac{y^2}{4} = 1$

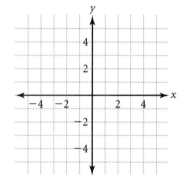

5. $\dfrac{x^2}{1} + \dfrac{y^2}{9} = 1$

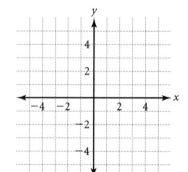

6. $\dfrac{(x-2)^2}{9} + \dfrac{(y-1)^2}{4} = 1$

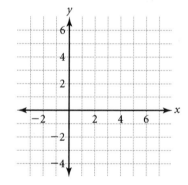

Write the standard equation for the ellipse with the given characteristics.

7. vertices: $(-9, 0)$ and $(9, 0)$; co-vertices: $(0, -7)$ and $(0, 7)$ _____

8. vertices: $(0, -12)$ and $(0, 12)$; co-vertices: $(-5, 0)$ and $(5, 0)$ _____

9. foci: $(-10, 0)$ and $(0, 10)$; vertices: $(-12, 0)$ and $(0, 12)$ _____

10. co-vertices: $(-4, 0)$ and $(0, 4)$; foci: $(0, -9)$ and $(0, 9)$ _____

172 Practice Masters Levels A, B, and C Algebra 2

Practice Masters Level B

9.4 Ellipses

Write the standard equation for each ellipse.

1.
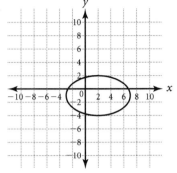

2.
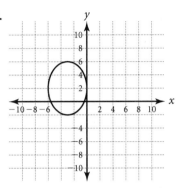

3.
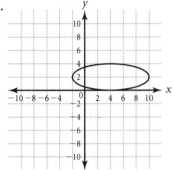

Sketch the graph of each ellipse. Label the center, foci, vertices, and co-vertices.

4. $\dfrac{(x+3)^2}{16} + \dfrac{(y+4)^2}{10} = 1$

5. $\dfrac{(x-2)^2}{6} + \dfrac{(y-4)^2}{9} = 1$

6. $\dfrac{x^2}{12} + \dfrac{(y-2)^2}{4} = 1$

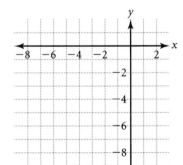
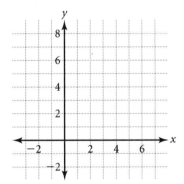
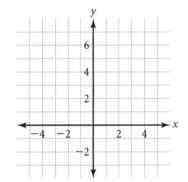

Write the standard equation for each ellipse.

7. $x^2 + 3y^2 + 4x - 6y = 5$ _____

8. $x^2 + 4y^2 + 4x - 24y = 60$ _____

9. $9x^2 + 4y^2 - 18x + 8y = 23$ _____

10. $4x^2 + 16y^2 - 8x + 64y = 28$ _____

Algebra 2   Practice Masters Levels A, B, and C

Practice Masters Level C
9.4 Ellipses

Write the standard equation for each ellipse.

1.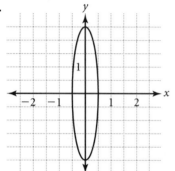

2.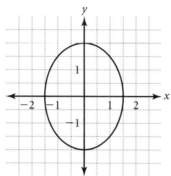

3.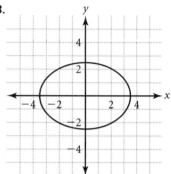

_____   _____   _____

Sketch the graph of each ellipse. Label the center, foci, vertices, and co-vertices.

4. $\dfrac{x^2}{4} + \dfrac{y^2}{2.25} = 1$

5. $\dfrac{x^2}{6.25} + \dfrac{y^2}{0.25} = 1$

6. $\dfrac{x^2}{2.25} + \dfrac{y^2}{12.25} = 1$

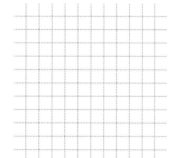 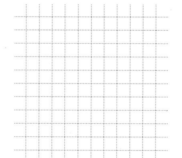 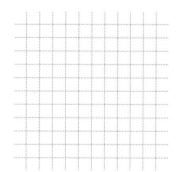

Write the standard equation for each ellipse.

7. $x^2 + 4y^2 + 3x - 8y = 13\dfrac{3}{4}$ _____

8. $2x^2 + y^2 - 6x - 7y = \dfrac{-35}{4}$ _____

9. $3x^2 + 5y^2 + 18x - 20y + 2 = 0$ _____

10. $5x^2 + 3y^2 - 15x + 9y + 16 = 0$ _____

Practice Masters Level A

9.5 Hyperbolas

Write the standard equation for each hyperbola.

1.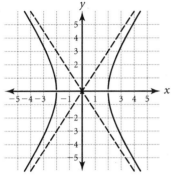

2.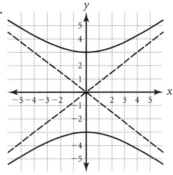

Graph each hyperbola. Label the center, vertices, co-vertices, foci and asymptotes.

3. $\dfrac{x^2}{4} + \dfrac{y^2}{16} = 1$

4. $\dfrac{y^2}{9} + \dfrac{x^2}{25} = 1$

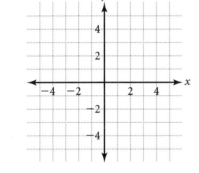

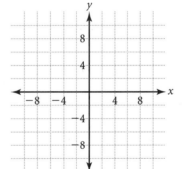

Write the standard equation for the hyperbola with the given characteristics.

5. vertices: $(-5, 0)$ and $(5, 0)$; co-vertices: $(0, -6)$ and $(0, 6)$ _____

6. foci: $(0, -6)$ and $(0, 6)$; vertices: $(0, -4)$ and $(0, 4)$ _____

7. co-vertices: $(-7, 0)$ and $(7, 0)$; foci: $(0, -8)$ and $(0, 8)$ _____

8. vertices: $(-10, 0)$ and $(10, 0)$; foci: $(-12, 0)$ and $(12, 0)$ _____

Algebra 2   Practice Masters Levels A, B, and C

Practice Masters Level B
9.5 Hyperbolas

Write the standard equation for each hyperbola.

1.
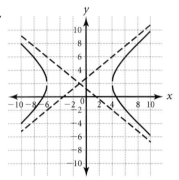

2.
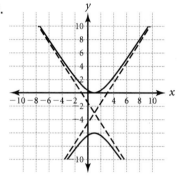

Graph each hyperbola. Label the center, vertices, co-vertices, foci and asymptotes.

3. $(x + 2)^2 - \dfrac{(y - 1)^2}{4} = 1$

4. $\dfrac{(y + 1)^2}{16} - \dfrac{(x + 2)^2}{9} = 1$

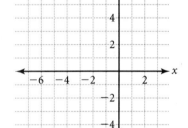

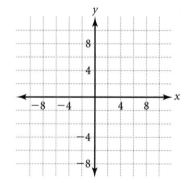

Write the standard equation for the hyperbola with the given characteristics.

5. $x^2 - 4y^2 + 6x + 16y = 11$ _____

6. $4x^2 - y^2 - 8x + 10y = 33$ _____

7. $y^2 - 2x^2 + 12x - 8y = 12$ _____

8. $8y^2 - 3x^2 - 12x - 32y = 4$ _____

Practice Masters Level C

9.5 Hyperbolas

Write the standard equation for each hyperbola.

1.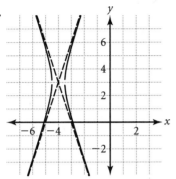

2.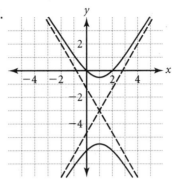

Graph each hyperbola. Label the center, vertices, co-vertices, foci and asymptotes.

3. $\dfrac{(x-3)^2}{2.25} - \dfrac{(y-1)^2}{6.25} = 1$

4. $\dfrac{(y+1)^2}{2.25} - \dfrac{(x-2)^2}{0.25} = 1$

Write the standard equation for the hyperbola with the given characteristics.

5. $2x^2 - 4y^2 + 12x + 24y - 26 = 0$ _____

6. $5y^2 - 3x^2 - 30x - 60y + 60 = 0$ _____

7. $6x^2 - 5y^2 + 18x + 25y - 47.75 = 0$ _____

8. $4y^2 - 9x^2 - 27x - 28y - 43.25 = 0$ _____

Practice Masters Level A
9.6 Solving Nonlinear Systems

Use the substitution method to solve each system. If there are no real-number solutions, write *none*.

1. $\begin{cases} y = 4x \\ y = x^2 \end{cases}$

2. $\begin{cases} y = 2 - x \\ x = y^2 \end{cases}$

3. $\begin{cases} y = x^2 \\ x^2 + y^2 = 20 \end{cases}$

Use the elimination method to solve each system. If there are no real-number solutions, write *none*.

4. $\begin{cases} 2x^2 - 4y^2 = 4 \\ 4x^2 + 9y^2 = 25 \end{cases}$

5. $\begin{cases} 4x^2 - 8y^2 = -68 \\ x^2 + y^2 = 10 \end{cases}$

6. $\begin{cases} x^2 - y^2 = 5 \\ 2x^2 + 9y^2 = 54 \end{cases}$

Solve each system by graphing. If there are no real-number solutions, write *none*.

7. $\begin{cases} x^2 + y^2 = 4 \\ x^2 - y^2 = 4 \end{cases}$

8. $\begin{cases} x^2 + 5y^2 = 70 \\ 3x^2 - 5y^2 = 30 \end{cases}$

9. $\begin{cases} y = x^2 \\ y = -x^2 + 4x \end{cases}$

Classify the conic section defined by each equation.

10. $x^2 + y^2 - 6x + 3y - 12 = 0$

11. $x - y^2 + 2x + y = 0$

12. $3x^2 - 2y^2 + 6x - 10y + 1 = 0$

13. $4x^2 + 9y^2 - 16x + 18y - 5 = 0$

14. $x^2 + 3y - 12y + 2 = 0$

15. $x^2 + 2y^2 + 2x - 15y + 8 = 0$

Practice Masters Level B
9.6 Solving Nonlinear Systems

Use the substitution method to solve each system. If there are no real-number solutions, write *none*.

1. $\begin{cases} y = 2x^2 - 14 \\ x^2 + y^2 = 25 \end{cases}$

2. $\begin{cases} x = -5y^2 + 81 \\ 2x^2 + 4y^2 = 66 \end{cases}$

3. $\begin{cases} -x^2 + y = 1 \\ x + y^2 = 7 \end{cases}$

Use the elimination method to solve each system. If there are no real-number solutions, write *none*.

4. $\begin{cases} 5x^2 + 4y^2 = 145 \\ -3x^2 + 2y^2 = 23 \end{cases}$

5. $\begin{cases} x^2 + y^2 = 26 \\ 9x^2 - 4y^2 = -91 \end{cases}$

6. $\begin{cases} 3x^2 + 5y^2 = 75 \\ 4x^2 - 2y^2 = 100 \end{cases}$

Solve each system by graphing. If there are no real-number solutions, write *none*.

7. $\begin{cases} x^2 - y^2 = 4 \\ 9x^2 + 4y^2 = 36 \end{cases}$

8. $\begin{cases} x^2 + y^2 = 9 \\ y = x^2 + 3 \end{cases}$

9. $\begin{cases} 16x^2 - 4y^2 = 64 \\ 3x^2 - y = -1 \end{cases}$

Classify the conic section defined by each equation. Write the standard equation of the conic section.

10. $x^2 + y - 2x + 3y = 0$

11. $9x^2 - 4y^2 + 18x + 16y = 23$

12. $16x^2 + 25y^2 + 80x + 625y = 125$

13. $4x^2 + 4y^2 - 64 = 0$

14. $x - 2y^2 + 6y - 4 = 0$

15. $6x^2 - 2y^2 + 18x + 4y = 20$

Algebra 1　　　Practice Masters Levels A, B, and C　　　179

NAME _____ CLASS _____ DATE _____

Practice Masters Level C
9.6 Solving Nonlinear Systems

Use any method to solve each system. If there are no real-number solutions, write *none*.

1. $\begin{cases} x = -3y^2 + 29 \\ y^2 - x^2 = 5 \end{cases}$

2. $\begin{cases} 3x^2 + 8y^2 = 140 \\ 5x^2 - 3y^2 = -28 \end{cases}$

3. $\begin{cases} x^2 + y = 3 \\ y = x^2 + 2.5 \end{cases}$

4. $\begin{cases} x = y^2 - 2 \\ x^2 + y^2 = \dfrac{37}{16} \end{cases}$

5. $\begin{cases} 3x^2 - 5y^2 = \dfrac{-1}{2} \\ 6x^2 + 8y^2 = 3\dfrac{1}{2} \end{cases}$

6. $\begin{cases} 9x^2 - 4y^2 = 81 \\ x^2 + y^2 = 9 \end{cases}$

Use a graphics calculator to solve the systems by graphing. Round answers to the nearest hundredth, if necessary.

7. $\begin{cases} y = x^2 + 2x \\ x^2 + y^2 = 9 \end{cases}$

8. $\begin{cases} 4x^2 - 25y^2 = 100 \\ 4y^2 + 25x^2 = 100 \end{cases}$

9. $\begin{cases} x = y^2 + 4y \\ 5x^2 + 3y^2 = 45 \end{cases}$

Classify the conic section defined by each equation. Write the standard equation of the conic section.

10. $3x^2 - 2y^2 + 12x - 8y + 2 = 0$

11. $6x^2 + 6y^2 - 3x + 12y = 10$

12. $x^2 + y - 6x - 2y + 8 = 0$

13. $4x^2 + 3y^2 = 24$

14. $4x^2 - 2y^2 + 3y - 10 = 0$

15. $x + 3y^2 - 2x - y = 11$

Practice Masters Level A
10.1 Introduction to Probability

A box contains 5 red chips, 8 white chips, and 7 blue chips. Find the probability of each event for one draw.

1. a white chip

2. a blue chip

3. a red chip

A spinner is divided into 12 equal regions, numbered 1 through 12. Find the probability of each event for one spin.

4. 2

5. 10

6. an even number

7. an odd number

8. a prime number

9. a number greater than 3

10. a number less than 8

11. a number greater than 1

12. a number ≤ 12

A phone call will come to George's house between 6:00 P.M. and 7:00 P.M. Calculate the probability that George will be home to receive the call for each given time that George arrives at his house.

13. 6:20 P.M.

14. 6:27 P.M.

15. 6:59 P.M.

Find the number of different stereo setups that can be formed given each list of the number of components from which to choose.

16. 5 CD players
    8 sets of speakers
    6 tuners
    4 graphic equalizers

17. 8 CD players
    3 sets of speakers
    4 tuners
    1 graphic equalizer

18. 3 CD players
    5 sets of speakers
    6 tuners
    5 graphic equalizers

Algebra 2 — Practice Masters Levels A, B, and C — 181

Practice Masters Level B
10.1 Introduction to Probability

A card is drawn from a standard 52-card deck. Find the probability of each event for one draw.

1. heart

2. red

3. face card

4. 3

5. 5 of clubs

6. black 6

7. not a 10

8. a jack or an ace

9. not a queen or a king

A hat contains names according to the table below. For Exercises 10–12, use the table to find the probability of each given event if one name is randomly selected from the hat.

Age	Male	Female
Under 12	12	6
13–30	40	52
31 or older	58	42

10. male

11. under 12

12. female 31 or older

13. What is the probability that a randomly thrown dart that hits the square board will land in the circular region?

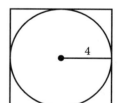

14. A car has 12 choices for outside color, 5 choices for interior design, 3 different tire packages, and 3 different engine sizes. How many different cars can be produced?

15. How many different license plates can be produced using 4 digits and 2 letters if the first digit cannot be the number 0 and the letters O and I cannot be used as letters?

NAME _____ CLASS _____ DATE _____

Practice Masters Level C
10.1 Introduction to Probability

A letter is chosen from the alphabet. Find the probability of each event.

1. not a vowel

2. a letter that comes after P

3. 6

4. a vowel or consonant

5. a letter in the word *cat*

6. a letter in the word *buzzard*

The following table gives the ages of people in the audience at a movie. For Exercises 7–12, use the table to find the probability of each event if one person is randomly selected from the audience.

Age	Male	Female
Under 2	3	5
3–10	24	35
11–16	42	53
17 or older	121	97

7. 10 years old or younger

8. 11 years old or older

9. not a male

10. an 11–16 year old

11. 3 years old or older

12. female over 17 years old

13. What is the probability that a randomly thrown dart that hits the square board will not land in the circular region?

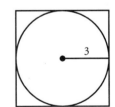

Find the number of possible passwords (numbers 0 and 1 and letters O and L excluded) for each given condition.

14. 3 letters followed by 2 digits and one letter

15. 1 letter followed by 3 digits and 2 letters

Algebra 2   Practice Masters Levels A, B, and C

NAME _____ CLASS _____ DATE _____

Practice Masters Level A
10.2 Permutations

If 10 children are sitting in their classroom ready to line up for lunch, find the number of permutations for each situation.

1. all 10 children line up

2. 5 children line up

3. 9 children line up

4. 3 children line up

If a shelf is large enough to hold 7 trophies, how many ways can the trophies be arranged if the number of trophies available to choose from are:

5. 7 trophies

6. 10 trophies

7. 11 trophies

8. 30 trophies

Find the number of permutations of the letters in each word.

9. math

10. money

11. chosen

12. objects

13. catering

14. circular

15. A lock contains 3 dials, each with ten digits. How many possible sequences of numbers exist?

16. Twelve people would like to sit at a circular table. In how many ways can this be accomplished?

17. The same 12 people in Exercise 16 decide to sit in a straight line. In how many ways can this be accomplished?

18. Four students are to be chosen from a group of 10 to fill the positions of president, vice-president, treasurer and secretary. In how many ways can this be accomplished?

---

184   Practice Masters Levels A, B, and C   Algebra 2

Practice Masters Level B
10.2 Permutations

Find the number of permutations of the digits 0 through 9 for each situation.

1. 2-digit numbers

2. 4-digit numbers

3. 6-digit numbers

4. 10-digit numbers

A display case has room for 10 statues. Find the number of ways 10 statues can be displayed if each number of statues is available.

5. 10 statues

6. 9 statues

7. 11 statues

8. 20 statues

Find the number of permutations of the letters in each word.

9. factorial

10. combination

11. permutation

12. mathematician

13. applications

14. Mississippi

If a class has 30 students, find the number of permutations for each situation.

15. 4 students for student judiciary

16. 7 students for the debate team

17. 12 students for the basketball team

18. 25 students for the baseball team

19. 1 student for President

20. first, second, and third place in the art show

Algebra 2

Practice Masters Level C
10.2 Permutations

Find the number of permutations of the letters of the alphabet for each situation.

1. 4-letter words

2. 6-letter words

3. 13-letter words

4. 26-letter words

A band contains 20 members. In how many ways can members be selected to carry colored flags for each situation?

5. 1 red flag, 1 white flag, and 1 blue flag

6. 3 green flags, 1 yellow flag, and 1 white flag

7. 2 orange flags, 2 brown flags, and 1 yellow

8. 2 blue flags, 2 yellow flags, and 2 white flags

9. a. If 4 delegates are to speak at a convention, in how many ways can they speak?

   b. Find the probability that Bob, a delegate, speaks last.

   c. Find the probability that Bob speaks first or second.

10. Six boys and 6 girls are sitting such that they alternate seats.

    a. In how many ways can they sit in a row?

    b. In how many ways can they sit at a circular table?

11. A form of an identification number contains 9 digits. The first digit cannot be a 0. How many such arrangements can be made?

12. A football team takes the huddle in a circular form with the quarterback in the center. How many ways can the 10 players form the huddle?

13. In how many ways can the expression $w^2x^5y^2z^4$ be written without exponents?

14. In how many ways can the expression $(w^3x^4y^6)^2$ be written without exponents?

NAME _____ CLASS _____ DATE _____

Practice Masters Level A
10.3 Combinations

Find the value of each expression.

1. $_6C_3$

2. $_8C_5$

3. $_7C_2$

4. $_9C_9$

5. $_{10}C_1$

6. $_6C_5 \times _{12}C_4$

Find the number of ways in which each committee can be selected.

7. a committee of 7 from a group of 10 _____

8. a committee of 3 from a group of 8 _____

9. a committee of 9 from a group of 9 _____

10. a committee of 6 from a group of 12 _____

A restaurant offers entrees with a choice of side dishes. Available are 5 different vegetables, 3 types of salad, and 3 types of bread. In how many ways can the following items be chosen?

11. 2 vegetables, 2 salads, and 2 breads

12. 4 vegetables, 2 salads, and 3 breads

13. 2 vegetables, 3 salads, and 1 bread

14. 5 vegetables, and 3 salads

A box contains 5 red chips, 12 white chips, and 3 blue chips. Find the number of ways of selecting each combination.

15. 2 red and 2 white chips

16. 3 white and 2 blue chips

17. 4 red and 3 blue chips

18. 5 red and 6 white chips

Determine if the situation involves a permutation or a combination.

19. In how many ways can 3 numbers be dialed on a locker to open the lock?

20. In how many ways can 5 students out of 25 be given superiors on their science projects?

Algebra 2   Practice Masters Levels A, B, and C   187

NAME _____ CLASS _____ DATE _____

Practice Masters Level B
10.3 Combinations

Find the value of each expression.

1. $_8C_3 \times {}_{10}C_4$

2. $\dfrac{_6C_2}{_4C_3}$

3. $\dfrac{_{12}C_3}{_3C_2}$

4. $\dfrac{_9C_5 \times {}_9C_4}{3!2!}$

5. $\dfrac{_{15}C_3 \times {}_{20}C_4}{_{16}C_2}$

6. $\dfrac{_{14}C_8 \times {}_{14}C_6}{_{14}C_{14}}$

Find the number of ways in which a French test can be made.

7. 20 questions from a test bank of 100 questions _____

8. 16 questions from a test bank of 50 questions _____

9. 4 questions from a test bank of 12 questions _____

10. 32 questions from a test bank of 200 questions _____

Pizzas can be topped with 4 different sauces, 5 different meats and 4 different cheeses. In how many ways can a pizza be made with the following ingredients?

11. 2 meats and 2 cheeses

12. 2 sauces and 3 meats

13. 3 cheeses and 4 meats

14. 3 sauces, 4 meats, and 2 cheeses

A bowl contains 24 candies. Six are red, 10 are yellow, and the rest are green. Find the probability of selecting each combination in a random handful.

15. 4 yellow and 3 green

16. 3 red and 6 green

17. 8 yellow and 2 red

18. 3 red, 5 yellow, and 4 green

Determine if the situation involves a permutation or a combination.

19. In how many ways can 12 members of a jury be selected from a jury pool of 150?

20. In how many ways can a foreman, assistant foreman, and secretary be selected from a 12-member group?

188   Practice Masters Levels A, B, and C   Algebra 2

Practice Masters Level C
10.3 Combinations

A standard deck of cards contain 52 cards.

1. Find the number of 2-card hands that can be dealt. _____

2. Find the probability that both cards in a random 2-card hand are jacks. _____

3. Find the number of 5-card hands that can be dealt. _____

4. Find the probability that 4 aces and a king are randomly dealt. _____

5. Find the probability that 3 tens and 2 sevens are randomly dealt. _____

6. Find the number of 7-card hands that can be dealt. _____

A state lottery game contains numbered balls numbered 1 through 40. Five balls are drawn randomly. If a person matches all 5 numbers they win the jackpot.

7. How many possible number sequences are there if the numbers must be matched in the order that they were drawn? _____

8. How many possible number sequences are there if the numbers can be drawn in any order? _____

9. Find the probability, if the order does not matter, that any single number combination will be drawn. _____

10. Evaluate.

   a. $_{100}C_3$ _____   b. $_{100}C_{97}$ _____

   c. Explain your answers to parts a and b. _____

Determine if the situation involves a permutation or a combination.

11. In how many ways can a 7-digit phone number be formed? _____

12. Find the number of ways that 100 people can be selected for a study on high cholesterol. _____

13. In how many ways can the President select the members of the cabinet from a pool of 20 politicians? _____

14. In how many ways can 5 players be selected to play basketball? _____

15. In how many ways can 10 out of 12 respondents in a survey pick a certain brand of toothpaste? _____

NAME _____ CLASS _____ DATE _____

Practice Masters Level A
10.4 Using Addition with Probability

Two number cubes are rolled. The table shows the possible outcomes. Use the table to state whether the events in each pair below are inclusive or mutually exclusive. Then find the probability of each pair of events.

(1, 1)	(1, 2)	(1, 3)	(1, 4)	(1, 5)	(1, 6)
(2, 1)	(2, 2)	(2, 3)	(2, 4)	(2, 5)	(2, 6)
(3, 1)	(3, 2)	(3, 3)	(3, 4)	(3, 5)	(3, 6)
(4, 1)	(4, 2)	(4, 3)	(4, 4)	(4, 5)	(4, 6)
(5, 1)	(5, 2)	(5, 3)	(5, 4)	(5, 5)	(5, 6)
(6, 1)	(6, 2)	(6, 3)	(6, 4)	(6, 5)	(6, 6)

1. a sum of 6 or a sum of 10

2. a sum of 3 or a sum of 7

3. a sum of 9 or a 5

4. a product of 20 or a 6

5. a sum less than 10 or a sum greater than 8

6. a sum of 3 or less, or double ones

7. a product of 5 or less, or a 6

8. an even number or a sum of 7

9. an odd number or a product greater than 25

10. a product greater than 20 or a product less than 15

A swim team with 25 members has 8 swimmers who swim freestyle, 5 swimmers who swim backstroke, and 7 swimmers who swim breaststroke. Some swimmers participate in more than one event according to the Venn diagram. Find the probability of each event if a swimmer is selected at random.

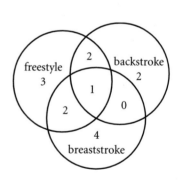

11. swims freestyle

12. swims exactly 2 events

13. swims breaststroke or backstroke

14. does not swim backstroke

15. swims freestyle and backstroke

16. does not swim freestyle, breaststroke, or backstroke

190   Practice Masters Levels A, B, and C   Algebra 2

Practice Masters Level B
10.4 Using Addition with Probability

A card is drawn from a standard 52-card deck. Tell whether the events A and B are inclusive or mutually exclusive. Then find P(A or B).

1. A: The card is red.
   B: The card is a 4.

2. A: The card is a face card.
   B: The card is a club.

3. A: The card is black.
   B: The card is red.

4. A: The card is a heart or a spade.
   B: The card is not a heart.

5. A: The card is less than 10.
   B: The card is a red face card.

6. A: The card is not a face card.
   B: The card is an ace.

7. A: The card is an ace of clubs.
   B: The card is red.

8. A: The card is red.
   B: The card is not a diamond or a heart.

The spinner shown is spun once. Find the probability of each event.

9. The number is odd or blue.

10. The number is prime or white.

11. The number is a six or red.

12. The number is less than 5 or blue.

13. The number is a multiple of 2 or red.

14. The number is greater than 3 or not red.

15. The number has a factor of 3 or not white.

16. The number has more than two factors or is blue.

A coin is tossed 3 times. Find the probability of each event.

17. 1 tail or 3 heads

18. at least 2 heads or no more than 2 tails

Algebra 2 — Practice Masters Levels A, B, and C — 191

Practice Masters Level C
10.4 Using Addition with Probability

A survey of families with 4 children is taken. Tell whether the events are inclusive or mutually exclusive. Then find the probability of each event.

1. 1 boy or 1 girl

2. at least 2 boys or 2 girls

3. 3 boys or 2 girls

4. no more than 1 boy or at least 2 girls

Five coins are tossed. Find the probability of each event.

5. 3 heads or 2 tails

6. no more than 2 heads or 5 tails

7. less than 4 heads or less than 4 tails

8. 2 or 3 tails or 1 head

A survey was taken to find out what newspapers people read. The results of 125 respondents are listed below.

60 read newspaper A 22 read A and B 15 read A, B, and C
50 read newspaper B 35 read A and C
75 read newspaper C 28 read B and C

Make a Venn diagram to help find the probability of each event if a person is selected at random.

9. reads newspaper A only

10. reads B only

11. reads C only

12. reads at least two newspapers

13. reads fewer than 2 newspapers

14. reads no newspapers

A committee of 3 is selected from 6 people: *A, B, C, D, E,* and *F.* Find the probability of each event.

15. A is selected.

16. B or C is selected.

17. D is *not* selected.

18. B or D is *not* selected.

192   Practice Masters Levels A, B, and C   Algebra 2

Practice Masters Level A
10.5 Independent Events

Events A, B, C, and D are independent, and P(A) = 0.3, P(B) = 0.5, P(C) = 0.4, and P(D) = 0.1. Find each probability.

1. P(A and B) _____
2. P(C and B) _____
3. P(A and D) _____
4. P(B and D) _____
5. P(C and A) _____
6. P(C and D) _____

Use the definition of independent events to determine whether the events from rolling a number cube are independent or dependent.

7. the event odd and the event 3 or 6 _____
8. the event less than 3 and the event more than 5 _____
9. the event even and the event prime number _____
10. the event more than 4 and the event 2 _____

A bag contains 6 red chips, 9 white chips, and 5 blue chips. A chip is selected and then replaced. Then a second chip is selected. Find probability of each event.

11. Both chips are white. _____
12. Neither chip is blue. _____
13. The first chip is red and the second chip is white. _____
14. The first chip is blue and the second chip is not blue. _____
15. Both chips are blue. _____
16. The first chip is not red and the second chip is not white. _____
17. The first chip is yellow and the second chip is blue. _____
18. Both chips are red, white or blue. _____
19. The first chip is green and the second chip is green. _____
20. The first chip is blue and the second chip is white. _____

Algebra 2 Practice Masters Levels A, B, and C

NAME _____ CLASS _____ DATE _____

Practice Masters Level B
10.5 Independent Events

Events *W, X, Y,* and *Z* are independent, and $P(W) = 0.75$, $P(X) = 0.25$, $P(Y) = 0.2$, and $P(Z) = 0.3$. Find each probability.

1. $P(W \text{ and } X)$ _____ 2. $P(W \text{ and } Y)$ _____

3. $P(W \text{ and } Z)$ _____ 4. $P(X \text{ and } Y)$ _____

5. $P(X \text{ and } Z)$ _____ 6. $P(Y \text{ and } Z)$ _____

Use the definition of independent events to determine whether the events from rolling a number cube are independent or dependent.

7. The event multiple of 2 and the event factor of 12. _____

8. The event composite number and the event multiple of 3. _____

9. The event factor of 15 and the event even. _____

10. The event factor of 30 and the event odd. _____

Two red flower bulbs are mixed with 3 white ones to form a packet. In the fall, the bulbs are planted in a row by selecting one bulb from a packet for the first plant, one bulb from another packet for the second plant and so on. Find the probability of each event.

11. The first two bulbs are red. _____

12. The first two bulbs are white. _____

13. The first bulb is red and the second bulb is white. _____

14. The first bulb is white and the second bulb is red. _____

15. The first and third bulbs are red and the second bulb is white. _____

16. The first 3 bulbs are white. _____

17. The colors alternate white, then red for the first 5 bulbs. _____

18. The second, third, and fourth bulbs in a row of 8 are red and the remainder are white. _____

Practice Masters Level C
10.5 Independent Events

Three cards are drawn from a standard deck of 52 cards, one at a time, with each card replaced before the next card is drawn. Find the probability of each event.

1. All 3 cards are red. _____
2. All 3 cards are spades. _____
3. All 3 cards are sevens. _____
4. All 3 cards are less than 6. _____
5. The first 2 cards are face cards, the third card is an ace. _____
6. The first card is a 10 of hearts, the second is a diamond, and the third is a 5. _____
7. The first 2 cards are less than 10 and the third card is a face card. _____
8. The first card is a diamond, the second is a heart, and the third is a club. _____
9. The first card is a black ace, the second is an ace of spades, and the third is an ace of clubs. _____
10. The first 2 cards are not face cards, and the third card is a face card. _____

Ping pong balls numbered 1 through 20 are placed in a hopper. One ball is drawn and replaced before the next ball is drawn. A total of 4 balls are to be drawn. Find the probability of each event.

11. All 4 balls have even numbers. _____
12. The first ball is 10 or less, while the other 3 balls are greater than 5. _____
13. All 4 balls have the same number. _____
14. All 4 balls have different numbers. _____
15. The first 2 balls are the same number and the other 2 balls are different numbers. _____
16. All 4 balls are factors of 100. _____
17. All 4 balls are factors of 100 but different from each other. _____
18. Three balls are greater than 15, and 1 ball is not greater than 15. _____
19. All 4 balls are prime numbers. _____
20. Two of the 4 balls are multiples of 5, and the other 2 balls are not multiples of 5. _____

Algebra 2   Practice Masters Levels A, B, and C   195

NAME _____ CLASS _____ DATE _____

Practice Masters Level A
10.6 Dependent Events and Conditional Probability

A box contains 5 purple marbles, 3 green marbles, and 2 orange marbles. Two consecutive draws are made from the box without replacement of the first draw. Find the probability of each event.

1. purple first, orange second _____
2. green first, purple second _____
3. green first, green second _____
4. orange first, green second _____
5. orange first, purple second _____
6. orange first, orange second _____
7. purple first, purple second _____
8. purple first, blue second _____

Let A and B represent events.

9. Given $P(A \text{ and } B) = \frac{1}{2}$ and $P(A) = \frac{2}{3}$, find $P(B|A)$. _____

10. Given $P(A \text{ and } B) = 0.12$ and $P(A) = 0.2$, find $P(B|A)$. _____

11. Given $P(A) = \frac{1}{4}$ and $P(B|A) = \frac{1}{3}$, find $P(A \text{ and } B)$. _____

12. Given $P(A) = 0.37$ and $P(B|A) = 0.42$, find $P(A \text{ and } B)$. _____

13. Given $P(B|A) = \frac{2}{3}$ and, $P(A \text{ and } B) = \frac{1}{5}$, find $P(A)$. _____

14. Given $P(B|A) = 0.63$ and, $P(A \text{ and } B) = 0.27$, find $P(A)$. _____

Two number cubes are rolled and the first cube shows a 3. Find the probability of each event.

15. Both numbers are 3s.

16. A sum of 7.

17. The numbers are both odd.

18. A sum of 2.

For one roll of a number cube, let A be the event "multiple of 2" and let B be the event "factor of 12." Find each probability.

19. $P(A)$ _____
20. $P(A \text{ and } B)$ _____
21. $P(B|A)$ _____
22. $P(A|B)$ _____

196   Practice Masters Levels A, B, and C   Algebra 2

Practice Masters Level B

10.6 Dependent Events and Conditional Probability

A bag contains 6 red chips, 9 white chips, and 5 blue chips. Three consecutive chips are drawn from the bag without replacement. Find the probability of each event.

1. 3 red _____

2. 3 white _____

3. white first, white second and blue third _____

4. blue first, red second and blue third _____

5. blue first, white second and red third _____

6. red first, white second and white third _____

7. Given $P(A \text{ and } B) = 0.1$ and $P(A) = 0.3$, find $P(B|A)$. _____

8. Given $P(A) = 0.24$ and $P(B|A) = 0.72$, find $P(A \text{ and } B)$. _____

9. Given $P(A \text{ and } B) = \frac{1}{7}$ and $P(B|A) = \frac{4}{9}$, find $P(A)$. _____

10. Given $P(A) = \frac{1}{4}$ and, $P(A \text{ and } B) = \frac{1}{9}$ find $P(B|A)$. _____

11. Given $P(B|A) = \frac{7}{12}$ and, $P(A \text{ and } B) = \frac{2}{13}$, find $P(A)$. _____

12. Given $P(B|A) = \frac{2}{5}$ and, $P(A) = \frac{1}{11}$, find $P(A \text{ and } B)$. _____

Two pink tulip bulbs are accidentally mixed with 9 white tulip bulbs. In the fall, 4 bulbs are planted in a row. Find the probability of each event.

13. The first 2 are pink. _____

14. The first 2 are white. _____

15. The first 2 bulbs consist of 1 pink and 1 white. _____

16. All 4 bulbs are white. _____

17. The first bulb is white and the other 3 bulbs are pink. _____

18. The order is white, pink, white, pink. _____

19. There are exactly 2 white bulbs planted. _____

Practice Masters Level C
10.6 Dependent Events and Conditional Probability

Five cards are dealt from a standard deck of cards without replacement. Find each probability.

1. All 5 cards are red.

2. All 5 cards are the same suit (a flush).

3. 5 cards in sequence from two to six (a straight)

4. a pair of sevens

5. 3 queens

6. 4 aces

7. 3 tens and 2 eights (a full house)

8. 4 clubs and 1 heart

9. 4 fives and 1 two

10. 5-card sequence from 10 to ace in the same suit (a royal flush)

A lottery game draws 6 numbers from 1 through 44. The jackpot goes to the person who matches all 6 numbers. Use this information for Exercises 11–13.

11. If a number cannot be chosen more than once, how many possible number combinations are there?

12. What is the probability of winning the jackpot?

13. Suppose that choosing 5 out of 6 correct numbers wins a smaller jackpot. What is the probability of winning this smaller jackpot?

NAME _____ CLASS _____ DATE _____

Practice Masters Level A

10.7 Experimental Probability and Simulation

Use a simulation with 10 trials to estimate the probability of each event. Record your results in the table.

1. 2 consecutive tails

Trial	Results
1	
2	
3	
4	
5	
6	
7	
8	
9	
10	

2. tails followed by heads

Trial	Results
1	
2	
3	
4	
5	
6	
7	
8	
9	
10	

3. heads followed by tails

Trial	Results
1	
2	
3	
4	
5	
6	
7	
8	
9	
10	

4. In 3 tosses of the coin, tails appears exactly once.

Trial	Results
1	
2	
3	
4	
5	
6	
7	
8	
9	
10	

5. In 4 tosses of the coin, heads appears at least 3 times.

Trial	Results
1	
2	
3	
4	
5	
6	
7	
8	
9	
10	

6. In 3 rolls of a number cube, the number 5 appears exactly once.

Trial	Results
1	
2	
3	
4	
5	
6	
7	
8	
9	
10	

Algebra 2   Practice Masters Levels A, B, and C   199

Practice Masters Level B
10.7 Experimental Probability and Simulation

Use a simulation with 10 trials to estimate the probability of each event. Record your results in the table.

1. In 4 tosses of the coin, the number of tails is greater than the number of heads.

Trial	Results
1	
2	
3	
4	
5	
6	
7	
8	
9	
10	

2. In 4 rolls of a number cube, the sum of the numbers is greater than 10.

Trial	Results
1	
2	
3	
4	
5	
6	
7	
8	
9	
10	

3. In 5 rolls of a number cube, at least 3 consecutive numbers will be rolled. (ex: 7, 2, 3, 4, 1)

Trial	Results
1	
2	
3	
4	
5	
6	
7	
8	
9	
10	

A multiple-choice test consists of 20 questions. Use a simulation to estimate the probability of each event given the number of possible answers for each question. Assume that each answer is a guess.

4. Each question has 2 possible answers. Find the probability of answering at least 5 correctly.

Trial	Results
1	
2	
3	
4	
5	
6	
7	
8	
9	
10	

5. Each question has 3 possible answers. Find the probability of answering at least 5 correctly.

Trial	Results
1	
2	
3	
4	
5	
6	
7	
8	
9	
10	

6. Each question has 4 possible answers. Find the probability of answering at least 5 correctly.

Trial	Results
1	
2	
3	
4	
5	
6	
7	
8	
9	
10	

Practice Masters Level C

10.7 Experimental Probability and Simulation

Of 180 motorists observed at an intersection, 57 went straight, 82 turned left, and 41 turned right. Use a simulation with 10 trials to estimate the probability of each event.

1. Exactly 3 of every 4 motorists turns left.

Trial	Results
1	
2	
3	
4	
5	
6	
7	
8	
9	
10	

2. At most 2 of every 4 motorists go straight.

Trial	Results
1	
2	
3	
4	
5	
6	
7	
8	
9	
10	

3. At least 1 motorist turns right.

Trial	Results
1	
2	
3	
4	
5	
6	
7	
8	
9	
10	

A fast food restaurant is giving away 1 out of 5 buttons with the purchase of a kid's meal. Assume that each button is equally likely to be given. Use a simulation with 10 trials to find the probability of each event.

4. All 5 buttons will be acquired after 5 kid's meals.

Trial	Results
1	
2	
3	
4	
5	
6	
7	
8	
9	
10	

5. All 5 buttons will be acquired after 7 kid's meals.

Trial	Results
1	
2	
3	
4	
5	
6	
7	
8	
9	
10	

6. All 5 buttons will be acquired after 10 kid's meals.

Trial	Results
1	
2	
3	
4	
5	
6	
7	
8	
9	
10	

Algebra 2

Practice Masters Level A
11.1 Sequences and Series

Write the first five terms of each sequence defined by the given explicit formula.

1. $t_n = 3n + 1$

2. $t_n = 5n - 2$

3. $t_n = 2n - 7$

4. $t_n = 3 - 2n$

5. $t_n = -4n + 6$

6. $t_n = -7n - 3$

7. $t_n = n^2$

8. $t_n = 3n^2$

9. $t_n = n^2 + 4$

10. $t_n = n^2 - 1$

Write the first five terms of each sequence defined by the given recursive formula.

11. $t_1 = 2$
    $t_n = t_{n-1} + 3$

12. $t_1 = 0$
    $t_n = t_{n-1} - 2$

13. $t_1 = 1$
    $t_n = 4t_{n-1}$

14. $t_1 = 2$
    $t_n = -3t_{n-1}$

15. $t_1 = 0$
    $t_n = 3 - t_{n-1}$

16. $t_1 = 2$
    $t_n = (t_{n-1})^2$

Write the terms of each series. Then evaluate.

17. $\sum_{n=1}^{4} 2n$

18. $\sum_{n=1}^{4} n + 1$

19. $\sum_{n=1}^{5} n^2$

20. $\sum_{n=1}^{5} 2n - 1$

CLASS _____ DATE _____

Practice Masters Level B

11.1 Sequences and Series

Write the first five terms of each sequence defined by the given explicit formula.

1. $t_n = n^2 + 7$

2. $t_n = 4n^2$

3. $t_n = -3n^2$

4. $t_n = -2n^2 - 1$

5. $t_n = n^3$

6. $t_n = 3n^3$

7. $t_n = -2n^3$

8. $t_n = n^3 + 2$

Write the first five terms of each sequence defined by the given recursive formula.

9. $t_1 = 1$
   $t_n = 4t_{n-1} + 3$

10. $t_1 = 1$
    $t_n = -2t_{n-1} + 2$

11. $t_1 = 1$
    $t_n = -3t_{n-1} - 1$

12. $t_1 = 1$
    $t_n = (2t_{n-1})^2$

13. $t_1 = 0$
    $t_n = 4 - 3t_{n-1}$

14. $t_1 = 0$
    $t_n = (t_n - 1)^3$

Write the terms of each series. Then evaluate.

15. $\sum_{n=1}^{4} n^2 + 5$ _____

16. $\sum_{n=1}^{4} 4n^2$ _____

17. $\sum_{n=1}^{4} 3 - 2n^2$ _____

18. $\sum_{n=1}^{4} n^3$ _____

Evaluate the sum.

19. $\sum_{n=1}^{5} 4$ _____

20. $\sum_{n=1}^{5} n^2 - 2n + 1$ _____

Algebra 2   Practice Masters Levels A, B, and C

Practice Masters Level C

11.1 Sequences and Series

Write the first four terms of each sequence defined by the given explicit formula.

1. $t_n = \dfrac{1}{5}n + 3$

2. $t_n = \dfrac{-1}{4}n - 1$

3. $t_n = \left(\dfrac{1}{2}n\right)^2$

4. $t_n = 2^n$

5. $t_n = 3^{-n}$

6. $t_n = \left(\dfrac{1}{3}\right)^{-n} + 1$

For each sequence, write a recursive formula and find the next two terms.

7. 1, 3, 7, 15 _____

8. 2, 6, 38, 1446 _____

Write the first four terms of each sequence defined by the given recursive formula.

9. $t_1 = 1$
$t_n = (t_{n-1})^2 + 3t_{n-1}$

10. $t_1 = 0$
$t_n = \dfrac{1}{2}(t_{n-1}) + 2$

11. $t_1 = 1$
$t_n = -(t_{n-1})^3 + 3(t_{n-2})^2$

12. $t_1 = 2$
$t_n = (t_{n-1})^4$

Evaluate the sum.

13. $\displaystyle\sum_{n=1}^{4}(3-n)^2$

14. $\displaystyle\sum_{n=1}^{5}n^3 + 3$

15. $\displaystyle\sum_{n=1}^{4}(n-2)^3$

16. $\displaystyle\sum_{n=1}^{4}n^n$

Practice Masters Level A
11.2 Arithmetic Sequences

Based on the terms given, state whether or not each sequence is arithmetic. If it is, identify the common difference, d.

1. 5, 8, 11, 14, 17, ...

2. −7, −2, 3, 8, 13, ...

3. 2, 4, 8, 10, 12, ...

4. 9, 7, 5, 3, 1, ...

5. 4, −1, −6, −11, ...

6. 0, 3, 5, 7, 9, ...

Use the recursive formula given to find the first four terms of each arithmetic sequence.

7. $t_1 = 0$
   $t_n = t_{n-1} + 3$

8. $t_1 = -1$
   $t_n = t_{n-1} - 10$

9. $t_n = -2 + 3(n + 1)$

10. $t_n = \dfrac{-1}{2}n - 2$

Find the indicated term given two other terms.

11. 5th term; $t_4 = 7$ and $t_7 = 22$

12. 5th term; $t_3 = -5$ and $t_7 = 11$

13. 1st term; $t_2 = 3$ and $t_6 = 27$

14. 8th term; $t_1 = 20$ and $t_3 = 14$

Find the indicated arithmetic means.

15. 2 arithmetic means between 18 and 33

16. 2 arithmetic means between 10 and −2

17. 3 arithmetic means between 3 and −5

18. 3 arithmetic means between −20 and 24

19. 3 arithmetic means between 37 and −3

20. 3 arithmetic means between −60 and 28

NAME _____ CLASS _____ DATE _____

Practice Masters Level B
11.2 Arithmetic Sequences

Based on the terms given, state whether or not each sequence is arithmetic. If it is, identify the common difference, d.

1. $-11, -18, -25, -32, \ldots$

2. $1, 3, 6, 9, 12, \ldots$

3. $\dfrac{3}{2}, \dfrac{5}{2}, \dfrac{7}{2}, \dfrac{9}{2}, \ldots$

4. $1, 4, 9, 16, \ldots$

5. $\dfrac{-1}{3}, \dfrac{1}{3}, 1, \dfrac{5}{3}, \ldots$

6. $\dfrac{7}{4}, \dfrac{3}{2}, \dfrac{5}{4}, 1, \ldots$

List the first four terms of each arithmetic sequence.

7. $t_n = 4(n - 3) + 1$

8. $t_n = \dfrac{1}{3}n + \dfrac{2}{3}$

9. $t_1 = 4$
   $t_n = t_{n-1} + 1$

10. $t_1 = -2$
    $t_n = t_{n-1} - 5$

Find the indicated term given two other terms.

11. 4th term; $t_1 = \dfrac{3}{2}$ and $t_7 = \dfrac{21}{2}$

12. 3rd term; $t_1 = \dfrac{4}{3}$ and $t_7 = \dfrac{-8}{3}$

13. 6th term; $t_2 = \dfrac{-4}{5}$ and $t_5 = 1$

14. 1st term; $t_3 = \dfrac{-2}{3}$ and $t_5 = 2$

Find the indicated arithmetic means.

15. 3 arithmetic means between $\dfrac{3}{4}$ and $\dfrac{-1}{4}$

16. 3 arithmetic means between $\dfrac{-7}{5}$ and $\dfrac{9}{5}$

17. 3 arithmetic means between $\dfrac{-5}{2}$ and $\dfrac{-1}{2}$

18. 3 arithmetic means between -1 and $\dfrac{7}{5}$

Practice Masters Level C

11.2 Arithmetic Sequences

Use the recursive formula given to find the first four terms of each arithmetic sequence.

1. $t_1 = \dfrac{-1}{4}; \quad t_n = t_{n-1} + \dfrac{1}{2}$

2. $t_1 = \dfrac{3}{2}; \quad t_n = t_{n-1} - \dfrac{1}{3}$

3. $t_1 = \dfrac{-1}{2}; \quad t_n = t_{n-1} + \dfrac{2}{3}$

4. $t_1 = \dfrac{1}{3}; \quad t_n = t_{n-1} - \dfrac{3}{4}$

5. $t_1 = -3.1; \quad t_n = t_{n-1} - 7.4$

6. $t_1 = -5.7; \quad t_n = t_{n-1} + 11.3$

Write an explicit formula for the nth term of each arithmetic sequence.

7. $-3, 0, 3, 6, 9, \ldots$

8. $7, 12, 17, 22, 27$

9. $47, 34, 21, 8, -5, \ldots$

10. $-7, 1, 9, 17, 25, \ldots$

11. $-500, -488, -476, -464, -452, \ldots$

12. $\dfrac{3}{5}, \dfrac{1}{5}, \dfrac{-1}{5}, \dfrac{-3}{5}, -1, \ldots$

13. $-1, \dfrac{1}{3}, \dfrac{5}{3}, 3, \dfrac{13}{3}, \ldots$

14. $-7, \dfrac{-11}{2}, -4, \dfrac{-5}{2}, -1, \ldots$

Find the indicated arithmetic means.

15. 3 arithmetic means between 1 and $\dfrac{19}{3}$

16. 4 arithmetic means between -3 and $\dfrac{-1}{2}$

17. 4 arithmetic means between 1 and $\dfrac{8}{3}$

18. 4 arithmetic means between $\dfrac{1}{3}$ and -3

19. 4 arithmetic means between 50 and 115

20. 4 arithmetic means between -47 and -104.5

NAME _____ CLASS _____ DATE _____

Practice Masters Level A
11.3 Arithmetic Series

Use the formula for an arithmetic series to find each sum.

1. $1 + 4 + 7 + 10 + 13$

2. $-3 + (-5) + (-7) + (-9) + (-11)$

3. $4 + 16 + 28 + 40 + 52$

4. $-1 + (-7) + (-13) + (-19) + (-25)$

5. $15 + 18 + 21 + 24 + 27$

6. $-5 + (-9) + (-13) + (-17) + (-21)$

7. Find the sum of the first 100 natural numbers. _____

8. Find the sum of the first 50 even natural numbers. _____

9. Find the sum of the first 12 multiples of 10. _____

10. Find the sum of the multiples of 6 from 18 to 96, inclusive. _____

For each arithmetic series, find S_{10}.

11. $7 + 13 + 19 + 25 + \ldots$

12. $6 + 11 + 16 + 21 + \ldots$

13. $-20 + (-17) + (-14) + (-11) + \ldots$

14. $-8 + (-5) + (-2) + 1 + \ldots$

15. $-3 + (-1) + 1 + 3 + \ldots$

16. $10 + 21 + 32 + 43 + \ldots$

Evaluate.

17. $\sum_{n=1}^{5} (3n + 2)$

18. $\sum_{n=1}^{6} (5n - 3)$

19. $\sum_{n=1}^{8} (7 + 2n)$

20. $\sum_{n=1}^{10} (10 - 4n)$

Practice Masters Level B
11.3 Arithmetic Series

Use the formula for an arithmetic series to find each sum.

1. $-10 + (-17) + (-24) + (-31) + (-38)$ _____

2. $101 + 149 + 197 + 246 + 293$ _____

3. $-12 + (-5) + 2 + 9 + 16$ _____

4. $-96 + (-61) + (-36) + 9 + 44$ _____

5. $2.3 + 5.7 + 9.1 + 12.5 + 15.9$ _____

6. $-11.3 + (-15.9) + (-20.5) + (-25.1) + (-29.7)$ _____

7. Find the sum of the first 500 natural numbers. _____

8. Find the sum of the first 100 odd natural numbers. _____

9. Find the sum of the first 20 multiples of 15. _____

10. Find the sum of the multiples of 8 from 32 to 64, inclusive. _____

For each arithmetic series, find S_{20}.

11. $-2 + 5 + 12 + 19 + \ldots$ _____

12. $7 + (-6) + (-19) + (-32) + \ldots$ _____

13. $\dfrac{5}{2} + 4 + \dfrac{11}{2} + 7 + \ldots$ _____

14. $18 + 14.5 + 11 + 7.5 + \ldots$ _____

Evaluate.

15. $\sum\limits_{n=1}^{5} (100 + 10n)$ _____

16. $\sum\limits_{n=1}^{5} (55 - 2n)$ _____

17. $\sum\limits_{n=1}^{5} \left(\dfrac{1}{2} - \dfrac{1}{2}n\right)$ _____

18. $\sum\limits_{n=1}^{12} \left(\dfrac{2}{3}n + \dfrac{1}{3}\right)$ _____

19. $\sum\limits_{n=1}^{15} (3.1n + 6.7)$ _____

20. $\sum\limits_{n=1}^{20} (10.9 - 2.5n)$ _____

Algebra 2

NAME _____ CLASS _____ DATE _____

Practice Masters Level C
11.3 Arithmetic Series

Use the formula for an arithmetic series to find each sum.

1. Find the sum of the first 1000 natural numbers. _____

2. Find the sum of the first 10,000 natural numbers. _____

3. Find the sum of the first 100 multiples of 3. _____

4. Find the sum of the first 100 multiples of $\frac{1}{2}$. _____

5. Find the sum of the first 200 multiples of $\frac{1}{3}$. _____

6. Find the sum of the multiples of $\frac{2}{3}$ between 4 and 24. _____

For each arithmetic series, find S_{50}.

7. $50 + 47 + 44 + 41 + \cdots$

8. $82 + (-91) + (-100) + (-109) + \cdots$

9. $\frac{7}{3} + 4 + \frac{17}{3} + \frac{22}{3} + \cdots$

10. $\frac{1}{8} + \left(\frac{-3}{8}\right) + \left(\frac{-7}{8}\right) + \left(1\frac{3}{8}\right) + \cdots$

11. $6 + 7\frac{1}{3} + 8\frac{2}{3} + 10 + \cdots$

12. $13.7 + 19.1 + 24.5 + 29.9 + \cdots$

Evaluate.

13. $\sum_{n=1}^{6}\left(7 + \frac{1}{2}n\right)$

14. $\sum_{n=1}^{8}\left(5n - \frac{3}{2}\right)$

15. $\sum_{n=1}^{10}\left(\frac{5}{3}n - \frac{2}{3}\right)$

16. $\sum_{n=1}^{20}(100n - 25)$

17. $\sum_{n=1}^{50}(300 + 6n)$

18. $\sum_{n=1}^{100}(1000 - 100n)$

19. $\sum_{n=1}^{100}\left(\frac{1}{2}n - \frac{1}{4}\right)$

20. $\sum_{n=1}^{500}(1000n + 2)$

Practice Masters Level A
11.4 Geometric Sequences

Determine whether each sequence is geometric. If so, identify the common ratio, r, and give the next three terms.

1. 2, 6, 18, 54, ...

2. 1, 4, 16, 64, ...

3. 2, 4, 6, 8, ...

4. 1, −3, 9, −27, ...

5. 1, 4, 9, 16, ...

6. 2, −4, 8, −16, ...

List the first four terms of each geometric sequence.

7. $t_1 = 2$
   $t_n = 4t_{n-1}$

8. $t_1 = 3$
   $t_n = 3t_{n-1}$

9. $t_1 = -2$
   $t_n = -t_{n-1}$

10. $t_1 = 1$
    $t_n = -5t_{n-1}$

11. $t_1 = 2$
    $t_n = -10t_{n-1}$

12. $t_1 = 1$
    $t_n = 0.5t_{n-1}$

Find the indicated geometric means.

13. Find 2 geometric means between 12 and 324.

14. Find 2 geometric means between 30 and 6480.

15. Find 3 geometric means between −45 and −3645.

16. Find 2 geometric means between −18 and 144.

Algebra 2

Practice Masters Levels A, B, and C

Practice Masters Level B
11.4 Geometric Sequences

Determine whether each sequence is geometric. If so, identify the common ratio, r, and give the next three terms.

1. $4, -12, 36, -108, \ldots$

2. $1, 3, 5, 7, \ldots$

3. $1, \dfrac{1}{2}, \dfrac{1}{4}, \dfrac{1}{8}, \ldots$

4. $1, 0.2, 0.04, 0.008, \ldots$

5. $-2, -4, 8, 16, \ldots$

6. $0.6, 0.06, 0.006, 0.0006, \ldots$

List the first four terms of each geometric sequence.

7. $t_1 = 2$
$t_n = -30 t_{n-1}$

8. $t_1 = 3$
$t_n = \dfrac{1}{3} t_{n-1}$

9. $t_1 = -1$
$t_n = \dfrac{2}{5} t_{n-1}$

10. $t_1 = 2$
$t_n = \dfrac{-3}{4} t_{n-1}$

11. $t_1 = -2$
$t_n = 0.7 t_{n-1}$

12. $t_1 = 1$
$t_n = -0.2 t_{n-1}$

Find the indicated geometric means.

13. Find 2 geometric means between 9 and -1944.

14. Find 3 geometric means between 1.2 and 110.82252.

15. Find 3 geometric means between $\dfrac{5}{4}$ and $\dfrac{5}{1024}$.

16. Find 4 geometric means between 2 and $\dfrac{-1}{16}$.

Practice Masters Level C
11.4 Geometric Sequences

1. In a geometric sequence, $t_1 = 4$ and $r = -2$. Find t_{10}.

2. In a geometric sequence, $t_1 = -3$ and $r = \frac{1}{2}$. Find t_{20}.

Find the fifth term in the geometric sequence that includes each pair of terms.

3. $t_3 = 750$, $t_7 = 468{,}750$

4. $t_1 = -6$, $t_4 = 48{,}000$

5. $t_2 = 0.15$, $t_6 = 0.001215$

6. $t_3 = 5.625$, $t_6 = 87.890625$

7. $t_3 = \dfrac{400}{9}$, $t_7 = \dfrac{6400}{729}$

8. $t_7 = 640$, $t_{10} = -5120$

Write an explicit formula for the nth term of each geometric sequence.

9. $81, -243, 729, -2187, \ldots$

10. $20, \dfrac{40}{3}, \dfrac{80}{9}, \dfrac{160}{27}, \ldots$

11. $2700, 8100, 24300, 72900, \ldots$

12. $50, 25, 12.5, 6.25, \ldots$

13. $30, 12, 4.8, 1.92, \ldots$

14. $1.8, -0.54, 0.162, -0.0486, \ldots$

Find the indicated geometric means.

15. Find 4 geometric means between 10 and $-1{,}000{,}000$.

16. Find 3 geometric means between 8 and 0.03125.

17. Find 3 geometric means between -160 and $-40{,}960$.

18. Find 4 geometric means between 1 and $\dfrac{1}{32}$.

Algebra 2

Practice Masters Level A
11.5 Geometric Series and Mathematical Induction

Find each indicated sum of the geometric series $1 + (-2) + 4 + (-8) + \ldots$

1. S_4 _____

2. S_5 _____

3. S_{10} _____

4. S_{15} _____

Find each indicated sum of the geometric series $2 + 6 + 18 + 54 + \ldots$

5. S_4 _____

6. S_6 _____

7. S_{10} _____

8. S_{12} _____

Use the formula for the sum of the first n terms of a geometric series to find each sum.

9. $1 + 4 + 16 + 64 + 128$

10. $-2 + (-4) + (-8) + (-16) + (-32)$

11. $-3 + 6 + (-12) + 24 + (-48)$

12. $-5 + (-15) + (-45) + (-135) + (-405)$

Evaluate.

13. $\sum_{n=1}^{5} 3(4^{n-1})$

14. $\sum_{n=1}^{8} -6(3^{n-1})$

Use mathematical induction to prove that the statement is true for every natural number, n.

15. $3 \leq n + 2$ _____

214   Practice Masters Levels A, B, and C   Algebra 2

NAME _____ CLASS _____ DATE _____

Practice Masters Level B
11.5 Geometric Series and Mathematical Induction

Find each indicated sum of the geometric series $4 + 12 + 36 + 108 + \cdots$

1. S_4 _____ 2. S_{10} _____

3. S_{20} _____ 4. S_{25} _____

Find each indicated sum of the geometric series $10 + 5 + 2.5 + 1.25 + \cdots$

5. S_5 _____ 6. S_{10} _____

7. S_{20} _____ 8. S_{100} _____

Use the formula for the sum of the first n terms of a geometric series to find each sum. Give answers to the nearest tenth, if necessary.

9. $5 + (-10) + 20 + (-40) + 80$

10. $18 + (-9) + \dfrac{9}{2} + \left(\dfrac{-9}{4}\right) + \dfrac{9}{8}$

11. $\dfrac{1}{3} + \dfrac{2}{3} + \dfrac{4}{9} + \dfrac{8}{27} + \dfrac{16}{81}$

12. $\dfrac{1}{4} + \left(\dfrac{-3}{16}\right) + \dfrac{9}{64} + \left(\dfrac{-27}{256}\right) + \dfrac{81}{1024}$

Evaluate. Round answers to the nearest tenth, if necessary.

13. $\displaystyle\sum_{n=1}^{10} \dfrac{1}{2}(-2)^{n-1}$

14. $\displaystyle\sum_{n=1}^{12} -2\left(\dfrac{1}{4}\right)^{n-1}$

Use mathematical induction to prove that the statement is true for every natural number, n.

15. $1^2 + 2^2 + 3^2 + \ldots + n^2 = \dfrac{n(n + 1)(n + 2)}{6}$

Algebra 2   Practice Masters Levels A, B, and C

Practice Masters Level C

11.5 Geometric Series and Mathematical Induction

Find each indicated sum of the geometric series $5 + \frac{5}{3} + \frac{5}{9} + \frac{5}{27} + \cdots$

1. S_4 _____

2. S_5 _____

3. S_{10} _____

4. S_{12} _____

Use the formula for the sum of the first n terms of a geometric series to find each sum. Give answers to the nearest tenth, if necessary.

5. $9 + 2.7 + 0.81 + 0.243 + 0.0729$

6. $-6 - \frac{9}{2} - \frac{27}{8} - \frac{81}{32} - \frac{243}{128}$

7. $100 + 150 + 225 + 337.5 + 506.25$

8. $0.12 + 0.072 + 0.0432 + 0.02592 + 0.015552$

Evaluate. Round answers to the nearest tenth, if necessary.

9. $\sum_{n=1}^{6} \frac{1}{2}\left(\frac{1}{3}\right)^{n-1}$

10. $\sum_{n=1}^{8} \frac{-1}{4}(2)^{n-1}$

11. $\sum_{n=1}^{10} \frac{-2}{3}\left(\frac{3}{4}\right)^{n-1}$

12. $\sum_{n=1}^{10} 0.7(-0.6)^{n-1}$

13. $\sum_{n=1}^{12} 1000(0.3)^{n-1}$

14. $\sum_{n=1}^{20} 0.3(1000)^{n-1}$

Use mathematical induction to prove that the statement is true for every natural number, n.

15. $1^3 + 2^3 + 3^3 + \cdots + n^3 = \frac{n^2(n+1)^2}{4}$

Practice Masters Level A
11.6 Infinite Geometric Series

Find the sum of each infinite geometric series, if it exists.

1. $10 + 5 + 2.5 + 1.25 + \cdots$

2. $20 + 5 + \dfrac{5}{4} + \dfrac{5}{16} + \cdots$

3. $2 + 4 + 8 + 16 + \cdots$

4. $30 + 6 + \dfrac{6}{5} + \dfrac{6}{25} + \cdots$

5. $7 + 5 + 3 + 1 + \cdots$

6. $100 + 10 + 1 + 0.1 + \cdots$

Find the sum of each infinite geometric series, if it exists.

7. $\sum_{n=1}^{\infty} \left(\dfrac{1}{7}\right)^n$

8. $\sum_{n=1}^{\infty} \left(\dfrac{1}{4}\right)^n$

9. $\sum_{n=1}^{\infty} (-0.3)^n$

10. $\sum_{n=1}^{\infty} 6^n$

11. $\sum_{n=1}^{\infty} \dfrac{4}{10^n}$

12. $\sum_{n=1}^{\infty} \left(\dfrac{-1}{2}\right)^n$

Write an infinite geometric series that converges to the given number.

13. $0.444444444\ldots$

14. $0.32323232\ldots$

15. $65.65865865\ldots$

16. $0.90909090\ldots$

Write each decimal as a fraction in simplest form.

17. $0.\overline{3}$

18. $0.\overline{6}$

19. $0.\overline{8}$

20. $0.\overline{29}$

Algebra 2

NAME _____ CLASS _____ DATE _____

Practice Masters Level B
11.6 Infinite Geometric Series

Find the sum of each infinite geometric series, if it exists.

1. $-10 + (-5) + (-2.5) + (-1.25) + \cdots$

2. $1 + \dfrac{1}{5} + \dfrac{1}{10} + \dfrac{1}{15} + \cdots$

3. $-20 + 10 + (-5) + 2.5 + \cdots$

4. $160 + (-40) + 10 + \left(\dfrac{-5}{2}\right) + \cdots$

5. $-2 + (-4) + (-8) + (-16) + \cdots$

6. $1000 + 125 + \dfrac{125}{8} + \dfrac{125}{64} + \cdots$

Find the sum of each infinite geometric series, if it exists.

7. $\sum_{n=1}^{\infty} \left(\dfrac{1}{3}\right)^n$

8. $\sum_{n=1}^{\infty} \left(\dfrac{-1}{10}\right)^n$

9. $\sum_{n=0}^{\infty} (0.9)^n$

10. $\sum_{n=0}^{\infty} \dfrac{-3}{10^n}$

11. $\sum_{n=1}^{\infty} 5(0.2)^n$

12. $\sum_{n=0}^{\infty} -4(2)^n$

Write an infinite geometric series that converges to the given number.

13. $0.63636363\ldots$

14. $72.727272\ldots$

15. $0.134134134\ldots$

16. $0.703703703\ldots$

Write each decimal as a fraction in simplest form.

17. $0.\overline{47}$

18. $0.\overline{19}$

19. $0.\overline{362}$

20. $0.\overline{540}$

218   Practice Masters Levels A, B, and C   Algebra 2

Practice Masters Level C
11.6 Infinite Geometric Series

Find the sum of each infinite geometric series, if it exists.

1. $\sum_{n=1}^{\infty} \frac{1}{4}\left(\frac{1}{2}\right)^n$

2. $\sum_{n=0}^{\infty} \frac{-1}{3}\left(\frac{1}{4}\right)^n$

3. $\sum_{n=0}^{\infty} \frac{2}{3}\left(\frac{-1}{3}\right)^n$

4. $\sum_{n=1}^{\infty} \frac{1}{4}(2)^n$

5. $\sum_{n=1}^{\infty} \frac{-3}{4}\left(\frac{3}{4}\right)^n$

6. $\sum_{n=0}^{\infty} 0.65(0.3)^n$

7. $\sum_{n=0}^{\infty} 0.2(-0.1)^n$

8. $\sum_{n=1}^{\infty} 0.6(1.1)^n$

9. $\sum_{n=1}^{\infty} 4.7\left(\frac{1}{100}\right)^n$

10. $\sum_{n=0}^{\infty} \frac{1}{8}\left(\frac{1}{100}\right)^n$

11. $\sum_{n=0}^{\infty} 4(-2)^n$

12. $\sum_{n=1}^{\infty} \frac{1}{10}\left(\frac{1}{1000}\right)^n$

Write an infinite geometric series that converges to the given number.

13. $0.24302430\ldots$

14. $9299.92999299\ldots$

Write each decimal as a fraction in simplest form.

15. $0.\overline{114}$

16. $0.\overline{252}$

17. $0.\overline{1473}$

18. $0.\overline{8199}$

Algebra 2

Practice Masters Level A
11.7 Pascal's Triangle

State the location of each entry in Pascal's triangle. Then give the value of each entry.

1. $_4C_2$

2. $_5C_3$

3. $_6C_1$

4. $_7C_5$

5. $_8C_8$

6. $_{10}C_5$

Find the indicated entries in Pascal's triangle.

7. 2nd entry, row 3

8. 7th entry, row 6

9. 3rd entry, row 9

10. 4th entry, row 4

A fair coin is tossed the indicated number of times. Find the probability of each event.

11. 4 tosses; exactly 2 heads

12. 5 tosses; exactly 4 heads

13. 5 tosses; exactly 3 heads

14. 7 tosses; 4 or 5 heads

15. 8 tosses; 2 or 3 heads

16. 6 tosses; 3 or fewer heads

A family has 4 children. Find the probability that the children are the following:

17. exactly 4 girls

18. exactly 3 boys

19. at most 2 boys

20. at least 1 girl

NAME _____ CLASS _____ DATE _____

Practice Masters Level B
11.7 Pascal's Triangle

State the location of each entry in Pascal's triangle. Then give the value of each entry.

1. $_5C_3$

2. $_{10}C_6$

3. $_{10}C_7$

4. $_{20}C_4$

5. $_{15}C_{15}$

6. $_9C_3$

Find the indicated entries in Pascal's triangle.

7. 4th entry, row 6

8. 9th entry, row 11

9. 5th entry, row 12

10. 3rd entry, row 20

A family has the indicated number of children. Find the probability of each event.

11. 5 children; exactly 3 boys _____

12. 6 children; 4 or 5 girls _____

13. 6 children; at least 2 boys _____

14. 7 children; at most 2 boys _____

15. 7 children; 4, 5 or 6 girls _____

A student guesses the answers to 10 questions on a true-false quiz. Find the probability that the indicated number of guesses are correct.

16. exactly 3

17. exactly 5

18. 6 or 7

19. at least 6

Algebra 2   Practice Masters Levels A, B, and C   **221**

NAME _____ CLASS _____ DATE _____

Practice Masters Level C
11.7 Pascal's Triangle

1. List the entries to the 10th row of Pascal's triangle. _____

2. List the entries in the 15th row of Pascal's triangle. _____

Find the probability of each event. Give answers to the nearest percent.

3. at least 5 heads appear in 7 tosses of a coin _____

4. at most 8 heads appear in 10 tosses of a coin _____

5. at least 5 heads but no more than 7 heads appear in 10 tosses of a coin _____

Find the probability of each event. Give answers to the nearest percent.

6. at least 7 correct guesses out of 10 true-false questions _____

7. at least 5 correct guesses out of 15 true-false questions _____

8. at most 10 correct guesses out of 20 true-false questions _____

9. at most 2 correct guesses out of 20 true-false questions _____

10. between 4 and 8 correct guesses, inclusive, out of 20 true-false questions _____

Cards are drawn at random from a standard deck of 52 cards, one at a time with replacement, the indicated number of times. Find the probability of each event. Give answers to the nearest percent.

11. 20 red cards out of 50 draws _____

12. 50 black cards out of 100 draws _____

13. at least 5 red cards out of 20 draws _____

14. at least half the cards black out of 10 draws _____

15. between 10 and 15 red cards, inclusive, out of 20 draws _____

Practice Masters Level A
11.8 The Binomial Theorem

Expand each binomial raised to a power.

1. $(x + y)^3$

2. $(a + b)^6$

3. $(x + y)^7$

4. $(a - b)^8$

5. $(2x + 3y)^5$

Write each summation as a binomial raised to a power. Then write it in expanded form.

6. $\sum_{n=1}^{5} \binom{5}{n} x^{5-n} y^n$ _____

7. $\sum_{n=1}^{8} \binom{8}{n} a^{8-n} b^n$ _____

For the expansion of $(a + b)^{10}$, find the indicated terms.

8. 4th term

9. 7th term

A 10-question test has 4 multiple-choice answers per question that are all equally likely to be selected. Find the probability of each event. Give answers to the nearest percent.

10. exactly 3 correct _____

11. exactly 5 correct _____

12. at least 2 correct _____

13. none correct _____

Algebra 2

Practice Masters Level B

11.8 The Binomial Theorem

Expand each binomial raised to a power.

1. $(x - y)^5$

2. $(a + b)^{10}$

3. $(x - y)^7$

4. $(3x + 4y)^6$

5. $(2x - 5y)^5$

Write each summation as a binomial raised to a power. Then write it in expanded form.

6. $\sum_{n=0}^{7} \binom{7}{n} p^{7-n} q^n$

7. $\sum_{n=0}^{8} \binom{8}{0} a^{8-n} y^n$

For the expansion of $(x - y)^{12}$, find the indicated terms.

8. 9th term

9. 4th term

A 20-question test has 4 multiple-choice answers per question that are all equally likely to be selected. Find the probability of each event. Give answers to the nearest percent.

10. exactly 10 correct

11. 2 or 3 correct

12. between 4 and 8 correct, inclusive

13. none correct

Practice Masters Level C
11.8 The Binomial Theorem

Expand each binomial raised to a power.

1. $(x - 5y)^5$

2. $(4x - 4y)^6$

3. $(-x + 3y)^5$

4. $\left(\dfrac{1}{2}x + \dfrac{1}{4}y\right)^5$

5. $\left(\dfrac{2}{3}x - \dfrac{1}{3}y\right)^6$

6. Find the 6th term in the expansion $(a + b)^{20}$. _____

7. Find the 10th term in the expansion $(a + b)^{20}$. _____

8. How many terms are there in the expansion of $(a + b)^{20}$? _____

9. A basketball player has an 80% free-throw percentage. What is the probability that he will make 5 of his next 6 free-throws? _____

A standard deck of 52 cards is shuffled and the indicated number of cards is drawn. Find the probability of each event. Give answers to the nearest percent.

10. exactly 5 clubs out of 10 cards _____

11. exactly 1 ten out of 4 cards _____

12. at least 2 hearts out of 8 cards _____

13. at most 3 spades out of 12 cards _____

Algebra 2   Practice Masters Levels A, B, and C

NAME _____ CLASS _____ DATE _____

Practice Masters Level A
12.1 Measures of Central Tendency

Find the mean, median, and mode of each data set. Round answers to the nearest thousandth, when necessary.

1. 4, 12, 8, 10, 8, 4, 10, 4, 4
   mean _____ median _____ mode _____

2. 32, 36, 32, 35, 37, 39, 30
   mean _____ median _____ mode _____

3. 20, 14, 18, 10, 16
   mean _____ median _____ mode _____

4. 12, 14, 18, 10, 16
   mean _____ median _____ mode _____

5. −6, −7, −5, −6, −6, −8
   mean _____ median _____ mode _____

Find the mean, median, and mode of the data, and compare them.

6. The average mortgage rate of the years 1992 through 1999:
   8.11, 7.16, 7.47, 7.85, 7.71, 7.68, 7.10, 7.26

_____

_____

Make a frequency table for the data and find the mean.

7. Shoe size of members of the basketball team:
   10, 12, 11, 13, 12, 10, 12, 9, 14, 11, 10, 12

Shoe size	Tally	Frequency
9		
10		
11		
12		
13		
14		

   mean: _____

Make a grouped frequency table for the data and use the table to estimate the mean.

8. Number of days absent for students for a year.
   5, 7, 2, 4, 5, 6, 2, 1, 10, 3, 0, 12, 10, 8, 6, 13, 11, 10, 5, 5

Number of days	Class mean	Frequency	Product
0 – 2			
3 – 5			
6 – 8			
9 – 11			
12 – 14			

   estimated mean: _____

226   Practice Masters Levels A, B, and C   Algebra 2

NAME _____ CLASS _____ DATE _____

Practice Masters Level B
12.1 Measures of Central Tendency

Find the mean, median, and mode of each data set. Round answers to the nearest thousandth, when necessary.

1. 26, 27, 22, 22, 24, 26, 29, 20, 42, 24, 22
   mean _____ median _____ mode _____

2. −11, −12, −9, −16, −15, −11, −10, −15
   mean _____ median _____ mode _____

3. 2.9, 3.1, 2.4, 2.3, 2.6, 2.7
   mean _____ median _____ mode _____

4. 14.2, 14.8, 14.7, 14.8, 15.6, 14.2
   mean _____ median _____ mode _____

5. $\frac{2}{3}, \frac{1}{4}, \frac{1}{4}, \frac{1}{2}, \frac{1}{3}, \frac{2}{3}, \frac{1}{2}, \frac{2}{3}$
   mean _____ median _____ mode _____

Find the mean, median, and mode of the data, and compare them.

6. The average ACT scores of the years 1990 through 1999:
   20.6, 20.6, 20.6, 20.7, 20.8, 20.8, 20.9, 21.0, 21.0, 21.0

_____

_____

Make a frequency table for the data and find the mean.

7. Number of family members in each student's family:
   5, 5, 6, 5, 4, 6, 5, 6, 4, 7, 8, 8, 5, 3, 2

   mean: _____

Number of members	Tally	Frequency
2		
3		
4		
5		
6		
7		
8		

Make a grouped frequency table for the data and use the table to estimate the mean.

8. Number of days of lead time to fill an order at a wholesale warehouse:
   3, 5, 10, 12, 7, 6, 20, 18, 6, 8, 13, 15, 10, 4, 17, 19, 7, 17, 15, 15, 14, 11, 10, 13, 14

   estimated mean: _____

Number of days	Class mean	Frequency	Product
3 – 5			
6 – 8			
9 – 11			
12 – 14			
15 – 17			
18 – 20			

Algebra 2   Practice Masters Levels A, B, and C   227

NAME _____ CLASS _____ DATE _____

Practice Masters Level C
12.1 Measures of Central Tendency

Make a frequency table for the data. Find the mean, median, and mode.
Round answers to the nearest thousandth.

1. The number of home runs hit by each team on August 6 during the 2000 season.

National League Team	Home Runs
Arizona	130
Atlanta	130
Chicago	137
Cincinnati	133
Colorado	107
Florida	112
Houston	161
Los Angeles	151
Milwaukee	115
Montreal	121
New York	130
Philadelphia	108
Pittsburgh	123
St. Louis	179
San Diego	119
San Francisco	154

Number of home runs	Class mean	Frequency	Product
101 – 110			
111 – 120			
121 – 130			
131 – 140			
141 – 150			
151 – 160			
161 – 170			
171 – 180			
Total			

Mean _____

Median _____

Mode _____

2. The batting average for each team on August 6 during the 2000 season.

American League Team	Average
Anaheim	0.283
Baltimore	0.278
Boston	0.270
Chicago	0.289
Cleveland	0.280
Detroit	0.267
Kansas City	0.282
Minnesota	0.270
New York	0.281
Oakland	0.268
Seattle	0.269
Tampa Bay	0.265
Texas	0.289
Toronto	0.275

Batting averages	Class mean	Frequency	Product
0.261 – 0.265			
0.266 – 0.270			
0.271 – 0.275			
0.276 – 0.280			
0.281 – 0.285			
0.286 – 0.290			
Total			

Mean _____

Median _____

Mode _____

NAME _____ CLASS _____ DATE _____

Practice Masters Level A

12.2 Stem-and-Leaf Plots, Histograms, and Circle Graphs

Make a stem-and-leaf plot for each data set.

1. 32, 46, 38, 42, 51, 33, 32, 42, 46, 32, 52

Stem	Leaf

2. 13, 21, 17, 9, 22, 17, 26, 13, 6, 26, 21, 17, 29

Stem	Leaf

Make frequency table and a histogram for the data.

3. 1, 3, 8, 2, 7, 8, 2, 3, 1, 8, 1, 1, 5, 7, 8, 3, 3, 1, 1, 6

Number	Frequency

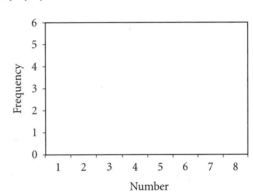

Make a relative frequency table for the data.

4. 1, 3, 8, 2, 7, 8, 2, 3, 1, 8, 1, 1, 5, 7, 8, 3, 3, 1, 1, 6

Number	Frequency	Relative frequency
1 – 2		
3 – 4		
5 – 6		
7 – 8		
9 – 10		

5. Find the probability that a number randomly selected from the data set in Exercise 4 will be *at least* 6. _____

Algebra 2

NAME _____ CLASS _____ DATE _____

Practice Masters Level B

12.2 Stem-and-Leaf Plots, Histograms, and Circle Graphs

Make a stem-and-leaf plot for each data set.

1. 8.5, 5.7, 5.9, 6.0, 6.7, 5.1, 6.9, 7.8, 8.2, 7.3, 8.1, 5.5, 7.3, 8.5, 8.1, 6.6, 7.8, 6.0

2. 0.59, 0.44, 0.37, 0.41, 0.56, 0.43, 0.49, 0.37, 0.26, 0.20, 0.41, 0.54, 0.16, 0.32, 0.33, 0.24, 0.50, 0.59, 0.34, 0.30

Stem	Leaf

Stem	Leaf

Make a circle graph for the data.

3.

Most Popular Colors for Sports Cars 1998 Model Year					
Dark Green	Black	White	Silver	Red	
16%	15%	14%	10%	16%	
Lite Brown	Teal	Purple	Blue	Others	
7%	4%	3%	5%	10%	

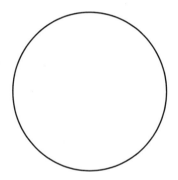

Make a relative frequency table and a relative frequency histogram for the data.

4. 1.3, 1.7, 1.5, 1.5, 1.2, 1.1, 1.4, 1.6, 1.1, 1.7, 1.0, 1.6, 1.3, 1.0, 1.9, 1.8, 1.3, 1.7, 1.2, 1.1

Number	Frequency	Relative frequency

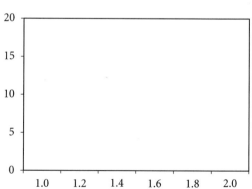

5. Find the probability that a number randomly selected from the data set in Exercise 4 will be *at least* 1.6. _____

Practice Masters Level C

12.2 Stem-and-Leaf Plots, Histograms, and Circle Graphs

1. A police officer has written tickets for the month according to the data in the table.

Type of Ticket	Number of Tickets
Speeding	29
Failing to stop	14
Running red lights	6
Illegal parking	5
No seat belt	10
Other	20

   a. Make a circle graph for the data.

   b. What is the probability that a person was not ticketed for running a red light if the person was ticketed by this officer?

   _____

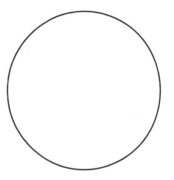

2. a. Make a frequency table for the data set.
   1.6, 2.6, 1.4, 3.3, 2.9, 0.1, 1.6, 1.5, 0.9, 0.7, 0.9, 3.1, 1.8, 1.3, 2.9, 3.8, 2.9, 0.5, 1.5, 2.3, 2.3, 1.4, 0.9, 1.1

Number	Frequency	Relative frequency

   b. Make a histogram for the data set.

   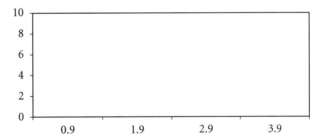

   c. Estimate the mean of the data set. _____

   d. Find the probability that a number randomly selected will be *at most* 2.9. _____

Algebra 2   Practice Masters Levels A, B, and C   231

NAME _____ CLASS _____ DATE _____

Practice Masters Level A
12.3 Box-and-Whisker Plots

Find the quartiles, the range, and the interquartile range for each data set.

1. 7, 1, 2, 8, 4, 6, 5, 10, 4, 2, 5, 10, 3, 5, 1, 10, 4, 6, 5, 3

2. 12, 20, 13, 12, 20, 14, 14, 17, 13, 16, 13, 18, 14, 18, 15, 16, 12, 12, 18, 15

3. 18, 55, 34, 103, 62, 37, 92, 35, 95, 81, 11, 94, 45, 32, 90, 24, 10, 109, 20, 17

Find the minimum and maximum values, quartiles, range and interquartile range for each data set. Then make a box-and-whisker plot for each data set.

4. 12, 14, 16, 15, 20, 17, 20, 2, 9, 18, 11, 8, 6, 19, 11

5. 39, 31, 39, 40, 32, 31, 37, 35, 33, 33, 37, 32, 35, 40, 32

6. 65, 75, 65, 76, 65, 68, 67, 68, 69, 72, 56, 72, 72, 73, 67, 76, 82

7. $1.09, $0.99, $1.29, $1.59, $0.99, $1.42, $1.33, $1.25, $1.10, $1.19, $1.12

NAME _____ CLASS _____ DATE _____

Practice Masters Level B
12.3 Box-and-Whisker Plots

Find the quartiles, the range, and the interquartile range for each data set.

1. 3.5, 0.7, 8.6, 7.1, 2.7, 4.1, 7.7, 2.1, 0.3, 5.6, 3.2, 6.6, 8.9, 6.5, 0.5, 3.8, 4.5, 2.5, 1.5, 8.9

2. 886, 592, 522, 738, 343, 650, 530, 92, 981, 557, 374, 899, 271, 424, 907

3. The table below gives the ages of the Presidents of the United States when they died. Find the minimum, and maximum values, quartiles, range, and interquartile range. Then make a box-and-whisker-plot.

President	Age	President	Age	President	Age
Washington	67	Johnson	66	Truman	88
Adams	90	Grant	63	Eisenhower	78
Jefferson	83	Hayes	70	Kennedy	46
Madison	85	Garfield	49	Johnson	64
Monroe	73	Arthur	56	Nixon	81
Q. Adams	80	Cleveland	71		
Jackson	78	Harrison	67		
Van Buren	79	Cleveland	71		
Harrison	68	McKinley	58		
Tyler	71	Roosevelt	60		
Polk	53	Taft	72		
Taylor	65	Wilson	67		
Fillmore	74	Harding	57		
Pierce	64	Coolidge	60		
Buchanan	77	Hoover	90		
Lincoln	56	Roosevelt	63		

Minimum _____ Maximum _____

$Q_1 = $ _____ $Q_2 = $ _____ $Q_3 = $ _____

Range _____   IQR _____

Box-and-whisker plot:

Algebra 2   Practice Masters Levels A, B, and C   233

Practice Masters Level C
12.3 Box-and-Whisker-Plots

The table gives the number of points scored by the teams in the first 33 Superbowls.

Super Bowl	Winners score	Losers score	Super Bowl	Winners score	Losers score	Super Bowl	Winners score	Losers score
1	35	10	12	27	10	23	20	16
2	33	14	13	35	31	24	55	10
3	16	7	14	31	19	25	20	19
4	23	7	15	27	10	26	37	24
5	16	13	16	26	21	27	52	17
6	24	3	17	27	17	28	30	13
7	14	7	18	38	9	29	49	26
8	24	7	19	38	16	30	27	17
9	16	6	20	46	10	31	35	21
10	21	17	21	39	20	32	31	24
11	32	14	22	42	10	33	34	19

1. For the data set that includes winners only, find the minimum, and maximum values, quartiles, range and interquartile range. _____

2. For the data set that includes losers only, find the minimum, and maximum values, quartiles, range and interquartile range. _____

3. Make a box-and-whisker plot using the information from both Exercises 1 and 2.

NAME _____ CLASS _____ DATE _____

Practice Masters Level A
12.4 Measures of Dispersion

Find the range and mean deviation for each data set.

1. 7, 5, 2, 8, 5, 6

2. 12, 13, 17, 14, 12, 15

3. 47, 51, 42, 68, 55, 70

4. 110, 108, 112, 104, 111, 105

5. −4, −10, −11, −7, −12, −8

6. 3, 5, −2, 7, −3, 0

Find the variance and standard deviation for each data set.

7. 7, 5, 2, 8, 5, 6

8. 12, 13, 17, 14, 12, 15

9. 47, 51, 42, 68, 55, 70

10. 110, 108, 112, 104, 111, 105

11. −4, −10, −11, −7, −12, −8

12. 3, 5, −2, 7, −3, 0

The winning times in seconds for the women's 100-meter dash for several Olympics are shown in the table.

Year	1960	'64	'68	'72	'76	'80	'84	'88	'92	'96
Winning time	11.0	11.4	11.0	11.07	11.08	11.6	10.97	10.54	10.82	10.94

13. Find the mean and median winning times.

   Mean _____   Median _____

14. Find the range and the mean deviation.

   Range _____   Mean deviation _____

15. Find the standard deviation. _____

Algebra 2   Practice Masters Levels A, B, and C   **235**

Practice Masters Level B
12.4 Measures of Dispersion

Find the range and mean deviation for each data set.

1. 27, 36, 42, 35, 22, 51, 41, 47, 37

2. $-14, -17, -10, -20, -11, -12, -15$

3. 6, -7, 3, -8, 5, -7, -10, -12

4. 4.2, 5.7, 3.1, 5.2, 4.9, 3.2, 4.3, 4.2

5. $\dfrac{1}{2}, \dfrac{2}{3}, \dfrac{1}{10}, \dfrac{1}{5}, \dfrac{1}{3}, \dfrac{3}{4}, \dfrac{1}{4}, \dfrac{1}{6}$

6. $\dfrac{-2}{5}, \dfrac{3}{4}, \dfrac{3}{5}, \dfrac{-1}{2}, \dfrac{-1}{3}, \dfrac{1}{4}, \dfrac{1}{2}, \dfrac{2}{5}$

Find the variance and standard deviation for each data set.

7. 27, 36, 42, 35, 22, 51, 41, 47, 37

8. $-14, -17, -10, -20, -11, -12, -15$

9. 6, -7, 3, -8, 5, -7, -10, -12

10. 4.2, 5.7, 3.1, 5.2, 4.9, 3.2, 4.3, 4.2

11. $\dfrac{1}{2}, \dfrac{2}{3}, \dfrac{1}{10}, \dfrac{1}{5}, \dfrac{1}{3}, \dfrac{3}{4}, \dfrac{1}{4}, \dfrac{1}{6}$

12. $\dfrac{-2}{5}, \dfrac{3}{4}, \dfrac{3}{5}, \dfrac{-1}{2}, \dfrac{-1}{3}, \dfrac{1}{4}, \dfrac{1}{2}, \dfrac{2}{5}$

The winning times (minutes : seconds . hundredths of a second) for the men's single luge event in several Olympics are shown in the table.

Year	1964	'68	'72	'76	'80	'84	'88	'92	'94	'98
Winning time	3:26.77	2:52.48	3:27.58	3:27.69	2:54.80	3:04.26	3:05.55	3:02.36	3:21.57	3:18.44

13. Find the mean and median winning times.

    Mean _____

    Median _____

14. Find the range and the mean deviation.

    Range _____

    Mean deviation _____

15. Find the standard deviation. _____

NAME _____ CLASS _____ DATE _____

Practice Masters Level C
12.4 Measures of Dispersion

Find the mean deviation and standard deviation for each data set.

1. 91, 64, 15, 90, 46, 10, 43, 10, 42, 69

2. 6.2, 3.8, 3.2, 3.7, 2.4, 5.8, 5.4, 2.7, 6.9, 4.2

3. −48, −56, −27, −39, −47, −86, −89, −28, −46, −68

4. −27, 68, 28, −37, −36, 85, −12, 61, 90, −24

5. $\frac{3}{5}, \frac{1}{2}, \frac{3}{4}, \frac{1}{5}, \frac{1}{4}, \frac{2}{5}, \frac{1}{3}, \frac{1}{5}, \frac{1}{4}, \frac{3}{5}$

6. 193, 434, 771, 921, 235, 797, 369, 848, 467, 818

7. $\frac{-3}{4}, \frac{2}{3}, \frac{3}{5}, \frac{-1}{2}, \frac{-1}{4}, \frac{1}{3}, \frac{2}{3}, \frac{-3}{5}, \frac{4}{5}, \frac{1}{4}$

8. 79.94, 37.01, 13.27, 39.58, 27.72, 30.23, 66.15, 63.65, 57.19, 69.96

Which measure of variation, mean deviation or standard deviation, is less affected by an extreme value in the following data sets?

9. 12, 215, 11, 16, 14

10. 166, 150, 214, 180, 0, 152

Name any values that lie outside one standard deviation of the mean.

11. 7, 10, 9, 4, 8, 11, 12, 10, 8, 10

12. −6, −8, −9, −5, −10, −13, −7, −8, −4, −9

The data below is the number of executions by state:
16, 8, 16, 4, 1, 8, 39, 22, 1, 10, 5, 1, 24, 2,
4, 29, 1, 3, 6, 8, 9, 2, 2, 13, 144, 5, 46, 2, 1
and twenty-one states had zero executions

13. Find the mean and median of the data.
    Mean _____ Median _____

14. Find the standard deviation. _____

Algebra 2   Practice Masters Levels A, B, and C   237

Practice Masters Level A
12.5 Binomial Distributions

A coin is flipped 5 times. Find the probability of each event.

1. Exactly 2 are heads.

2. Exactly 4 are tails.

3. More than 2 are heads.

4. At least 2 are heads.

A number cube is rolled 5 times. Find the probability of each event.

5. Exactly 5 are sixes.

6. Exactly 3 are fours.

7. More than 3 are ones.

8. At least 2 are threes.

9. At most 3 are threes.

10. No more than 4 are fives.

A multiple-choice test has 8 questions. Each question has 1 correct choice and 3 incorrect choices. Find the probability of each event if a person randomly guesses at each question.

11. Exactly 2 are correct.

12. Exactly 4 are correct.

13. Exactly 0 are correct.

14. At least 1 is correct.

15. At most 2 are correct.

16. No more than 6 are correct.

A family has 5 children. If the probabilities of having a boy and having a girl are equal, find the probability of each event.

17. Exactly 3 are boys.

18. At least 1 is a girl.

238   Practice Masters Levels A, B, and C   Algebra 2

NAME _____ CLASS _____ DATE _____

Practice Masters Level B
12.5 Binomial Distributions

Suppose that a new medicine treatment has a 70% cure rate. Find the probability of each event for the next 20 people who undergo the treatment.

1. Exactly 5 people are cured.

2. Exactly 10 people cured.

3. All 20 people are cured.

4. At least 4 persons cured.

5. At most 15 people are cured.

6. Between 15 and 18 people, inclusive, are cured.

In a certain geographical location, 75% of the respondents to a questionnaire stated that they were Republicans. Find the following probabilities based on 10 people who are selected at random.

7. Exactly 7 are Republican.

8. Exactly 8 are Republican.

9. At least 1 is Democrat.

10. At most 1 is Democrat.

11. Between 1 and 9, inclusive, are Republican.

12. Between 5 and 10, inclusive, are Republican.

Find the probability that a batter with the following batting averages will get a hit in 4 of the next 10 at-bats.

13. 0.300

14. 0.400

15. 0.350

16. 0.250

17. 0.273

18. 0.311

Algebra 2   Practice Masters Levels A, B, and C   **239**

Practice Masters Level C
12.5 Binomial Distributions

Suppose a test to determine pregnancy is advertised as 90% accurate. (The test is positive when pregnant and negative when not pregnant.) Find the probability of each event.

1. 5 out of 10 will receive the correct result. _____

2. 9 out of 10 will receive the correct result. _____

3. At least 8 out of 10 will receive the correct result. _____

4. At most 7 out of 10 will receive the correct result. _____

A shipment of light bulbs is known to have 5 defective bulbs for every 200 light bulbs. Find the probability of each event for a shipment of 1000 light bulbs.

5. Exactly 5 bulbs are defective.

6. Exactly 20 bulbs are defective.

7. Exactly 25 bulbs are defective.

8. Exactly 30 bulbs are defective.

A certain brand of bottled sports drink has placed a cash prize under the cap of 1 out of 6 bottles. Find the probability of each event for a sample of 24 bottles.

9. Exactly 1 has a cash prize.

10. Exactly 4 have cash prizes.

11. Exactly 6 have cash prizes.

12. At most 2 have cash prizes.

13. At least 1 has a cash prize.

14. At least 3 have cash prizes.

Practice Masters Level A
12.6 Normal Distributions

Let x be a random variable with a standard normal distribution. Use the area table for a standard normal curve, given on page 807 of the textbook, to find each probability.

1. $P(x \geq 0.2)$

2. $P(x \geq 1.8)$

3. $P(x \leq 1.2)$

4. $P(x \leq 1.0)$

5. $P(x \geq -1.4)$

6. $P(x \leq -0.4)$

7. $P(x \geq -1)$

8. $P(x \leq -0.6)$

A light bulb company advertises that a 75-watt light bulb has an average life of 500 hours with a standard deviation of 30 hours.

9. What percent of the light bulbs have an average life of less than 425 hours?

10. What percent of the light bulbs have an average life of more than 550 hours?

11. What percent of the light bulbs have an average life of more than 450 hours?

12. What percent of the light bulbs have an average life of more than 500 hours and less than 550 hours?

13. What percent of the light bulbs have an average life of more than 425 hours and less than 575 hours?

The weights of a box of cereal are normally distributed with a mean of 12.2 ounces and a standard deviation of 0.3 ounces. Find the probability of each event.

14. a box weighs less than 12 ounces

15. a box weighs more than 12.5 ounces

16. a box weighs between 12.2 and 12.6 ounces

Algebra 2   Practice Masters Levels A, B, and C   241

NAME _____ CLASS _____ DATE _____

Practice Masters Level B
12.6 Normal Distributions

Let *x* be a random variable with a standard normal distribution. Use the area table for a standard normal curve, given on page 807 of the textbook, to find each probability.

1. $P(x \geq 0.6)$

2. $P(x \leq -0.8)$

3. $P(x \geq -1.0)$

4. $P(x \leq 2.0)$

5. $P(1.4 \leq x \leq 1.8)$

6. $P(-2.0 \leq x \leq -1.0)$

7. $P(-1.2 \leq x \leq 1.0)$

8. $P(-2.0 \leq x \leq 1.8)$

A poll by a television advertising company found that the time a person will watch television uninterrupted is normally distributed with a mean of 97 minutes with a standard deviation of 21 minutes. What is the probability that a viewer will watch television uninterrupted for each given time interval?

9. more than 100 minutes

10. less than 60 minutes

11. more than 45 minutes

12. less than 2 hours

13. 1 to 1.5 hours

14. 90 minutes to 2 hours

If the weight of a population is normally distributed with a mean of 185 pounds and a standard deviation of 100 pounds, find the following probability of each event.

15. A person weighs less than twice the average weight. _____

16. A person weighs more than half the average weight. _____

242   Practice Masters Levels A, B, and C   Algebra 2

NAME _____ CLASS _____ DATE _____

Practice Masters Level C
12.6 Normal Distributions

The waiting time in a hospital emergency room is normally distributed with a mean of 25 minutes and a standard deviation of 8 minutes.

1. What is the probability that a person will wait less than 10 minutes? _____

2. What is the probability that a person will wait more than 30 minutes? _____

3. What is the probability that a person will wait between 10 and 40 minutes? _____

4. What is the probability that a person will wait between 28 and 32 minutes? _____

5. What is the probability that a person will wait less than 15 minutes or more than 32 minutes? _____

6. Approximately 68% of the people will wait between what two times? _____

At a high school, classes begin at 7:40. The average arrival time is 7:35 with a standard deviation of 3.5 minutes.

7. On an ordinary day, what percent of the students are late? _____

8. What percent of the students arrive before 7:30? _____

9. What percent of the students arrive later than 7:42? _____

10. What percent of the students arrive between 7:35 and 7:40? _____

11. What percent of the students arrive between 7:38 and 7:43? _____

12. Approximately 99% of the students arrive between what two times? _____

A soft drink dispenser fills cups, on average, 8 ounces with a standard deviation of 0.3 ounces. The volumes are normally distributed. Find the probability of each event.

13. A cup contains more than 8.5 ounces. _____

14. A cup contains less than 7.8 ounces. _____

15. A cup contains more than 7.9 ounces. _____

16. A cup contains more than 7.7 ounces and less than 8.4 ounces. _____

17. A cup contains less than 7.6 ounces or more than 8.3 ounces. _____

18. Approximately 95% of the cups contain between what two values? _____

Algebra 2   Practice Masters Levels A, B, and C

Practice Masters Level A
13.1 Right-Triangle Trigonometry

Refer to △ABC at the right to find each value listed. Give exact answers and answers to the nearest tenth.

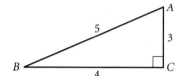

1. sin A

2. cos A

3. tan A

4. sin B

5. cos B

6. tan B

7. sec A

8. csc B

Find m∠A to the nearest tenth by using inverse trigonometric functions.

9.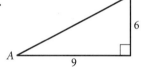

10.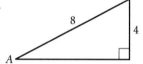

11.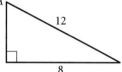

12.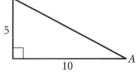

13.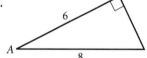

14.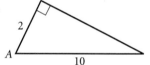

Solve each triangle. Give angle measures to the nearest degree and side lengths to the nearest tenth.

15.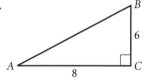

AB = _____

m∠A = _____

m∠B = _____

16.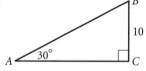

AB = _____

AC = _____

m∠B = _____

17.

AB = _____

AC = _____

m∠A = _____

NAME _____ CLASS _____ DATE _____

Practice Masters Level B
13.1 Right-Triangle Trigonometry

Refer to △ABC at the right to find each value listed. Give exact answers and answers to the nearest ten-thousandth.

1. sin A

2. cos A

3. tan A

4. sin B

5. cos B

6. tan B

7. sec A

8. csc B

Find m∠A to the nearest tenth by using inverse trigonometric functions.

9.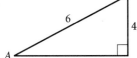

10.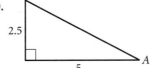

11.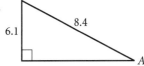

12.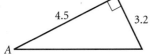

13.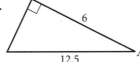

14.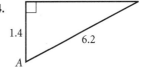

Solve each triangle. Give angle measures to the nearest degree and side lengths to the nearest tenth.

15.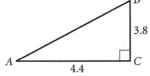

16.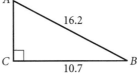

17.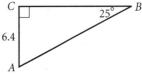

AB = _____   AC = _____   AB = _____

m∠A = _____   m∠A = _____   BC = _____

m∠B = _____   m∠B = _____   m∠A = _____

Algebra 2   Practice Masters Levels A, B, and C   245

NAME _____ CLASS _____ DATE _____

Practice Masters Level C
13.1 Right-Triangle Trigonometry

Solve each triangle. Give angle measures to the nearest degree and side lengths to the nearest tenth.

1.

AB = _____

m∠A = _____

m∠B = _____

2.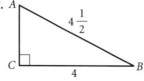

AC = _____

m∠A = _____

m∠B = _____

3.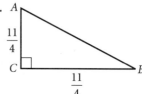

AB = _____

m∠A = _____

m∠B = _____

4.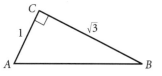

AB = _____

m∠A = _____

m∠B = _____

5.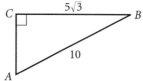

AC = _____

m∠A = _____

m∠B = _____

6.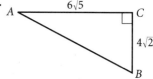

AB = _____

m∠A = _____

m∠B = _____

7.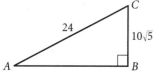

AB = _____

m∠A = _____

m∠B = _____

8.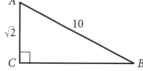

CB = _____

m∠A = _____

m∠B = _____

9.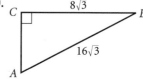

AC = _____

m∠A = _____

m∠B = _____

10. The base of a 12-foot ladder is placed 4 feet from the base of a building. What angle of elevation does the ladder make with the ground? _____

11. A cement patio is to be made in the shape of a right triangle with a hypotenuse of 20 feet. If the two legs of the right triangle are to be equal, find all the side and angle measures for the patio. _____

NAME _____ CLASS _____ DATE _____

Practice Masters Level A
13.2 Angles of Rotation

For each angle, find all coterminal angles such that $-360° < \theta < 360°$.

1. 20° _____   2. 55° _____   3. 100° _____

4. 179° _____   5. 240° _____   6. 270° _____

7. 330° _____   8. 400° _____   9. −80° _____

Find the reference angle.

10. 130° _____   11. 65° _____   12. 200° _____

13. 290° _____   14. 355° _____ 15. 112° _____

16. −40° _____   17. −100° _____ 18. 400° _____

Find the exact value of the six trigonometric functions of θ given each point on the terminal side of θ in standard position.

19. (2, 2)

sin θ = _____
cos θ = _____
tan θ = _____
csc θ = _____
sec θ = _____
cot θ = _____

20. (3, 5)

sin θ = _____
cos θ = _____
tan θ = _____
csc θ = _____
sec θ = _____
cot θ = _____

21. (−4, 2)

sin θ = _____
cos θ = _____
tan θ = _____
csc θ = _____
sec θ = _____
cot θ = _____

Given the quadrant of θ in standard position and a trigonometric function value of θ, find exact values for the indicated functions.

22. II, $\sin \theta = \dfrac{1}{2}$; $\cos \theta$

23. II, $\tan \theta = \dfrac{-2}{5}$; $\sin \theta$

24. IV, $\cos \theta = \dfrac{2}{3}$; $\sin \theta$

25. I, $\tan \theta = \dfrac{5}{8}$; $\cos \theta$

26. III, $\sin \theta = \dfrac{-4}{9}$; $\tan \theta$

27. IV, $\tan \theta = \dfrac{-5}{11}$; $\cos \theta$

Algebra 2

NAME _____ CLASS _____ DATE _____

Practice Masters Level B
13.2 Angles of Rotation

For each angle, find all coterminal angles such that $-360° < \theta < 360°$.

1. 77° _____   2. 112° _____   3. −60° _____

4. −120° _____   5. 500° _____   6. 845° _____

Find the reference angle.

7. −120° _____   8. −200° _____   9. 450° _____

10. 600° _____   11. −295° _____   12. −340° _____

Find the exact value of the six trigonometric functions of θ given each point on the terminal side of θ in standard position.

13. $(3, -4)$

sin θ = _____
cos θ = _____
tan θ = _____
csc θ = _____
sec θ = _____
cot θ = _____

14. $(-2, 4)$

sin θ = _____
cos θ = _____
tan θ = _____
csc θ = _____
sec θ = _____
cot θ = _____

15. $(-4, -6)$

sin θ = _____
cos θ = _____
tan θ = _____
csc θ = _____
sec θ = _____
cot θ = _____

Given the quadrant of θ in standard position and a trigonometric function value of θ, find exact values for the indicated functions.

16. II, $\sin \theta = \frac{4}{7}$; $\cos \theta$

17. IV, $\cos \theta = \frac{1}{2}$; $\tan \theta$

18. III, $\tan \theta = \frac{2}{5}$; $\sin \theta$

19. III, $\tan \theta = \frac{4}{7}$; $\cos \theta$

20. II, $\cos \theta = \frac{-2}{3}$; $\tan \theta$

21. II, $\tan \theta = \frac{-5}{9}$; $\sin \theta$

Find the number of rotations or the fraction of a rotation represented by each angle.

22. 180° _____   23. 450° _____   24. 630° _____

248   Practice Masters Levels A, B, and C   Algebra 2

NAME _____ CLASS _____ DATE _____

Practice Masters Level C
13.2 Angles of Rotation

For each angle, find all coterminal angles such that
$-360° < \theta < 360°$. Then find the corresponding reference angle.

1. 550° _____   2. −350° _____   3. −740° _____

Find the number of rotations or the fraction of a rotation
represented by each angle. Then indicate the direction.

4. 1020° _____   5. −525° _____   6. −631° _____

Find the exact value of the six trigonometric functions of θ given
each point on the terminal side of θ in standard position.

7. $(-5, 6)$

sin θ = _____

cos θ = _____

tan θ = _____

csc θ = _____

sec θ = _____

cot θ = _____

8. $(2, -5)$

sin θ = _____

cos θ = _____

tan θ = _____

csc θ = _____

sec θ = _____

cot θ = _____

9. $\left(\dfrac{1}{2}, 2\right)$

sin θ = _____

cos θ = _____

tan θ = _____

csc θ = _____

sec θ = _____

cot θ = _____

Find the exact value of the six trigonometric functions of θ given
each point on the terminal side of θ in standard position.

10. IV, $\sin \theta = \dfrac{-2}{5}$

sin θ = _____

cos θ = _____

tan θ = _____

csc θ = _____

sec θ = _____

cot θ = _____

11. II, $\sin \theta = \dfrac{2}{9}$

sin θ = _____

cos θ = _____

tan θ = _____

csc θ = _____

sec θ = _____

cot θ = _____

12. III, $\tan \theta = \dfrac{4}{7}$

sin θ = _____

cos θ = _____

tan θ = _____

csc θ = _____

sec θ = _____

cot θ = _____

Algebra 2   Practice Masters Levels A, B, and C

NAME _____ CLASS _____ DATE _____

Practice Masters Level A
13.3 Trigonometric Functions of Any Angle

Find the exact values of sine, cosine, and tangent of each angle.

1. 45°
   sin: _____
   cos: _____
   tan: _____

2. 60°
   sin: _____
   cos: _____
   tan: _____

3. 90°
   sin: _____
   cos: _____
   tan: _____

4. −240°
   sin: _____
   cos: _____
   tan: _____

5. 270°
   sin: _____
   cos: _____
   tan: _____

6. −30°
   sin: _____
   cos: _____
   tan: _____

Point P is located at the intersection of the unit circle and the terminal side of angle θ in standard position. Find the coordinates of P to the nearest hundredth.

7. $\theta = 70°$

8. $\theta = 120°$

9. $\theta = 190°$

10. $\theta = 250°$

11. $\theta = -10°$

12. $\theta = -50°$

Find the exact values of sine, cosine, and tangent of each angle.

13. 540°
    sin: _____
    cos: _____
    tan: _____

14. 390°
    sin: _____
    cos: _____
    tan: _____

15. 405°
    sin: _____
    cos: _____
    tan: _____

Find each trigonometric function value. Give answers to the nearest thousandth.

16. sec 100°

17. csc 200°

18. cot 89°

250   Practice Masters Levels A, B, and C   Algebra 2

NAME _____ CLASS _____ DATE _____

Practice Masters Level B

13.3 Trigonometric Functions of Any Angle

Find the exact values of sine, cosine, and tangent of each angle.

1. 30°
   sin: _____
   cos: _____
   tan: _____

2. −135°
   sin: _____
   cos: _____
   tan: _____

3. 120°
   sin: _____
   cos: _____
   tan: _____

4. 330°
   sin: _____
   cos: _____
   tan: _____

5. 225°
   sin: _____
   cos: _____
   tan: _____

6. −60°
   sin: _____
   cos: _____
   tan: _____

Point P is located at the intersection of the unit circle and the terminal side of angle θ in standard position. Find the coordinates of P to the nearest hundredth.

7. $\theta = 110°$

8. $\theta = 205°$

9. $\theta = 41°$

10. $\theta = -37°$

11. $\theta = -58°$

12. $\theta = 199°$

Find the exact values of sine, cosine, and tangent of each angle.

13. 840°
    sin: _____
    cos: _____
    tan: _____

14. 1020°
    sin: _____
    cos: _____
    tan: _____

15. 1260°
    sin: _____
    cos: _____
    tan: _____

Find each trigonometric function value. Give answers to the nearest thousandth.

16. sec −300°

17. csc −400°

18. cot −510°

Algebra 2   Practice Masters Levels A, B, and C   251

NAME _____ CLASS _____ DATE _____

Practice Masters Level C
13.3 Trigonometric Functions of Any Angle

Point *P* is located at the intersection of a circle with a radius of *r*, and the terminal side of angle θ in the standard position. Find the exact coordinates of *P*.

1. $\theta = 300°; r = 4$

2. $\theta = -135°; r = 6$

3. $\theta = 495°; r = 10$

4. $\theta = -120°; r = 5$

5. $\theta = -330°; r = 8$

6. $\theta = -540°; r = 5$

Point *P* is located at the intersection of the unit circle and the terminal side of angle θ in standard position. Find the coordinates of *P* to the nearest hundredth.

7. $\theta = -97°$

8. $\theta = -32°$

9. $\theta = -205°$

Find the exact values of sine, cosine, and tangent of each angle.

10. $-315°$
   sin: _____
   cos: _____
   tan: _____

11. $-600°$
   sin: _____
   cos: _____
   tan: _____

12. $585°$
   sin: _____
   cos: _____
   tan: _____

13. $780°$
   sin: _____
   cos: _____
   tan: _____

14. $1350°$
   sin: _____
   cos: _____
   tan: _____

15. $1035°$
   sin: _____
   cos: _____
   tan: _____

Find each trigonometric function value. Give answers to the nearest thousandth.

16. $\sec 480°$

17. $\csc -600°$

18. $\cot -840°$

252   Practice Masters Levels A, B, and C   Algebra 2

NAME _____ CLASS _____ DATE _____

Practice Masters Level A
13.4 Radian Measure and Arc Length

Convert each degree measure to radian measure. Give exact answers.

1. 45°

2. 60°

3. 240°

4. 150°

5. −120°

6. −150°

7. 100°

8. −350°

9. 140°

Convert each radian measure to degree measure. Round answers to the nearest tenth of a degree.

10. 3π

11. $\dfrac{\pi}{4}$

12. $\dfrac{\pi}{3}$

13. $\dfrac{-\pi}{3}$

14. $\dfrac{5\pi}{6}$

15. $\dfrac{2\pi}{3}$

16. 1

17. −2

18. 3.1

A circle has a diameter of 12 inches. For each central angle measure, find the length in inches of the arc intercepted by the angle.

19. π radians

20. $\dfrac{\pi}{6}$ radians

21. $\dfrac{3\pi}{4}$ radians

22. 2 radians

23. 4.5 radians

24. $\dfrac{3\pi}{2}$ radians

Evaluate each expression. Give exact answers.

25. $\sin\dfrac{\pi}{2}$

26. $\cos\dfrac{\pi}{4}$

27. $\tan\dfrac{\pi}{3}$

Algebra 2   Practice Masters Levels A, B, and C   253

Practice Masters Level B
13.4 Radian Measure and Arc Length

Convert each degree measure to radian measure. Give exact answers.

1. 40°

2. −70°

3. 130°

4. 200°

5. −125°

6. 145°

7. 27°

8. 54°

9. −36°

Convert each radian measure to degree measure. Round answers to the nearest tenth of a degree.

10. $\dfrac{3\pi}{4}$

11. $\dfrac{3\pi}{5}$

12. $\dfrac{3\pi}{2}$

13. $\dfrac{7\pi}{4}$

14. $\dfrac{-5\pi}{3}$

15. $\dfrac{-\pi}{12}$

16. 3.5

17. −2.7

18. 1.2

A circle has a diameter of 15 inches. For each central angle measure, find the length in inches of the arc intercepted by the angle.

19. $\dfrac{\pi}{3}$ radians

20. $\dfrac{7\pi}{6}$ radians

21. $\dfrac{7\pi}{4}$ radians

22. 1 radian

23. 2.3 radians

24. 3.19 radians

Evaluate each expression. Give exact answers.

25. $\sin \dfrac{11\pi}{6}$

26. $\cos \dfrac{7\pi}{2}$

27. $\tan \dfrac{9\pi}{4}$

254   Practice Masters Levels A, B, and C   Algebra 2

NAME _____ CLASS _____ DATE _____

Practice Masters Level C
13.4 Radian Measure and Arc Length

Convert each degree measure to radian measure. Give exact answers.

1. 420°

2. −392°

3. 536°

4. −412°

5. 187.5°

6. −29°

Convert each radian measure to degree measure. Round answers to the nearest tenth of a degree.

7. $\dfrac{13\pi}{12}$

8. $\dfrac{23\pi}{4}$

9. $\dfrac{-17\pi}{2}$

A circle has a radius of 9.25 inches. For each central angle measure, find the length in inches of the arc intercepted by the angle.

10. $\dfrac{7\pi}{15}$ radians

11. 2.73 radians

12. 1.09 radians

Evaluate each expression. Give exact answers.

13. $\sin \dfrac{\pi}{3}$

14. $\cos \dfrac{-7\pi}{4}$

15. $\tan \dfrac{\pi}{6}$

16. $\sec \dfrac{-11\pi}{2}$

17. $\csc \dfrac{5\pi}{4}$

18. $\cot \dfrac{-13\pi}{4}$

19. A 20-inch diameter bicycle tire rotates 900 times. How many feet does the bicycle tire travel during this time?

20. What angle measure, in radians, does the hand on the clock form at 5:00?

Algebra 2   Practice Masters Levels A, B, and C   255

Practice Masters Level A

13.5 Graphing Trigonometric Functions

Identify the amplitude, if it exists, and the period of each function.

1. $y = 3 \sin \theta$
2. $y = 2.5 \cos \theta$
3. $y = 6 \sin 4\theta$

4. $y = 2 \tan \theta$
5. $y = -2 \cos 2\theta$
6. $y = 1.5 \sin \theta$

Identify the phase shift and the vertical translation of each function from its parent function.

7. $y = \sin(\theta + 90°) + 2$
8. $y = \sin(\theta - 30°) - 1$

9. $y = \cos(\theta - 60°) + 3$
10. $y = \cos(\theta + 45°) - 3$

Describe the transformation of each function from its parent function. Then graph at least one period of the function along with its parent function.

11. $y = 3 \sin 3\theta - 1$
12. $y = -\cos(\theta - 45°) + 2$

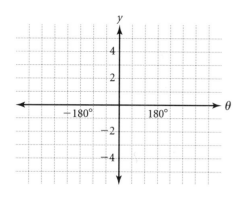
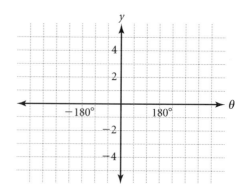

13. Write a function for a sound with a frequency of 100 hertz and an amplitude of 1.5.

NAME _____ CLASS _____ DATE _____

Practice Masters Level B
13.5 Graphing Trigonometric Functions

Identify the amplitude, if it exists, and the period for each function.

1. $y = \dfrac{1}{2} \sin \dfrac{1}{2}\theta$

2. $y = -3.5 \cos \dfrac{1}{4}\theta$

3. $y = -\sin(\theta - 30°)$

4. $y = \dfrac{-1}{4} \cos 4\theta$

5. $y = 2.5 \tan 2\theta$

6. $y = \dfrac{2}{5} \cos \dfrac{1}{3}\theta$

Identify the phase shift and the vertical translation of each function from its parent function.

7. $y = \cos(\theta - 180°)$

8. $y = 2\sin(\theta + 10°) - 2$

9. $y = 3 + \cos(\theta - 60°) + 2$

10. $y = 4 - \sin(\theta + 30°) + \dfrac{1}{2}$

11. $y = \dfrac{1}{2} \sin\left(\theta - \dfrac{\pi}{2}\right) + 1$

12. $y = -\sin(\theta + \pi) - 3$

Describe the transformation of each function from its parent function. Then graph at least one period of the function along with its parent function.

13. $y = 2 \cos\left(\theta - \dfrac{\pi}{2}\right) + 1$

14. $y = 2 - \sin\left(\theta + \dfrac{3\pi}{4}\right)$

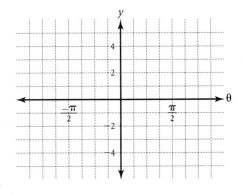
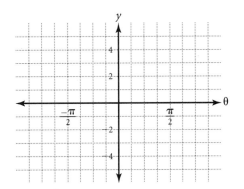

15. Write a function for a sound with a frequency of 75 hertz and an amplitude of 2 and a phase shift at one-third of a period to the right. _____

Algebra 2

Practice Masters Level C

13.5 Graphing Trigonometric Functions

1. Write a function for a sound wave with a frequency of 130 hertz, an amplitude of 3, and a phase shift of one-half of a period to the right. _____

Identify the amplitude, period, phase shift, and vertical shift of each function. Then graph at least one period of the function.

2. $y = -\tan(2\theta)$

3. $y = -2\cos\left(\theta + \dfrac{\pi}{4}\right) + 1$

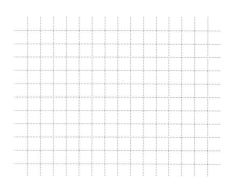

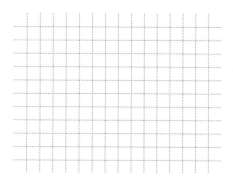

4. $y = 3\sin 2\left(\theta - \dfrac{\pi}{2}\right) - 1$

5. Write a function for the graph below.

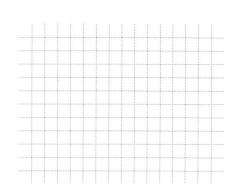

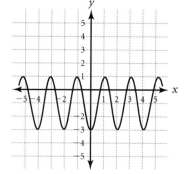

NAME _____ CLASS _____ DATE _____

Practice Masters Level A

13.6 Inverses of Trigonometric Functions

Evaluate each trigonometric expression. Give exact answers in degrees.

1. $\text{Sin}^{-1}\dfrac{1}{2}$

2. $\text{Sin}^{-1}(-1)$

3. $\text{Cos}^{-1}\dfrac{\sqrt{3}}{2}$

4. $\text{Cos}^{-1}\left(\dfrac{-\sqrt{2}}{2}\right)$

5. $\text{Tan}^{-1}\left(\dfrac{-\sqrt{3}}{3}\right)$

6. $\text{Cos}^{-1} 0$

Find all possible values for each expression.

7. $\sin^{-1}\dfrac{1}{2}$

8. $\cos^{-1} 1$

9. $\tan^{-1} 1$

10. $\sin^{-1}\left(\dfrac{-\sqrt{3}}{2}\right)$

11. $\cos^{-1}(-1)$

12. $\tan^{-1}\left(-\sqrt{3}\right)$

Evaluate each trigonometric expression.

13. $\sin\left(\text{Cos}^{-1}\dfrac{1}{2}\right)$

14. $\cos\left(\text{Sin}^{-1}\dfrac{\sqrt{2}}{2}\right)$

15. $\tan(\text{Sin}^{-1} 1)$

16. $\cos\left(\text{Cos}^{-1}(-1)\right)$

17. $\sin\left(\text{Tan}^{-1}\left(\dfrac{\sqrt{3}}{3}\right)\right)$

18. $\text{Sin}^{-1}(\cos 60°)$

Use inverse trigonometric functions to solve each problem.

19. A flag pole 30 feet tall casts a shadow of 50 feet on the ground. Find, to the nearest tenth of a degree, the angle of elevation of the sun.

20. A 20-foot extension ladder leans against a building. If the base of the ladder is placed 6 feet from the base of the building, what is the angle of elevation of the ladder?

21. An airplane flying at an altitude of 1000 feet spots a runway beacon 4000 feet away. What should the airplane's angle of descent be in order to land on the runway?

Algebra 2   Practice Masters Levels A, B, and C   259

NAME _____ CLASS _____ DATE _____

Practice Masters Level B
13.6 Inverses of Trigonometric Functions

Evaluate each trigonometric expression. Give exact answers in degrees.

1. $\text{Sin}^{-1}(0)$

2. $\text{Cos}^{-1}(-1)$

3. $\text{Tan}^{-1}(-1)$

4. $\text{Cos}^{-1}\left(\dfrac{\sqrt{2}}{2}\right)$

5. $\text{Sin}^{-1}\left(\dfrac{\sqrt{3}}{2}\right)$

6. $\text{Sin}^{-1}(-1)$

Find all possible values for each expression.

7. $\sin^{-1}(-1)$

8. $\cos^{-1}(0)$

9. $\sin^{-1}(0)$

10. $\tan^{-1}(0)$

11. $\cos^{-1}\left(\dfrac{-1}{2}\right)$

12. $\tan^{-1}\left(\dfrac{\sqrt{3}}{3}\right)$

Evaluate each trigonometric expression.

13. $\cos\left(\text{Sin}^{-1}\dfrac{\sqrt{3}}{2}\right)$

14. $\sin(\text{Tan}^{-1} 1)$

15. $\tan(\text{Cos}^{-1} 1)$

16. $\text{Sin}^{-1}\left(\sin\dfrac{\pi}{4}\right)$

17. $\text{Tan}^{-1}\left(\cos\dfrac{\pi}{6}\right)$

18. $\text{Cos}^{-1}\left(\tan\dfrac{\pi}{6}\right)$

Use inverse trigonometric functions to solve each problem.

19. A loading ramp rises 2 feet for every 10 horizontal feet.

   a. Find the angle of elevation of the ramp. _____

   b. If the overall length of the ramp is 22 feet, how much overall rise does the ramp have? _____

20. A person flying a kite lets out 350 feet of string. The person is holding the string 4 feet above the ground. The horizontal distance from the person to the point on the ground directly below the kite is 205 feet.

   a. What is the height of the kite? _____

   b. What is the angle of elevation of the string with the ground? _____

NAME _____ CLASS _____ DATE _____

Practice Masters Level C
13.6 Inverses of Trigonometric Functions

Evaluate each trigonometric expression. Give answers to the nearest tenth of a degree.

1. $\text{Sin}^{-1} 0.3$

2. $\text{Cos}^{-1} -0.2$

3. $\text{Tan}^{-1} 10.3$

Find all possible values for each expression. Give answers to the nearest tenth of a degree.

4. $\sin^{-1}(-0.4)$

5. $\cos^{-1} 0.8$

6. $\tan^{-1} 8.2$

Evaluate each trigonometric expression. Give answers to the nearest hundredth.

7. $\cos(\text{Sin}^{-1} 0.85)$

8. $\sin(\text{Tan}^{-1}(-4.2))$

9. $\text{Cos}^{-1}(\tan 10°)$

10. A 6-foot tall person spots a bird perched on a wire 30 feet in the air. The distance on the ground from the person to the point on the ground directly below the bird is 80 feet.

   a. How far is it from the person's eye to the bird, to the nearest foot? _____

   b. What is the angle of elevation from the person to the bird, to the nearest tenth of a degree? _____

11. Find the measure of the indicated sides and angles to the nearest tenth.

   a. $m\angle ABC =$ _____

   b. $m\angle CBD =$ _____

   c. $m\angle D =$ _____

   d. $BC =$ _____

   e. $BD =$ _____

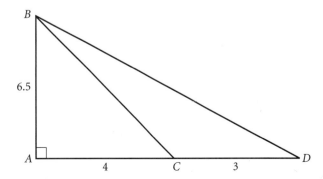

Algebra 2   Practice Masters Levels A, B, and C

NAME _____ CLASS _____ DATE _____

Practice Masters Level A
14.1 The Law of Sines

Find the indicated measures for each figure.

1. $B =$ _____

   $AB =$ _____

   $BC =$ _____

   Area $\triangle ABC =$ _____

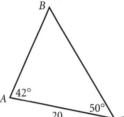

2. $B =$ _____

   $AC =$ _____

   $BC =$ _____

   Area $\triangle ABC =$ _____

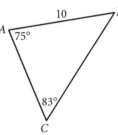

3. $B =$ _____

   $AB =$ _____

   $BC =$ _____

   Area $\triangle ABC =$ _____

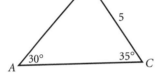

4. $C =$ _____

   $AB =$ _____

   $AC =$ _____

   Area $\triangle ABC =$ _____

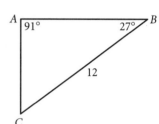

5. $B =$ _____

   $AC =$ _____

   $BC =$ _____

   Area $\triangle ABC =$ _____

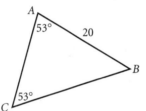

6. $C =$ _____

   $AC =$ _____

   $BC =$ _____

   Area $\triangle ABC =$ _____

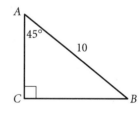

7. $A =$ _____

   $B =$ _____

   $AB =$ _____

   Area $\triangle ABC =$ _____

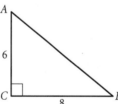

8. $B =$ _____

   $AC =$ _____

   $BC =$ _____

   Area $\triangle ABC =$ _____

9. Find the indicated measures for the triangular deck.

   $A =$ _____

   $C =$ _____

   $BC =$ _____

   Area of $\triangle ABC =$ _____

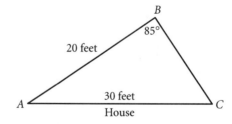

262   Practice Masters Levels A, B, and C   Algebra 2

NAME _____ CLASS _____ DATE _____

Practice Masters Level B
14.1 The Law of Sines

Find the area of △ABC to the nearest tenth of a square unit.

1. $a = 4$ in., $b = 7$ in., $C = 30°$

2. $b = 4$ m, $c = 10$ m, $A = 45°$

3. $b = 12$ cm, $c = 10$ cm, $A = 60°$

4. $a = 6$ in., $b = 9$ in., $C = 45°$

5. $a = 3$ ft, $c = 9$ ft, $B = 120°$

6. $b = 10$ m, $c = 8$ m, $A = 135°$

Use the given information to find the indicated side length in △ABC. Give answers to the nearest tenth.

7. Given $A = 30°$, $B = 62°$, and $a = 10$, find b. _____

8. Given $B = 50°$, $C = 38°$, and $c = 12$, find b. _____

9. Given $C = 41°$, $A = 54°$, and $a = 5$, find c. _____

10. Given $B = 68°$, $A = 52°$, and $b = 6$, find a. _____

Solve each triangle. Give answers to the nearest tenth, if necessary.

11. $A = 36°$, $B = 71°$, $b = 16$

12. $B = 50°$, $C = 100°$, $a = 14$

13. $A = 10°$, $C = 47°$, $c = 6$

14. $B = 55°$, $C = 75°$, $b = 8$

15. $A = 14°$, $B = 122°$, $a = 5$

16. $A = 97°$, $C = 33°$, $c = 4$

17. Two hunters, one at blind x, the other at blind y, 50 meters apart, spot a deer in the woods. How far from each blind is the deer if the angle at blind x is formed by the lines of sight to blind y, the deer is at 47°, and the angle formed by the lines of sight from blind y to blind x and the deer is 42°?

Algebra 2

Practice Masters Level C
14.1 The Law of Sines

Find the area of △ABC to the nearest tenth of a square unit.

1. $a = 5$ m, $b = 8$ m, $C = 50°$ _____

2. $b = 3$ cm, $c = 12$ cm, $A = 65°$ _____

3. $b = 15$ in., $c = 9$ in., $A = 70°$ _____

4. $a = 4$ m, $c = 11$ m, $B = 80°$ _____

Use the given information to find the indicated side length in △ABC. Give answers to the nearest tenth.

5. Given $A = 47°$, $B = 36°$, and $b = 10.2$, find a. _____

6. Given $B = 10°$, $C = 99°$, and $c = 11.4$, find b. _____

7. Given $C = 95°$, $A = 7°$, and $a = 9.2$, find c. _____

8. Given $B = 125°$, $A = 16°$, and $b = 8.6$, find a. _____

9. Given $A = 107°$, $C = 43°$, and $a = 4.3$, find c. _____

10. Given $C = 63°$, $B = 27°$, and $c = 5.1$, find b. _____

Solve each triangle. Give answers to the nearest tenth, if necessary.

11. $A = 37°$, $B = 62°$, $b = 4.1$

12. $B = 74°$, $C = 35°$, $c = 6.3$

13. $A = 54°$, $B = 26°$, $c = 5.5$

14. $A = 17°$, $C = 65°$, $a = 1.9$

State the number of triangles determined by the given information. If 1 or 2 triangles are formed, solve the triangle(s). Give answers to the nearest tenth, if necessary.

15. $A = 30°$, $B = 50°$, $c = 9$

16. $a = 8$, $b = 6$, $B = 40°$

17. Find the length, to the nearest tenth of a foot, of caution tape necessary to mark off the crime scene shown in the diagram.

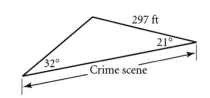

Practice Masters Levels A, B, and C

Practice Masters Level A

14.2 The Law of Cosines

Classify the type of information given, and then find the missing side length in each triangle. Give answers to the nearest tenth.

1.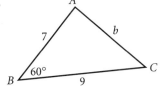

2.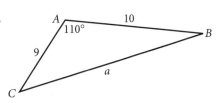

Classify the type of information given, and then find the measure of angle A in each triangle. Give answers to the nearest tenth.

3.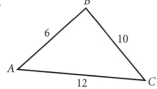

4.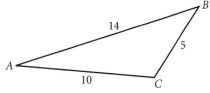

Solve each triangle. Give answers to the nearest tenth.

5. $a = 5, b = 6, c = 9$

6. $a = 4, b = 11, c = 10$

7. $a = 12, b = 9, c = 5$

8. $a = 10, b = 5, c = 7$

9. $A = 52°, b = 20, c = 13$

10. $B = 65°, a = 10, c = 14$

11. $C = 95°, a = 9, b = 6$

12. $A = 100°, a = 5, c = 4$

NAME _____ CLASS _____ DATE _____

Practice Masters Level B
14.2 The Law of Cosines

Classify the type of information given, and then use the law of cosines to find the missing side length in △ABC to the nearest tenth.

1. $A = 42°, b = 9, c = 12$ _____

2. $a = 20, B = 70°, c = 13$ _____

3. $a = 7, b = 3, C = 95°$ _____

4. $A = 120°, b = 6, c = 14$ _____

5. $a = 6.2, b = 8.5, C = 40°$ _____

6. $a = 9.1, B = 60°, c = 10.4$ _____

Solve each triangle. Give answers to the nearest tenth.

7. $a = 17, b = 6, c = 13$ _____

8. $a = 6, b = 5, c = 7$ _____

9. $A = 68°, b = 24, c = 20$ _____

10. $a = 4, B = 37°, c = 6$ _____

11. $a = 1.6, b = 2, C = 54°$ _____

12. $a = 6, A = 41°, B = 52°$ _____

Classify the type of information given, and then solve △ABC. Round answers to the nearest tenth. If no such triangle exists, write *not possible*.

13. $A = 38°, b = 6, c = 5$ _____

14. $a = 4, b = 3, c = 9$ _____

15. $a = 3, b = 5, C = 130°$ _____

16. $a = 4.0, B = 68°, c = 6$ _____

17. $a = 4, b = 3, C = 17°$ _____

18. $A = 47°, B = 129°, C = 4°$ _____

Practice Masters Level C
14.2 The Law of Cosines

Classify the type of information given, and then use the law of cosines to find the missing side length in △ABC to the nearest tenth.

1. $A = 42°, b = 6.5, c = 4.5$ _____

2. $a = 4.5, B = 62°, c = 3.7$ _____

3. $a = 6.3, b = 5.7, C = 40.5°$ _____

4. $a = 0.9, b = 0.6, C = 35.6°$ _____

5. $A = 80.2°, b = 13.5, c = 10.2$ _____

6. $a = 16, B = 103.4°, c = 12.2$ _____

Solve each triangle. Give answers to the nearest tenth.

7. $a = 10.5, b = 9.5, c = 11.5$ _____

8. $A = 72°, b = 6, c = 10.3$ _____

9. $a = 4.2, B = 36.2°, c = 5.1$ _____

10. $a = 13.6, B = 47.8°, c = 20.1$ _____

11. $a = 16.2, b = 10.8, c = 7.5$ _____

12. $a = 6.2, b = 5.9, C = 32.6°$ _____

Classify the type of information given, and then solve △ABC. Round answers to the nearest tenth. If no such triangle exists, write *not possible*.

13. $a = 16.1, b = 10.4, C = 47°$ _____

14. $A = 86.2°, b = 9, c = 4.7$ _____

15. $A = 32°, B = 61°, C = 87°$ _____

16. $a = 12.4, B = 36.4°, c = 9.8$ _____

17. $a = 3.6, b = 7.4, c = 11.1$ _____

18. $A = 106.2°, B = 18.3°, a = 9$ _____

Algebra 2

Practice Masters Level A

14.3 Fundamental Trigonometric Identities

Use definitions to prove each identity.

1. $\dfrac{\tan\theta}{\sec\theta} = \sin\theta$

2. $\sin^2\theta = 1 - \cos^2\theta$

3. $\dfrac{\csc\theta}{\cot\theta} = \sec\theta$

4. $\tan^2\theta = \sec^2\theta - 1$

Write each expression in terms of a single trigonometric function.

5. $\dfrac{\sin^2\theta + \cos^2\theta}{\cos\theta}$

6. $(\cot\theta)(\sec\theta)$

7. $\dfrac{\sin\theta}{1 - \cos^2\theta}$

8. $\tan\theta(\cot\theta + \tan\theta)$

9. $\csc\theta(1 - \cos^2\theta)$

10. $\dfrac{\csc^2\theta}{\sec^2\theta}$

Write each expression in terms of $\sin\theta$.

11. $2\tan\theta\cos\theta$

12. $\cos\theta(\sec\theta - \cos\theta)$

13. $\dfrac{\sec\theta}{\tan\theta + \cot\theta}$

NAME _____ CLASS _____ DATE _____

Practice Masters Level B
14.3 Fundamental Trigonometric Identities

Use definitions to prove each identity.

1. $\sin\theta \cot\theta \sec\theta = 1$

2. $\csc\theta = \sin\theta(1 + \cot^2\theta)$

3. $\dfrac{1 + \tan^2\theta}{\csc^2\theta} = \tan^2\theta$

4. $\csc^2\theta(1 - \cos^2\theta) = 1$

Write each expression in terms of a single trigonometric function.

5. $(\csc\theta - \cot\theta)(\csc\theta + \cot\theta)$

6. $\dfrac{\sec\theta}{\tan\theta \cot\theta}$

7. $\dfrac{\sec\theta}{\cos\theta} - 1$

8. $\dfrac{\cos\theta}{\csc\theta} - \dfrac{\sin\theta}{\sec\theta}$

9. $\dfrac{1}{\cos^2\theta} - 1$

10. $\dfrac{\tan^2\theta}{\sec\theta + 1} + 1$

Write each expression in terms of cosine θ.

11. $\dfrac{\sin\theta}{\csc\theta}$

12. $\dfrac{\sin^2\theta}{\sec^2\theta - 1}$

13. $\dfrac{3}{1 + \tan^2\theta}$

Algebra 2   Practice Masters Levels A, B, and C

Practice Masters Level C
14.3 Fundamental Trigonometric Identities

Use definitions to prove each identity.

1. $\cos^2\theta + \tan^2\theta + \sin^2\theta = \sec^2\theta$

2. $\dfrac{1+\cot^2\theta}{\csc^2\theta - 1} = \sec^2\theta$

3. $\tan\theta = \dfrac{\cos\theta}{\csc\theta - \sin\theta}$

4. $\dfrac{\sec\theta}{\csc\theta} - \dfrac{\sec\theta}{\sin\theta} = -\cot\theta$

Write each expression in terms of a single trigonometric function.

5. $\dfrac{\cos\theta \csc\theta}{\sin\theta \sec\theta}$

6. $\dfrac{1 + \tan^2\theta}{\tan^2\theta}$

7. $\dfrac{1}{1+\sin\theta} + \dfrac{1}{1-\sin\theta}$

8. $\cot\theta \csc\theta \sec\theta$

9. $\dfrac{1}{1+\csc\theta} + \dfrac{1}{1-\csc\theta}$

10. $\sec\theta \tan\theta \sin\theta \cos\theta \cot^2\theta$

Simplify each trigonometric expression.

11. $\dfrac{2}{\sec\theta - 1} + \dfrac{2}{\sec\theta + 1}$

12. $\dfrac{1}{\tan\theta \cot\theta}$

13. $\dfrac{\sec^2\theta + \csc^2\theta}{\csc^2\theta}$

NAME _____ CLASS _____ DATE _____

Practice Masters Level A
14.4 Sum and Difference Identities

Find the exact value of each expression.

1. $\sin(45° + 60°)$

2. $\sin(30° + 180°)$

3. $\cos(60° + 30°)$

4. $\cos(90° + 45°)$

5. $\sin(180° - 45°)$

6. $\cos(270° - 45°)$

7. $\sin\left(\dfrac{\pi}{4} + \dfrac{\pi}{6}\right)$

8. $\cos\left(\dfrac{\pi}{3} - \dfrac{\pi}{6}\right)$

Find the exact value of each expression.

9. $\sin(-240°)$

10. $\cos 150°$

11. $\sin 315°$

12. $\cos 330°$

13. $\cos 315°$

14. $\sin(-30°)$

Find the rotation matrix for each angle. Round entries to the nearest hundredth, if necessary.

15. $30°$

16. $-45°$

17. $-60°$

18. $330°$

19. $180°$

20. $150°$

21. Graph the function $y = \sin(90° - \theta)$.

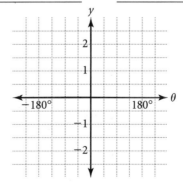

Algebra 2   Practice Masters Levels A, B, and C   271

NAME _____ CLASS _____ DATE _____

Practice Masters Level B
14.4 Sum and Difference Identities

Find the exact value of each expression.

1. $\sin(90° + 135°)$ 2. $\cos(210° + 45°)$ 3. $\sin(240° - 135°)$

4. $\cos(150° - 135°)$ 5. $\sin\left(\dfrac{5\pi}{6} + \dfrac{3\pi}{4}\right)$ 6. $\cos\left(\dfrac{11\pi}{6} - \dfrac{\pi}{3}\right)$

Find the exact value of each expression.

7. $\sin 330°$ 8. $\cos(-150°)$

9. $\cos 120°$ 10. $\sin(-225°)$

Find the rotation matrix for each angle. Round entries to the nearest hundredth, if necessary.

11. $120°$ 12. $-150°$ 13. $210°$

14. $35°$ 15. $50°$ 16. $-100°$

Graph each function.

17. $y = \sin(120° - \theta)$

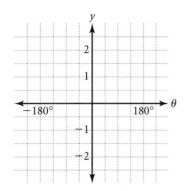

18. $y = \cos(150° + \theta)$

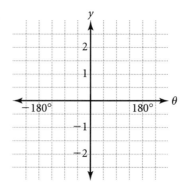

Find the coordinates of the image of each point after a 135° rotation.

19. $P(3, -4)$ 20. $P(-5, 2)$ 21. $P(2, 6)$

Practice Masters Level C

14.4 Sum and Difference Identities

Find the exact value of each expression.

1. $\sin(-15°)$

2. $\sin 240°$

3. $\sin(-165°)$

4. $\cos(-75°)$

5. $\sin 225°$

6. $\cos 15°$

7. $\cos(-315°)$

8. $\cos 225°$

9. $\sin(-225°)$

10. $\cos(-15°)$

Find the rotation matrix for each angle. Round entries to the nearest hundredth, if necessary.

11. $110°$

12. $-75°$

13. $155°$

14. $190°$

15. $-200°$

16. $130°$

Find the coordinates of the image of each point after a 120° rotation.

17. $P(4, 8)$

18. $P(-2, 5)$

19. $P(-4, -10)$

Graph each function.

20. $y = \sin(\theta - 120°)$

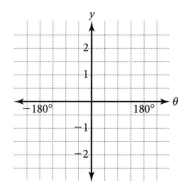

21. $y = \cos(135° + \theta)$

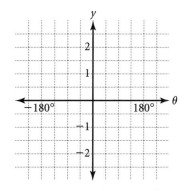

Algebra 2

NAME _____ CLASS _____ DATE _____

Practice Masters Level A
14.5 Double-Angle and Half-Angle Identities

Simplify.

1. $\dfrac{\sin 2\theta}{\sin \theta}$

2. $\cos 2\theta - \cos^2 \theta$

3. $\sqrt{\dfrac{1 - \cos 2\theta}{2}}$

4. $\dfrac{2\cos^3 \theta - \cos \theta}{\cos \theta}$

Write each expression in terms of trigonometric functions of θ rather than multiples of θ.

5. $\dfrac{\cos 2\theta}{\cos^2 \theta}$

6. $\cos 2\theta + \sin^2 \theta$

Find the exact value of each trigonometric expression.

7. $\sin 67.5°$

8. $\sin 22.5°$

9. $\cos 22.5°$

Use the information given to find the exact value of $\sin 2\theta$ and $\cos 2\theta$.

10. $0° \leq \theta \leq 90°$; $\sin \theta = \dfrac{3}{5}$

11. $270° \leq \theta \leq 360°$; $\cos \theta = \dfrac{3}{5}$

12. $90° \leq \theta \leq 180°$; $\sin \theta = \dfrac{15}{17}$

13. $180° \leq \theta \leq 270°$; $\cos \theta = \dfrac{-7}{25}$

Use the information given to find the exact value of $\sin \dfrac{\theta}{2}$ and $\cos \dfrac{\theta}{2}$.

14. $0° \leq \theta \leq 90°$; $\sin \theta = \dfrac{4}{5}$

15. $90° \leq \theta \leq 180°$; $\cos \theta = \dfrac{-3}{5}$

16. $180° \leq \theta \leq 270°$; $\sin \theta = \dfrac{-3}{5}$

17. $270° \leq \theta \leq 360°$; $\cos \theta = \dfrac{4}{5}$

Practice Masters Level B

14.5 Double-Angle and Half-Angle Identities

Simplify.

1. $\dfrac{1 - \tan^2 \theta}{\sec^2 \theta}$

2. $\sin \theta \tan \theta + \cos 2\theta \sec \theta$

3. $4 - 8 \sin^2 4\theta$

4. $\sqrt{\dfrac{1 + \cos 4\theta}{2}}$

Write each expression in terms of trigonometric functions of θ rather than multiples of θ.

5. $\dfrac{2 \sin \theta}{\sin 2\theta}$

6. $\dfrac{\cos 2\theta - 1}{2}$

Find the exact value of each trigonometric expression.

7. $\cos 67.5°$

8. $\sin 105°$

9. $\cos 105°$

Use the information given to find the exact value of $\sin 2\theta$ and $\cos 2\theta$.

10. $0° \leq \theta \leq 90°$; $\sin \theta = \dfrac{15}{17}$

11. $90° \leq \theta \leq 180°$; $\cos \theta = \dfrac{-8}{17}$

12. $180° \leq \theta \leq 270°$; $\sin \theta = \dfrac{-24}{25}$

13. $270° \leq \theta \leq 360°$; $\cos \theta = \dfrac{7}{25}$

Use the information given to find the exact value of $\sin \dfrac{\theta}{2}$ and $\cos \dfrac{\theta}{2}$.

14. $0° \leq \theta \leq 90°$; $\sin \theta = \dfrac{7}{25}$

15. $90° \leq \theta \leq 180°$; $\cos \theta = \dfrac{-24}{25}$

16. $180° \leq \theta \leq 270°$; $\sin \theta = \dfrac{-15}{17}$

17. $270° \leq \theta \leq 360°$; $\cos \theta = \dfrac{8}{17}$

Practice Masters Level C
14.5 Double-Angle and Half-Angle Identities

Simplify.

1. $\dfrac{\sin 2\theta}{1 + \cos 2\theta}$

2. $-\sqrt{\dfrac{9 - 9\cos 3\theta}{2}}$

3. $2\sin^2\theta + \dfrac{\sin 2\theta}{\tan\theta}$

4. $\dfrac{\cos 2\theta}{\cos\theta + \sin\theta}$

Write each expression in terms of trigonometric functions of θ rather than multiples of θ.

5. $4\cos^2 5\theta - 2$

6. $3 - 6\sin^2 3\theta$

Find the exact value of each trigonometric expression.

7. $\sin 112.5°$

8. $\cos 112.5°$

9. $\sin 165°$

Use the information given to find the exact value of $\sin 2\theta$ and $\cos 2\theta$.

10. $0° \leq \theta \leq 90°$; $\sin\theta = \dfrac{2}{5}$

11. $90° \leq \theta \leq 180°$; $\cos\theta = \dfrac{-5}{8}$

12. $180° \leq \theta \leq 270°$; $\sin\theta = \dfrac{-1}{3}$

13. $270° \leq \theta \leq 360°$; $\cos\theta = \dfrac{\sqrt{5}}{12}$

Use the information given to find the exact value of $\sin\dfrac{\theta}{2}$ and $\cos\dfrac{\theta}{2}$.

14. $0° \leq \theta \leq 90°$; $\cos\theta = \dfrac{4}{9}$

15. $90° \leq \theta \leq 180°$; $\sin\theta = \dfrac{1}{4}$

16. $180° \leq \theta \leq 270°$; $\cos\theta = \dfrac{-3}{7}$

17. $270° \leq \theta \leq 360°$; $\sin\theta = \dfrac{-3}{11}$

Practice Masters Level A
14.6 Solving Trigonometric Equations

Find all the solutions of each equation.

1. $2 \sin \theta + 1 = 0$

2. $2 \cos \theta - 1 = 0$

3. $\sin \theta - \sqrt{3} = -\sin \theta$

4. $3 \tan \theta = \sqrt{3}$

5. $2 \cos \theta + \sqrt{3} = 0$

6. $\tan \theta - 1 = 0$

Find the exact solution of each equation for $0° \le \theta \le 360°$.

7. $4 \sin^2 \theta = 3$

8. $\sec \theta = -1$

9. $4 \sin \theta \cos \theta = \sqrt{3}$

10. $\tan \theta = \sin \theta$

Find the exact solution of each equation for $0 \le \theta \le 2\pi$.

11. $2 \tan \theta - 2\sqrt{3} = 0$

12. $2 \cos \theta + 1 = 0$

13. $\sin^2 \theta = \dfrac{1}{2}$

14. $4 \cos^2 \theta = 3$

Solve each equation to the nearest tenth of a degree for $0° \le \theta \le 360°$.

15. $4 \sin \theta = 3$

16. $\sin 2\theta - \sin \theta = 1$

17. The position of a weight attached to an oscillating spring is given by $y = 5 \cos \pi t$, where t is the time in seconds and y is the vertical distance in centimeters. Rest position is at the point where $y = 0$. Find the times at which the weight is 2 centimeters above its rest position.

Algebra 2

NAME _____ CLASS _____ DATE _____

Practice Masters Level B
14.6 Solving Trigonometric Equations

Find all the solutions of each equation.

1. $\dfrac{2\sin\theta + 3}{3} = 1$

2. $2\sin\theta + 3 = \sin\theta + 2$

3. $5\cos\theta = \cos\theta + 2$

4. $\tan\theta = 4\tan\theta - \sqrt{3}$

Find the exact solution of each equation for $0° \le \theta \le 360°$.

5. $\cos^2\theta + \sin^2\theta = \sin\theta$

6. $2\cos^2\theta - 3\cos\theta + 1 = 0$

7. $2\sin^2\theta - 3\sin\theta = -1$

8. $2\cos^2\theta - \sqrt{3}\cos\theta = 0$

Find the exact solution of each equation for $0 \le \theta \le 2\pi$.

9. $\cos\theta - \sec\theta = 0$

10. $\sin\theta - \sin 2\theta = 0$

11. $2 + \csc\theta = 0$

12. $4\sin^2\theta - 1 = 0$

Solve each equation to the nearest tenth of a degree for $0° \le \theta \le 360°$.

13. $\sin^2\theta + 3\sin\theta - 1 = 0$

14. $3\cos^2\theta = 2$

15. The function $y(t) = 122\sin\theta - 16t^2$ models the altitude of a ball t seconds after it was hit at an angle of θ degrees. Determine, to the nearest tenth of a degree, the measure of the angle at which the ball was hit, if it had an altitude of 40 feet after 1.5 seconds.

NAME _____ CLASS _____ DATE _____

Practice Masters Level C
14.6 Solving Trigonometric Equations

Find all the solutions of each equation.

1. $\tan \theta = \cot \theta$

2. $\sin \theta = \csc \theta$

3. $\cos \theta = \sin \theta$

4. $\dfrac{2}{\cos \theta} = \tan^2 \theta + 2$

Find the exact solution of each equation for $0° \leq \theta \leq 360°$.

5. $\sec^2 \theta + 2 = 3 \sec \theta$

6. $\cos 2\theta = \sin \theta + 1$

7. $\sin^2 \theta + 3 \cos^2 \theta + \cos 2\theta = 1$

8. $4 \sin^2 \theta = 3$

Find the exact solution of each equation for $0 \leq \theta \leq 2\pi$.

9. $\cos^2 \theta + \cos 2\theta = 2$

10. $\sin^2 \theta - \cos^2 \theta + \cos \theta = 0$

11. $\tan^2 \theta - \sec^2 \theta + 2 = \cos \theta$

12. $\sin^2 \theta + \cos \theta = \sin \theta + \cos^2 \theta$

Solve each equation to the nearest tenth of a degree for $0° \leq \theta \leq 360°$.

13. $\sec \theta + \cot^2 \theta = \csc \theta$

14. $\sin 2\theta - \sin^2 \theta + 1 = 0$

15. Given that the index of refraction for water, n_{water}, is approximately 1.33, find each indicated angle, to the nearest tenth of a degree.

   a. θ_{water} if θ_{air} is $45°$

   b. θ_{air} if θ_{water} is $20°$

   c. θ_{air} if θ_{water} is $\dfrac{\pi}{10}$

   d. θ_{water} if θ_{air} is $\dfrac{3\pi}{4}$

Algebra 2   Practice Masters Levels A, B, and C

Answers

Lesson 1.1
Level A

1. Linear
2. Linear
3. Not linear
4. Linear
5. Not linear
6. Linear
7. Not linear
8. Linear
9. Not linear

10.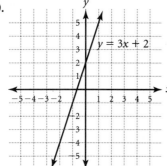

11.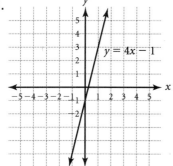

12.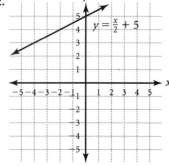

13.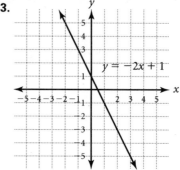

14.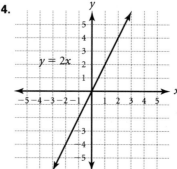

15.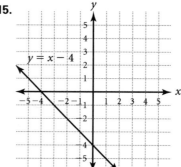

16. $(4, 8)$
17. Not linear
18. $(2, 5)$

Lesson 1.1
Level B

1. Linear
2. Linear
3. Not linear
4. Linear
5. Not linear

Answers

6. Linear
7. Not linear
8. Linear
9. Not linear

10.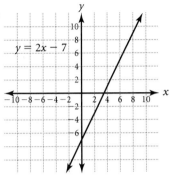

11.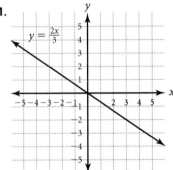

12.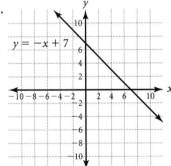

13.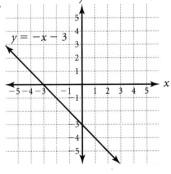

14.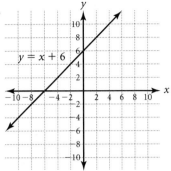

15.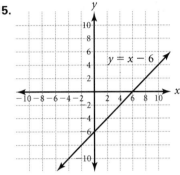

16. $(-10, 8)$

17. Not linear

18. $(0, -1)$

Lesson 1.1
Level C

1. Not linear

2.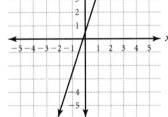

Answers

3.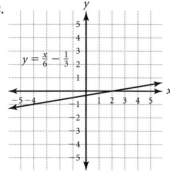

4. horizontal line

5. A straight line that intercepts the *x*- and *y*-axes at the origin.

6. a. $C = 5 + 1.5d$
   b. 9.50
   c. 5

7. $h = 15d$

8. a. $y = 6x + 58$
   b. 88°F

Lesson 1.2
Level A

1. $y = 2x - 3$
2. $y = \dfrac{1}{2}x + 3$
3. $y = -x + 4$
4. $y = \dfrac{-2}{3}x$
5. $y = -7$
6. $y = \dfrac{-1}{2}x + \dfrac{3}{4}$
7. 1
8. $\dfrac{-7}{4}$
9. 1
10. 1

11. $-\dfrac{1}{3}$

12. 3

13. $m = 5, b = -3$

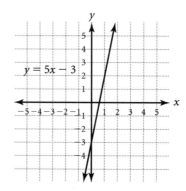

14. $m = -4, b = 1$

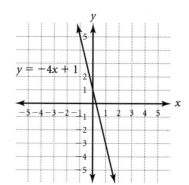

15. $m = \dfrac{1}{2}, b = 3$

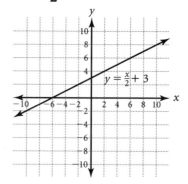

Answers

16. $m = 0, b = 3$

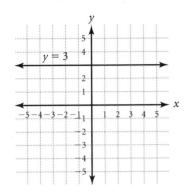

17. $m =$ undefined, $b =$ none

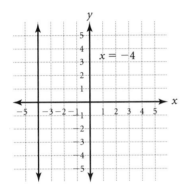

18. $m = 2, b = -3$

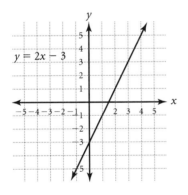

Lesson 1.2
Level B

1. $y = -\dfrac{2}{3}x - \dfrac{1}{3}$

2. $y = \dfrac{-6}{7}$

3. $y = \dfrac{-3}{5}x$

4. $y = -x$

5. $\dfrac{5}{7.7} \approx 0.65$

6. -2

7. 0

8. $m = \dfrac{-2}{3}; b = \dfrac{8}{3}$

9. $m = \dfrac{2}{5}; b = -1$

10. $m = 1; b = 0$

11. $m = 6; b = -44$

12. $m = 1.5; b = 0$

13. $m = -2; b = -2$

14. $y = 3x + 1$

15. $y = -4$

16. $x = -1$

17.

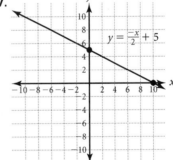

18.

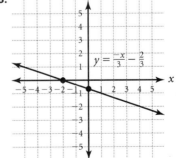

Algebra 2   Practice Masters Levels A, B, and C   **283**

Answers

Lesson 1.2
Level C

1. undefined
2. $-\dfrac{1}{2}$
3.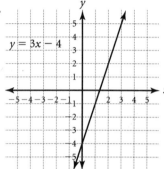
4.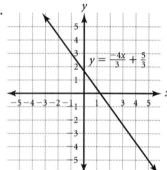
5.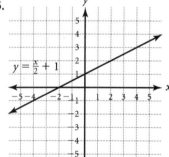
6.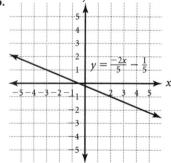

7. 6
8. 45.833
9. a. 0.133 pounds per day
   b. 65 days
   c. $y = 0.133x - 0.59$
   d. how much the chicken weighed at birth

Lesson 1.3
Level A

1. $y = 2x - 6$
2. $y = \dfrac{7}{2}x + 3$
3. $y = \dfrac{-1}{5}x + \dfrac{7}{5}$
4. $y = \dfrac{11}{4}x + \dfrac{23}{2}$
5. $y = -5$
6. $x = 2$
7. $y = 2x - 4$
8. $y = -3x + 3$
9. $y = \dfrac{1}{2}x - 6$
10. $y = -\dfrac{3}{4}x + 6$
11. $y = 4x$
12. $y = 2$
13. $y = x + 3$
14. $y = -5x - 14$
15. $y = \dfrac{5}{3}x - 9$
16. $y = x + 3$
17. $y = \dfrac{-1}{2}x + 3$

Answers

18. $y = -3$

19. $y = -3x + 4$

20. $y = 2x - 2$

21. $y = -9x + 51$

22. $y = x + \dfrac{9}{2}$

23. $y = \dfrac{-4}{3}x - 5$

24. $y = \dfrac{1}{3}x$

25. $y = x + 4$

26. $y = \dfrac{-1}{3}x - 4$

27. $y = 4x + 10$

28. $x = 7$

29. $y = -\dfrac{1}{6}x - 1$

30. $y = -\dfrac{1}{4}x - 9$

31. $y = -\dfrac{1}{2}x + 20$

32. $y = -\dfrac{2}{3}x$

Lesson 1.3
Level B

1. $y = \dfrac{-1}{2}x + \dfrac{5}{2}$

2. $y = -4x - 12$

3. $y = -8x + \dfrac{16}{3}$

4. $y = 0$

5. $y = -5x - 11$

6. $y = \dfrac{3}{4}x + \dfrac{65}{4}$

7. $y = -\dfrac{1}{2}x + \dfrac{1}{3}$

8. $y = \dfrac{-11}{4}$

9. $x = -2$

10. $x = 0$

11. $y = \dfrac{1}{3}x - \dfrac{5}{3}$; $y = -3x + 15$

12. $y = x - 4$; $y = -x + 2$

13. $y = -x + 3$; $y = x - 1$

14. $y = \dfrac{2}{3}x - \dfrac{1}{2}$; $y = \dfrac{-3}{2}x - \dfrac{1}{2}$

15. $y = \dfrac{2}{3}x + 4$

16. $y = \dfrac{-5}{2}x + 25$

17. parallel

18. neither

Lesson 1.3
Level C

1. $y = \dfrac{4}{5}x - 2$

2. $x = 2$

3. $y = 5x + 29$

4. $y = -6x + 2.25$

5. $y = \dfrac{4}{3}x - \dfrac{5}{3}$

Answers

6. $y = -6x + 6$
7. $(2, -5)$
8. $(-6, -5); (6, -3); (4, 7)$
9. perpendicular
10. parallel
11. perpendicular
12. parallel
13. a. $y = 25x + 50$
    b. 78 visits

Lesson 1.4
Level A

1. yes
2. yes
3. no
4. yes
5. no
6. yes
7. $k = 2; y = 2x$
8. $k = \dfrac{-7}{3}; y = \dfrac{-7}{3}x$
9. $k = \dfrac{3}{5}; y = \dfrac{3}{5}x$
10. $k = -1; y = -1x$
11. $k = -\dfrac{1}{6}; y = \dfrac{-1}{6}x$
12. $k = -20; y = -20x$
13. $k = \dfrac{1}{3}; y = \dfrac{1}{3}x$
14. $k = -200; y = -200x$
15. $x = 2$

16. $x = -12$
17. $x = \dfrac{13}{6}$
18. $x = 0.4$
19. $x = -3$
20. $x = 3$
21. $x = -2$
22. $x = 9$
23. $y = 10$
24. $x = 8$
25. $y = 3$
26. $y = -2$
27. $x = 2$

Lesson 1.4
Level B

1. $k = \dfrac{-3}{2}; y = \dfrac{-3}{2}x$
2. $k = 2.\overline{6}; y = 2.\overline{6}x$
3. $k = 2.\overline{2}; y = 2.\overline{2}x$
4. $k = \dfrac{-4}{3}; y = \dfrac{-4}{3}x$
5. $k = 2.6; y = 2.6x$
6. $k = 420; y = 420x$
7. yes, $y = 28x$
8. no
9. $x = 10$
10. $x = \dfrac{6}{5}$
11. $x = 5$

Answers

12. $x = \dfrac{-3}{4}$

13. $a = 8.4$

14. $b = -3$

15. $a = -10\dfrac{2}{3}$

16. $b = 90$

Lesson 1.4
Level C

1. $a = \dfrac{5}{8}$

2. $b = 4\dfrac{1}{2}$

3. $y = -2.5x$

4. no

5. a. 0.6
   b. $16.\overline{6}$ pounds
   c. 3.84 feet
   d. $0.41\overline{6}$ pounds
   e. because 0 pounds causes 0 stretch

6. a. $3\dfrac{1}{3}$ feet
   b. 18 feet

Lesson 1.5
Level A

1. D
2. A
3. B
4. E
5. F
6. C

7. $3364

8. Correlation–positive
   $y = 2x + 5$

9. Correlation–negative
   $y = -4.1x - 2.6$

10. Correlation–negative
    $y = -4.7x + 13.9$

Lesson 1.5
Level B

1. False
2. True
3. True
4. Positive
   $y = 1.8x - 0.45$
5. Weak positive correlation
   $y = 0.14x + 5$
6. a. $y = 0.5x + 22.5$
   b. 0.41
   c. 27.5
   d. It is reasonable but not accurate, because the correlation coefficient is weak to moderately positive. He averaged 24.3 points per game.

Lesson 1.5
Level C

1. $y = -1.8x + 72$
2. -0.33
3. 54 goals
4. It is reasonable but not accurate. The correlation coefficient indicates weak negative correlation.
5. $y = 4.9x + 18.5$
6. 0.68

Answers

7. 67

8. $y = 1.7x + 1$; 0.5

9. $6.1 million

Lesson 1.6
Level A

1. $x = 6$
2. $x = -11$
3. $x = \dfrac{3}{2}$
4. $x = 0$
5. $x = -12$
6. $x = 12$
7. $x = -6$
8. $x = -3$
9. $x = 4$
10. $x = -2$
11. $x = 12.69$
12. $x = 1.2$
13. $l = \dfrac{V}{wh}$
14. $b = \dfrac{2A}{h}$
15. $h = \dfrac{2A}{b_1 + b_2}$
16. $t = \dfrac{I}{pr}$
17. $r = \dfrac{C}{2\pi}$
18. $g = \dfrac{2V}{t^2}$

19. $10y + 2x = C$; $36
20. $125 + 7C = E$; $167

Lesson 1.6
Level B

1. $n = -2$
2. $y = -10$
3. $m = 5$
4. $x = \dfrac{3}{5}$
5. $a = \dfrac{1}{4}$
6. $x = \dfrac{15}{2}$
7. $m = \dfrac{4}{3}$
8. no solution
9. $x = -12.70$
10. $x = -46.4$
11. $x = 2.66$
12. $x = $ any real number
13. $b_1 = \dfrac{2A}{h} - b_2$
14. $l = \dfrac{P - 2w}{2}$
15. $T = \dfrac{A - P}{PR}$
16. $s = \dfrac{V - V_0}{b} - s_0$
17. $h = \dfrac{A}{2\pi r} - r$

Answers

18. $d_1 = \dfrac{2A}{d_2}$

19. $20 + 2x + x = 90; x = 23.3°$

20. $c = 30 + 18.75h$; 12 hours

Lesson 1.6
Level C

1. $x = -0.93$
2. $x = -1$
3. $x = 1$
4. $x = -8$
5. $x = -0.55$
6. $x = -1.59$
7. $x = -0.43$
8. $x = 1.29$
9. $x + \dfrac{4}{5}x = 90$, 40° and 50°
10. a. $y = 225x + 1200$
    b. $4800
    c. 24 months
11. a. Paul: $y = 5.75x + 150$
    Rene: $y = 5.25x + 175$
    b. 50 hours
    c. Rene

Lesson 1.7
Level A

1. $n \geq 1$
2. $n < -1$
3. $n \geq -2$
4. $n < \dfrac{3}{2}$

5. $x \geq -1$

6. $y \leq -2$

7. $z > 4$

8. $x \leq -3$

9. $x > 22$

10. $r \leq -3$

11. $x > -6$ and $x < 2$

12. $8x - 2 > 4$ or $3x + 6 < 12$;
    $x > \dfrac{3}{4}$ or $x < 3$

Lesson 1.7
Level B

1. $n < 0$
2. $n > 1$
3. $n > -2$ and $n \leq 4$
4. $n \leq -4$ or $n > -1$
5. $x < 10$

Answers

6. $x \geq -39$

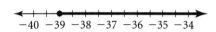

7. $y > -4$

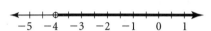

8. $x \geq -\dfrac{18}{5}$

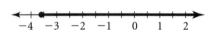

9. $x > 1$ or $x < -2$

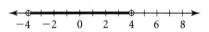

10. $z > -4$ and $z < 4$

11. $x < 5$ or $x < 1$

12. $n < 3$ and $n > 5$

**Lesson 1.7
Level C**

1. $t > \dfrac{1}{5}$

2. $m \geq 55$

3. $t \geq -\dfrac{29}{7}$

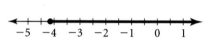

4. $m > -1.5$

5. all real numbers

6. $w < -4$ and $w > -47$

7. $z > -1.75$ and $z < 7\dfrac{1}{2}$

8. $x \leq -4$ or $x > 0$

9. 7 days

10. $1.8 < x < 13$

**Lesson 1.8
Level A**

1. e
2. b
3. d
4. a
5. c
6. $x = 6$ or $x = 2$

Answers

7. $x = 2$ or $x = -5$

8. $x = 7$ or $x = -4$

9. $x = 1$ or $x = -6$

10. $x > -5$ or $x < -7$

11. $x \leq 5$ and $x \geq \dfrac{5}{3}$

12. no solution

13. $x \geq 25$ or $x \leq -5$

14. $|2x| < 6$

15. $|10y - 12| > 10$

Lesson 1.8
Level B

1. $x = 4$ or $x = -5$

2. $m = 6$ or $m = -1$

3. $t = 3$ or $t = -3$

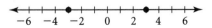

4. $x = 0$

5. $x = 2$ or $x = \dfrac{3}{2}$

6. all real numbers

7. $x \leq 2$ and $x \geq \dfrac{-2}{5}$

8. $b > -2$ and $b < 3$

9. $x > -2$ and $x < 14$

10. $x \geq 3$ or $x \leq \dfrac{-9}{2}$

11. $|d - 102| \leq 0.025$

Lesson 1.8
Level C

1. $x < 3$ or $x > 9$

2. $t \geq 5$ and $t \leq 55$

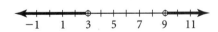

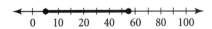

Answers

3. $x = \dfrac{-27}{5}$ or $x = \dfrac{21}{5}$

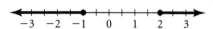

4. $m \geq 2$ and $m \leq -1$

5. $r = -8$ or $r = 28$

6. all real numbers

7. $x \geq 0.9$ and $x \leq 4.4$

8. $148.8 \leq x \leq 164.3$

9. a. $0.75m > 899$
   b. 1198.66 minutes; 19.98 hours

Answers

Lesson 2.1
Level A

1. rational, real
2. irrational, real
3. rational, real
4. rational, natural, whole, integer, real
5. rational, real
6. irrational, real
7. distributive
8. commutative for addition
9. inverse for multiplication
10. associative for addition
11. identity for multiplication
12. commutative for multiplication
13. inverse for addition
14. a. parenthesis
    b. exponents
    c. multiplication/division
    d. addition/subtraction
15. 54
16. 81
17. 58
18. 34
19. -3
20. $1\frac{1}{9}$

Lesson 2.1
Level B

1. rational; perfect square
2. rational; perfect square
3. irrational; non-terminating, non-repeating decimal
4. rational; perfect square
5. irrational; non-terminating, non-repeating decimal
6. irrational; non-terminating, non-repeating decimal
7. Distributive property; $3x + 18$
8. Distributive property;
   $6(c + g) - x(c + g)$
   Distributive property;
   $6c + 6g - xc - xg$
9. division; $2 + 1 - 1$
   addition; $3 - 1$
   subtraction; 2
10. multiplication; $\frac{2x}{16} \div 2$
    division; $\frac{x}{16}$
11. 50
12. -6
13. 12
14. $-\frac{31}{2}$
15. 34
16. 225
17. 2.35
18. $\frac{52}{15}$
19. -4
20. -5
21. 30
22. -8

Algebra 2   Practice Masters Levels A, B, and C   293

Answers

23. -5
24. -24
25. -4.5
26. 9
27. -10
28. -12

Lesson 2.1
Level C

1. $\dfrac{37}{5}$
2. 70
3. $\dfrac{15}{64}$
4. 10
5. 417
6. $\dfrac{21}{68}$
7. -6
8. -72
9. -16
10. $\dfrac{11}{9}$
11. $3 \times (6 + 2) \times (5 - 1) = 96$
12. $12 - 4 \times (5 + 3) = -20$
13. $(1.2 + 6.2) \times (3.4 - 1.5) \times 6 = 84.36$
14. $\dfrac{1}{4} \times \left(\dfrac{2}{3} + \dfrac{5}{6}\right) - \dfrac{3}{8}$
15. $\left(\dfrac{1}{2} + \dfrac{1}{3} \cdot \dfrac{1}{4}\right) - \dfrac{1}{8}$
16. $8^2 + 20 \div (5 + 5) \times 2$
17. It is easier to multiply by 6 and 1.5 to get 9, then multiply 9 by 14 to get 126.
18. It is easier to add 39 and -19 and get 20, then add 20 and 13 to get 33.

Lesson 2.2
Level A

1. $\dfrac{1}{25}$
2. 144
3. 1
4. 9
5. $\dfrac{27}{64}$
6. 256
7. 3
8. 16
9. 3125
10. 9
11. $\dfrac{1}{10}$
12. $\dfrac{1}{2}$
13. -5000
14. 1
15. m
16. x^{15}
17. $\dfrac{1}{x^4}$

Answers

18. $\dfrac{1}{x^6}$

19. r^3

20. $\dfrac{1}{p^4}$

21. w^{12}

22. $\dfrac{1}{w^2}$

23. $2w^8$

24. $\dfrac{x^{15}}{64}$

25. $x^2 y^6$

26. $-t^9$

27. $-16 x^{12} y^7$

28. $-108 a^{13} b^{18}$

29. $\dfrac{1}{2x^5}$

30. $-y^4$

Lesson 2.2
Level B

1. $\dfrac{16}{25}$

2. 100

3. 1

4. $\dfrac{1}{49}$

5. $-\dfrac{8}{27}$

6. 125

7. 216

8. 625

9. -400

10. $\dfrac{1}{216}$

11. $\dfrac{1}{16}$

12. $\dfrac{1}{100{,}000}$

13. 144

14. 125

15. $-a^{12}$

16. $\dfrac{1}{x^3 y^3}$

17. $15 r^5 s^7$

18. $x^{\frac{10}{9}}$

19. x^6

20. $\dfrac{1}{x^{\frac{1}{5}}}$

21. $\dfrac{1}{x^{\frac{8}{3}}}$

22. $\dfrac{2}{3} a^6 b^2$

23. $10{,}000 x^{10}$

24. $\dfrac{m^{\frac{2}{3}}}{n^{\frac{2}{3}}}$

25. $r^{14} s^{17}$

26. $6x^5 + 3 - x^3 + 10x^2$

27. $-\dfrac{1}{x^{18}}$

Answers

28. a^2y^2

29. $\dfrac{1}{400x^4}$

30. $\dfrac{1}{y^{26}}$

Lesson 2.2
Level C

1. y^{4+6m}
2. x^{m+4}
3. z^{3p+2}
4. x^{y^2}
5. w^3
6. r^{4x}

7. a. 100.1 per second
   b. 242.176 meters per second

8. a. 16.2 N
   b. 3.49×10^{-5}

Lesson 2.3
Level A

1. Yes
2. No
3. Yes
4. Yes
5. Yes
6. No
7. Yes
8. No
9. $D = \{-1, -0.5, 0, 0.5, 1\}$
   $R = \{5, 0, 1, 3, -4\}$

10. $D = \{-3, -2, -1, 0\}$
    $R = \{1, 0\}$

11. $D = \{5, 4, 3, 2\}$
    $R = \{-1\}$

12. $D = \{-2 \leq x \leq 1\}$
    $R = \{-1 \leq y \leq 3\}$

13. $-1, 11$

14. $-1, \dfrac{1}{2}$

15. $1, -6$

Lesson 2.3
Level B

1. Yes
2. Yes
3. No
4. No
5. No

6. $\left\{-8, -7\dfrac{1}{2}, -7, -6\dfrac{1}{2}, -6\right\}$

7. $\{13, 7, 1, -5, -11\}$

8. $\left\{\dfrac{-3}{4}, 5\dfrac{1}{4}, 6, \dfrac{-3}{4}, 5\dfrac{1}{4}\right\}$

9. $\left\{1\dfrac{2}{27}, 1\dfrac{2}{243}, 1, 1\dfrac{2}{243}, 1\dfrac{2}{27}\right\}$

10. $R = \{-9, -4, -1, 0\}$

11. -4

12. $D = \{\text{all real numbers}\}$
    $R = \{\text{all real numbers}\}$

13. $D = \{\text{all real numbers}\}$
    $R = \{f(x) \geq 0\}$

14. $D = \{\text{all real numbers}\}$
    $R = \{f(x) \leq 5\}$

Answers

Lesson 2.3
Level C

1. $\{-21, -7, -3, -3\}$
2. $\{-262, -195, -138, -91\}$
3. $\{0, 2.25, 17, 56.25, 132, 256.25\}$
4. $\left\{\dfrac{-7}{4}, \dfrac{-11}{8}, -1, \dfrac{-5}{8}, -\dfrac{1}{4}\right\}$
5. $\dfrac{11}{2}$
6. 5
7. 8
8. $\dfrac{1}{2}a + 4$
9. D = {all real numbers}
   R = $\{f(x) \leq -6\}$
10. D = {all real numbers}
    R = {all real numbers}
11. D = {all real numbers}
    R = $\{f(x) \geq -3\}$
12. D = {all real numbers}
    R = $\{f(x) \geq 4\}$
13. a. yes; $S(w) = 0.35(w) + w$
    b. $s(1.50) = \$2.03$
    c. $2.07
    d. $4.31

Lesson 2.4
Level A

1. $f + g = 2x + 2$
   $f - g = -8x + 16$
2. $f + g = x^2 + \dfrac{5}{2}x + 2$
   $f - g = -x^2 - \dfrac{3}{2}x + 2$
3. $f + g = 3x^2 - 9x + 7$
   $f - g = 3x^2 - x - 7$
4. $f \cdot g = x^3 - 9x^2 + 9x - 81$
   $\dfrac{f}{g} = \dfrac{x^2 + 9}{x - 9}; x \neq 9$
   $\dfrac{f}{g} = x - 2$
5. $f \cdot g = \dfrac{5}{2}x^3$
   $\dfrac{f}{g} = 10x; x \neq 0$
6. $5x + 10$
7. $4x^2 + 16x + 16$
8. $3x + 6$
9. $4; x \neq -2$
10. $-3x - 6$
11. $\dfrac{1}{4}; x \neq -2$
12. -15
13. 9
14. -6
15. 4
16. $1\dfrac{1}{2}$
17. 4
18. $f \cdot g = 24x - 48$
    $g \cdot f = 24x - 8$
19. $f \cdot g = -18x + 3$
    $g \cdot f = -18x + 18$

Algebra 2   Practice Masters Levels A, B, and C

Answers

Lesson 2.4
Level B

1. $f + g = 2x^2 + 4x$
   $f - g = 4x^2$

2. $f + g = 5x^2 + 4x - 1$
   $f - g = -x^2 + 4x - 1$

3. $f + g = x^2 + 1$
   $f - g = x^2 - 2x - 1$

4. $f \cdot g = -3x^2$
   $\dfrac{f}{g} = -12; x \neq 0$

5. $f \cdot g = -4x^3 + 48x$
   $\dfrac{f}{g} = -\dfrac{x}{36} + \dfrac{1}{3x}; x \neq 0$

6. $f \cdot g = 2x^4 - x^3 - x^2$
   $\dfrac{f}{g} = \dfrac{x^3 - x^2}{2x + 1}$ $x \neq -\dfrac{1}{2}$

7. $x^2 - \dfrac{1}{2}x - 5$

8. $x^2 + \dfrac{1}{2}x - 13$

9. $-x^2 + \dfrac{1}{2}x + 5$

10. $\dfrac{x^2 - 9}{\frac{1}{2}x - 4}; x \neq 8$

11. $f \cdot g = 6x - 3$
    $\dfrac{f}{g} = 6x - 9$

12. $f \cdot g = 25x^2$
    $\dfrac{f}{g} = 5x^2$

13. $f \cdot g = \dfrac{1}{8}x - 3$
    $\dfrac{f}{g} = \dfrac{1}{8}x - \dfrac{3}{4}$

14. 9

15. 45

16. -5

Lesson 2.4
Level C

1. $|10x - 4| + (-8x^2)$

2. $|-80x^2 - 4|$

3. $|10x - 4| + 8x^2$

4. $\dfrac{|10x - 4|}{-8x^2}; x \neq 0$

5. 5

6. -16

7. 63

8. 19.25

9. 3

10. $\dfrac{53}{3}$

11. answers will vary
    $f(x) = 4x - 16$
    $g(x) = x^2$

12. a. $C(t) = 1.35(20t) + 200 = 27t + 200$
    b. $C(8) = \$416$
    c. 160

13. a. $C = (p + 0.0625p) - 0.1(p + 0.625p)$
       $= 0.95625p;$
       $C = (p - 0.1p)0.625 + p = 1.05625p$
    b. a, $0.95625p < 1.05625p$

Answers

Lesson 2.5
Level A

1. Yes, {(1, 0), (4, 1), (9, 2), (16, 3)}, yes

2. Yes, $\left\{\left(2, \dfrac{1}{2}\right), (3, 1), \left(2, \dfrac{3}{2}\right), (1, 2)\right\}$, no

3. No, {(5, 4), (10, 5), (6, 4), (2, 3)}, yes

4. No, {(1, 2), (1, 3), (2, 4), (5, 2)}, no

5. Yes, {(6, −1), (5, −2), (4, −3), (−4, 3)}, yes

6. Yes

7. Yes

8. $y = \dfrac{1}{3}x + \dfrac{1}{3}$

9. $y = \dfrac{1}{7}x - \dfrac{3}{7}$

10. $y = -x$

11. $y = \dfrac{3}{2}x + \dfrac{9}{2}$

12. $y = \dfrac{4}{3}x - \dfrac{8}{3}$

13. $y = \dfrac{x + 3.2}{-6.4}$

14. No

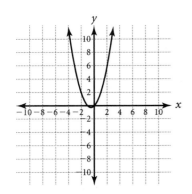

15. No

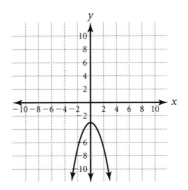

16. Yes

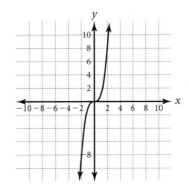

Lesson 2.5
Level B

1. Yes, {(3, 1), (3, 2), (5, 4), (5, 9)}, no

2. No, {(0, 0), (0, 1), (3, 2), (3, 1)}, no

3. Yes, {(−1, 2), (0, 3), (−2, 4), (−5, 5)}, yes

4. Yes, $\left\{(3, -5), \left(1, \dfrac{1}{2}\right), \left(0, -\dfrac{3}{2}\right), (2, 2)\right\}$, yes

5. Yes

6. No

7. $y = 3x + 5$

8. $y = \dfrac{4x - 7}{2}$

9. $y = \dfrac{8x + 2}{-3}$

Answers

10. $y = \dfrac{-2}{3}x + 4$

11. $y = \dfrac{1}{8}x - \dfrac{3}{2}$

12. $y = \dfrac{-2}{3}x - 1$

13. Yes

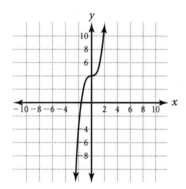

14. No

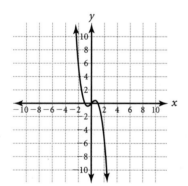

15. No

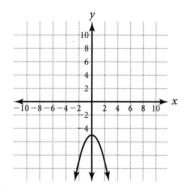

Lesson 2.5
Level C

1. Yes
2. Yes
3. No
4. Yes
5. No
6. $y = -4x - 8$
7. $y = \dfrac{54}{5}x + 2$
8. $y = \dfrac{-1}{3}x - \dfrac{1}{2}$
9. $y = \dfrac{4.6}{1.7} - \dfrac{x}{1.7}$
10. $y = x^2$
11. $y = \dfrac{2}{(x-1)}; x \neq 1$
12. No
13. Yes
14. No
15. No
16. a. $c = 6000 - 240(y - 1980)$

    b. $y = \dfrac{c - 6000}{-240} + 1980;$

    This gives the year of the computer given the cost of the computer.

Answers

Lesson 2.6
Level A

1.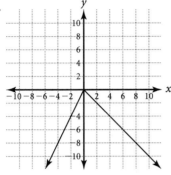

2.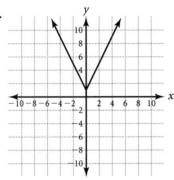

3.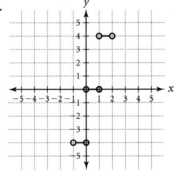

4.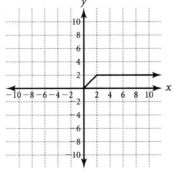

5.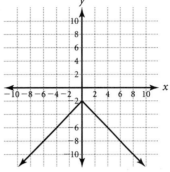

6.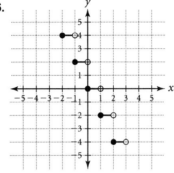

Answers to Exercises 7–9 may vary, depending on the inclusion strategy used by the student.

7. $y = \begin{cases} x + 4 & \text{if } x < 2 \\ 3x & \text{if } x \geq 2 \end{cases}$

8. $y = \begin{cases} -2x & \text{if } x < -2 \\ 4 & \text{if } -2 \leq x \leq 2 \\ 2x & \text{if } x > 2 \end{cases}$

9. $y = \begin{cases} -1 & \text{if } x < -2 \\ x + 1 & \text{if } -2 \leq x \leq 0 \\ 1 & \text{if } x > 0 \end{cases}$

Answers

**Lesson 2.6
Level B**

1.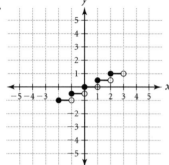

2.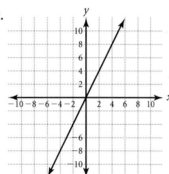

3.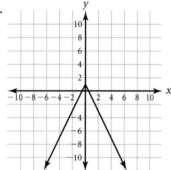

4. $y = 3|x| - 2$

5. $y = \dfrac{3}{2}[x]$

6. $y = \begin{cases} \dfrac{1}{2}x + 2 & \text{if } x < 0 \\ 0 & \text{if } x = 0 \\ \dfrac{1}{2}x - 2 & \text{if } x > 0 \end{cases}$

7. 4.7

8. -3.8

9. 16

10. 7

11. 0

12. -6

13. -14.2

14. 0.7

15. -1

16. 64

**Lesson 2.6
Level C**

1.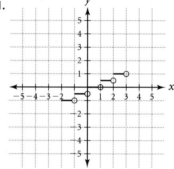

2.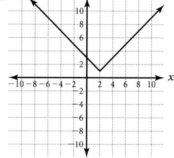

Answers

3.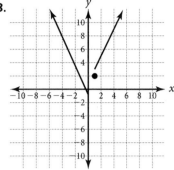

4. -4

5. -3.2

6. 729

7. 3.8

8. $m = [x]$

9. a. $550, $1000, $854.25, $1275

   b. $c = \begin{cases} 100 + 4.5b & \text{if } b \leq 200 \\ 4.25b & \text{if } b > 200 \end{cases}$

Lesson 2.7
Level A

1. vertical translation 2 units up

2. vertical stretch by a factor of 3

3. reflection across x-axis

4. reflection across x-axis and vertical stretch by a factor of 6

5. horizontal translation 5 units to the right

6. vertical translation 2 units up, reflection across x-axis, vertical stretch by a factor of 3

7. vertical translation 3 units up

8. reflection across x-axis

9. horizontal compression by a factor of $\frac{1}{6}$

10. vertical translation 6 units down, vertical stretch by a factor of 2

11. vertical compression by a factor of $\frac{1}{2}$

12. horizontal translation 2 units right

13. $g(x) = -x^2$

14. $g(x) = x^2$

15. $g(x) = x^4 + 5$

16. $g(x) = \left|\frac{1}{3}x\right|$

17. $g(x) = x^3 - 3$

18. $g(x) = (x - 4)^2$

Lesson 2.7
Level B

1. vertical translation 2 units down

2. horizontal translation 2 units right

3. horizontal compression by $\frac{1}{4}$ and reflection across x-axis

4. reflection about x-axis, vertical translation 1 unit up

5. vertical compression by a factor of $\frac{1}{3}$

6. vertical stretch by a factor of 2, horizontal translation $\frac{1}{2}$ unit to right

7. vertical translation 4 units up

8. horizontal stretch by a factor of $\frac{1}{5}$

9. vertical compression by a factor of $\frac{1}{2}$

10. reflection across x-axis, vertical translation 2 units down

11. horizontal translation 3 units right

Algebra 2 — Practice Masters Levels A, B, and C

Answers

12. horizontal translation 2 units right, vertical translation 4 units up
13. $f(x) = (-x)^3$
14. $f(x) = -(x+5)^2$
15. $f(x) = 4\sqrt{x}$
16. $f(x) = \sqrt{\frac{1}{5}x}$
17. $g(x) = \frac{-1}{3}x + 1$
18. $g(x) = \frac{1}{2}\sqrt{2x}$

14. $f(x) = \sqrt{\frac{1}{2}(x+4)}$
15. $f(x) = 2\sqrt{(x+10)}$
16. $f(x) = \sqrt{3(x-6)}$

Lesson 2.7
Level C

1. horizontal translation 2 units left
2. horizontal translation 2 units right
3. horizontal translation 1 unit left, reflected across x-axis
4. vertical translation 2 units up, horizontal translation 3 units right
5. vertical stretch by a factor of 3
6. horizontal compression by a factor of $\frac{1}{3}$
7. horizontal translation 3 units right
8. vertical stretch by a factor of 2
9. horizontal compression by a factor of $\frac{1}{2}$
10. vertical translation 2 units up
11. reflection across y-axis
12. reflection across x-axis and y-axis
13. $f(x) = (-x)^5$

Answers

Lesson 3.1
Level A

1. $(1, 1)$, independent

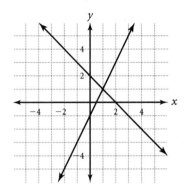

2. $(3, 3)$, independent

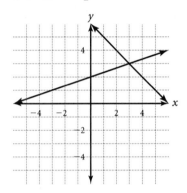

3. $(3, -3)$, independent

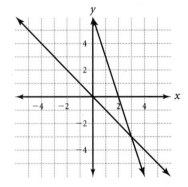

4. $(-2, 4)$, independent

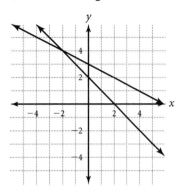

5. $(0, 0)$, independent

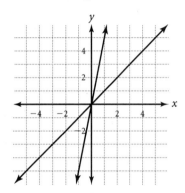

6. no solution, inconsistent

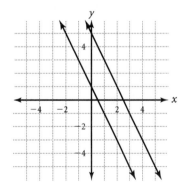

7. $(1, 4)$

8. $(2, 10)$

9. $(4, -1)$,

10. $(-3, -1)$

11. $(-5, 2)$

12. $(0, 3)$

Answers

Lesson 3.1
Level B

1. $(-3, -7)$, independent

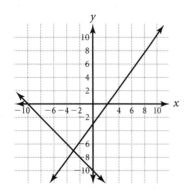

2. $(6, 0)$, independent

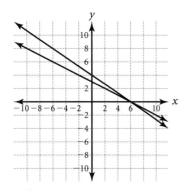

3. no solution, inconsistent

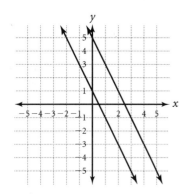

4. $(-1, 11)$

5. $(12, -4)$

6. all real numbers

7. $\left(\dfrac{1}{2}, \dfrac{2}{3}\right)$

8. no solution

9. $(-0.1, 0.4)$

10. 125 feet wide, 500 feet long

11. 75 mL acid solution
    25 mL pure acid

12. 8.5 pounds of pecans
    5.5 pounds of almonds

Lesson 3.1
Level C

1. $(1.73, -1.73)$

2. $(-0.68, 3.07)$

3. $(1, -2, -4)$

4. $\left(\dfrac{1}{2}, \dfrac{1}{4}, \dfrac{-27}{8}\right)$

5. $(2, -1, 1)$

6. $(-7, 5, -2.5)$

7. plane air speed is 280 mph
   wind speed is 40 mph

8. $6\dfrac{2}{3}$ mL chlorine solution
   $13\dfrac{1}{3}$ mL pure water

9. $444.44 at 11%
   $4555.56 at 6.5%

10. rate of boat $\dfrac{25d}{48}$ mph
    rate of current $\dfrac{7d}{48}$ mph

Lesson 3.2
Level A

1. $(24, -7)$

2. $(5, -3)$

3. $(5, 0)$

Answers

4. $(2, -3)$

5. $(-3, 2)$

6. no solution

7. $(0, -2)$

8. $(-10, 3)$

9. infinite number of solutions

10. $\left(\dfrac{1}{2}, \dfrac{-1}{2}\right)$

11. $(4, -1)$

12. $\left(\dfrac{-11}{42}, \dfrac{-1}{3}\right)$

13. $\left(\dfrac{-1}{3}, \dfrac{-1}{2}\right)$

14. $(-6, 3)$

15. no solution

16. $(5, 0)$

Lesson 3.2
Level B

1. $(10, 6)$

2. $(-12, 1)$

3. $(-1, -12)$

4. $\left(\dfrac{1}{6}, \dfrac{-1}{8}\right)$

5. $(5, 6)$

6. infinitely many solutions

7. no solution

8. $(-358, 269)$

9. $\left(4, \dfrac{7}{2}\right)$

10. $(20, -16)$

11. $\left(\dfrac{1}{2}, -3\right)$

12. $\left(\dfrac{1}{2}, \dfrac{-7}{2}\right)$

13. $(-6, -2)$

14. no solution

Lesson 3.2
Level C

1. no solution

2. $(4, -12)$

3. infinitely many solutions

4. $(2a + b, -3a - 2b)$

5. $(0.05, -0.08)$

6. $\left(\dfrac{5c + 2d}{7}, \dfrac{4c + 3d}{-7}\right)$

7. 9 skilled, 15 unskilled

8. 300 large, 240 small

9. $3695.65 at 19%
   $1304.35 at 7.5%

Lesson 3.3
Level A

1.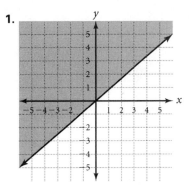

Algebra 2 Practice Masters Levels A, B, and C 307

Answers

2.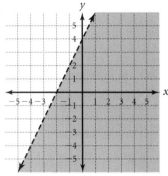

3.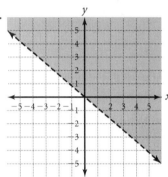

4.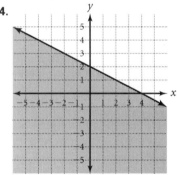

5.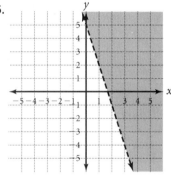

6.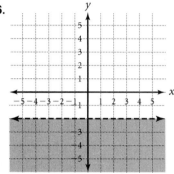

7. a. $1.0s + 2.0l \geq 300$

   b.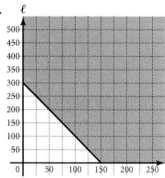

   c. answers may vary, sample answers: (134, 500), (1000, 1), (1, 577)

Lesson 3.3
Level B

1.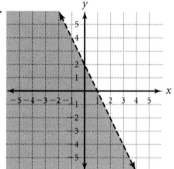

308 Practice Masters Levels A, B, and C Algebra 2

Answers

2.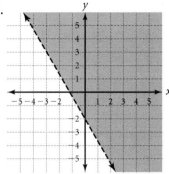

3.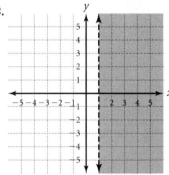

4.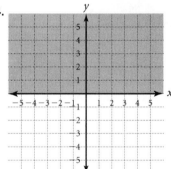

5.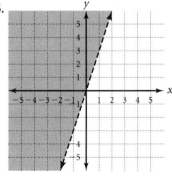

6.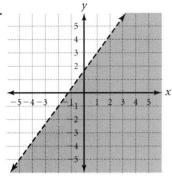

7. a. $400x + 250y > 800$
   b.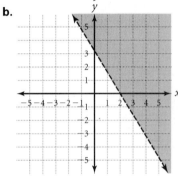
   c. at least $1.24 per shake

Lesson 3.3
Level C

1.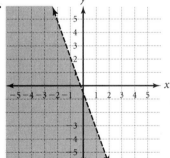

Answers

2.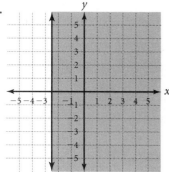

3.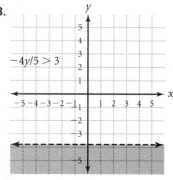
$-4y/5 > 3$

4. a. $D = \leq -1.5L + 5.8$
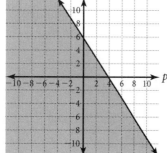

b. $S \leq -2.14L + 5.8$
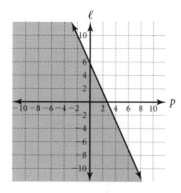

c. $D \leq -1.1P + 5.27$
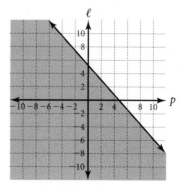

d. $S \leq -1.57P + 8.3$
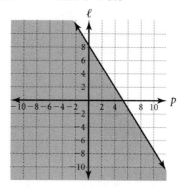

Lesson 3.4
Level A

1.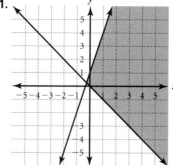

2.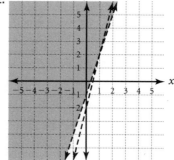

Answers

3.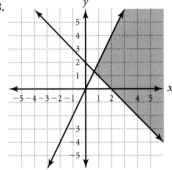

4. $\begin{cases} y < -2x + 4 \\ y > 3x - 1 \end{cases}$

5. $\begin{cases} y \leq -5x \\ y \geq x - 3 \end{cases}$

6. $\begin{cases} y < 2x - 3 \\ y \geq -2x + 2 \end{cases}$

7.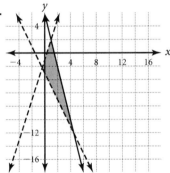

Lesson 3.4
Level B

1.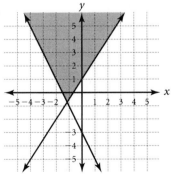

2.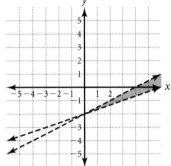

3.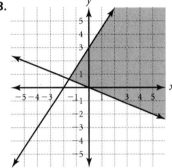

4. $\begin{cases} y \geq -2x + 3 \\ y \geq 2x - 3 \end{cases}$

5. $\begin{cases} y < 4x - 2 \\ y > \dfrac{1}{2}x - 2 \end{cases}$

6. $\begin{cases} y \leq 2x \\ y \geq \dfrac{3}{2}x - 3 \\ y \leq 0 \end{cases}$

7.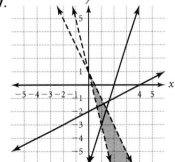

Algebra 2   Practice Masters Levels A, B, and C   311

Answers

Lesson 3.4
Level C

1. $\begin{cases} y < 2x - 5 \\ y > \dfrac{-1}{2}x + \dfrac{3}{2} \end{cases}$

2. $\begin{cases} y \geq \dfrac{2}{3}x - \dfrac{4}{3} \\ y \leq \dfrac{1}{2}x + 2 \\ y < 2x + 3 \end{cases}$

3. $\begin{cases} y \geq \dfrac{1}{4}x - 3 \\ y < \dfrac{-1}{4}x + 2 \\ y < 2x + \dfrac{5}{2} \end{cases}$

4. $35x + 25y \geq 3500$ and $x + y \leq 120$

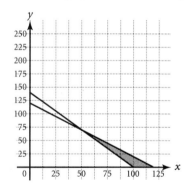

5. $x + y \leq 10{,}000$
$0.13x + 0.15y \geq 1400$

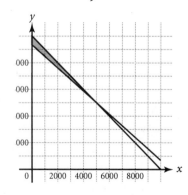

Lesson 3.5
Level A

1.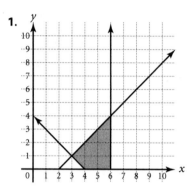

2. $\begin{cases} x + y \leq 500 \\ 100 \leq x \leq 400 \\ y \geq 80 \end{cases}$

3. Max: 15   Min: -12
4. Max: 19   Min: -6
5. Max: 6   Min: -26
6. Max: 11   Min: -3
7. Max: -7 Min: none exist
8. Max: 0   Min: -30

Lesson 3.5
Level B

1.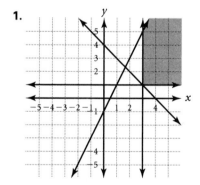

Answers

2. Max: 11   Min: -14

3. Max: $\dfrac{-2}{3}$ Min: -4

4. a. $\begin{cases} x + y \leq 400 \\ 50 \leq y \leq 250 \\ x \geq 100 \end{cases}$

b.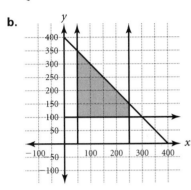

c. $R = 540(0.759)x + 34.9(6.8)y$
$R = 409.86x + 237.32y$

d. $126,197

Lesson 3.5
Level C

1.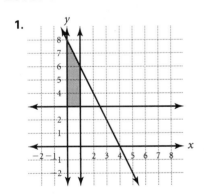

2. Max: 3   Min: -7

3. Max: 6   Min: -8.5

4. a. $1.75x + 2y \leq 200$
$4x + 6.5y \geq 600$
$x \leq 90$
$y \geq 30$

b.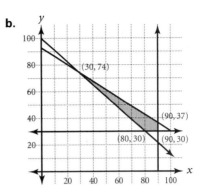

c. $P = 2.25x + 4.5y$

d. $P = \$400.50$

Lesson 3.6
Level A

1.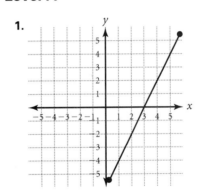

2.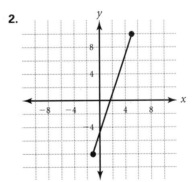

Answers

3.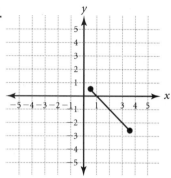

4. $y = 2x - 6$

5. $y = 3x - 5$

6. $y = 4 - x^2$

7. $y = \dfrac{-1}{2}x - 1$

8. $y = x^2 + 2x + 1$

9. $y = \dfrac{-3}{2}x - 12$

10. $y = \dfrac{-9}{8}x + \dfrac{5}{4}$

11. $y = -x + 1$

12. $y = 2x^2 - 8x + 1$

13. about 99 feet

Lesson 3.6
Level B

1.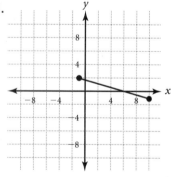

2.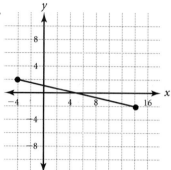

3.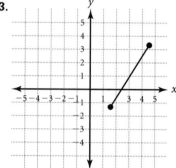

4. $y = \sqrt{x} - 2$

5. $y = \dfrac{2}{3}x - \dfrac{8}{3}$

6. $y = 2x^2 + 8x + 1$

7. $y = \sqrt{1 - x} - 3$

8.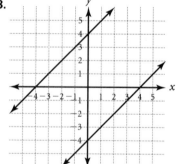

9. about 225 feet

Answers

Lesson 3.6
Level C

1. $y = \dfrac{3}{2}x - \dfrac{9}{2}$

2. $y = x^2 - 7x + 11$

3. $y = 60x^2 + 51x + 10.8$

4. $y = \dfrac{4}{3}x + 2\sqrt{\dfrac{x}{3}} - 1$

5. $y = 250x^2 - 15x + 0.425$

6. $y = \dfrac{-8}{9}x + \dfrac{107}{135}$

7.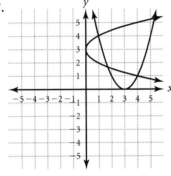

8.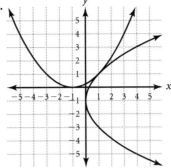

9. a. $x(t) = 12t + 30$
$y(t) = \dfrac{3}{4}t + \dfrac{1}{2}$

b. 47.5 years

c. $36\dfrac{1}{8}$ inches

Algebra 2   Practice Masters Levels A, B, and C   **315**

Answers

**Lesson 4.1
Level A**

1. 2×2
2. 2×2
3. 2×3
4. $\begin{bmatrix} 8 & -4 \\ -1 & -1 \end{bmatrix}$
5. $\begin{bmatrix} -1 & 1 & -2 \\ -3 & -4 & 5 \end{bmatrix}$
6. $\begin{bmatrix} 18 & -12 \\ -6 & 9 \end{bmatrix}$
7. $\begin{bmatrix} -4 & 4 \\ 3 & 7 \end{bmatrix}$
8. $\begin{bmatrix} \frac{1}{2} & \frac{-1}{2} & 1 \\ \frac{3}{2} & 2 & \frac{-5}{2} \end{bmatrix}$
9. $\begin{bmatrix} -6 & 0 \\ -3 & 12 \end{bmatrix}$
10. $\begin{bmatrix} 14 & -12 \\ -8 & 17 \end{bmatrix}$
11. $\begin{bmatrix} 0.01 & -0.01 & 0.02 \\ 0.03 & 0.04 & -0.05 \end{bmatrix}$
12. $\begin{bmatrix} 15 & -8 \\ \frac{-5}{2} & 0 \end{bmatrix}$
13. 3×3
14. 32
15. 21
16. 2; the number of TVs sold in February
17. $\begin{bmatrix} 13 & 11 & 15 & 2 \\ 14 & 8 & 10 & 6 \\ 10 & 2 & 5 & 9 \end{bmatrix}$

**Lesson 4.1
Level B**

1. 3×3
2. 2×3
3. 3×3
4. $\begin{bmatrix} -2 & 10 & 6 \\ 0 & -3 & -5 \\ -2 & -6 & 2 \end{bmatrix}$
5. $\begin{bmatrix} -1 & \frac{-4}{3} & \frac{-3}{2} \\ -3 & 1 & \frac{-8}{3} \end{bmatrix}$
6. $\begin{bmatrix} 8 & 20 & -4 \\ -4 & 18 & -2 \\ 4 & 4 & 12 \end{bmatrix}$
7. $\begin{bmatrix} 6 & 8 & 9 \\ 18 & -6 & -16 \end{bmatrix}$
8. $\begin{bmatrix} \frac{-1}{2} & -5 & -1 \\ \frac{1}{2} & \frac{-3}{2} & \frac{3}{2} \\ 0 & 1 & -2 \end{bmatrix}$
9. $\begin{bmatrix} \frac{-1}{2} & 10 & 4 \\ \frac{-1}{2} & 0 & -4 \\ -1 & -4 & 3 \end{bmatrix}$
10. $\begin{bmatrix} -3 & 30 & 14 \\ -1 & -3 & -13 \\ -4 & -14 & 8 \end{bmatrix}$
11. $\begin{bmatrix} 0.09 & 0 & -0.12 \\ -0.03 & 0.18 & 0.06 \\ 0.06 & 0.12 & 0.06 \end{bmatrix}$

Answers

12. $\begin{bmatrix} 2 & 5 & -1 \\ -1 & \frac{9}{2} & \frac{-1}{2} \\ 1 & 1 & 3 \end{bmatrix}$

13. 4×4

14. 1996; 920 trees

15. pecan

16. $\begin{bmatrix} 400 & 200 & 25 & 0 & 40 \\ 500 & 20 & 30 & 0 & 20 \\ 20 & 4 & 15 & 3 & 600 \\ 0 & 4 & 10 & 15 & 400 \\ 0 & 0 & 0 & 0 & 25 \end{bmatrix}$

Lesson 4.1
Level C

1. They do not have the same dimensions.

2. $\begin{bmatrix} -8 & 5 & -7 & \frac{-4}{3} \end{bmatrix}$

3. $\begin{bmatrix} 6.4 & -0.8 & -0.64 \\ -2.8 & 16.4 & 0.4 \end{bmatrix}$

4. $\begin{bmatrix} 16.9 & -3.55 & 2.22 \\ -16.45 & -46.25 & 75.95 \end{bmatrix}$

5. a. Fulton—February
   Heatherwood—January
   Darlington—June
   b. January
   c. $\begin{bmatrix} -275 & 5 & -20 & 0 & -10 & -35 \\ 25 & -10 & 40 & 30 & -35 & -65 \\ 10 & -5 & -35 & -20 & -15 & -30 \end{bmatrix}$

Lesson 4.2
Level A

1. $\begin{bmatrix} 0 & -4 \\ 14 & -8 \end{bmatrix}$

2. $\begin{bmatrix} -22 & -6 \\ -25 & -49 \end{bmatrix}$

3. $\begin{bmatrix} -3 & -6 & 30 \\ 0 & 0 & 0 \\ -1 & -2 & 10 \end{bmatrix}$

4. $[22]$

5. does not exist

6. $\begin{bmatrix} 2 \\ 3 \\ 10 \end{bmatrix}$

7. $\begin{bmatrix} 22 & 10 & 5 \\ 48 & 18 & 8 \\ -26 & -9 & 1 \end{bmatrix}$

8. $\begin{bmatrix} 2 \\ 20 \\ 10 \end{bmatrix}$

9. does not exist

10. $\begin{bmatrix} 24 & -10 & -5 \\ 3 & 5 & -5 \\ -2 & 0 & 25 \end{bmatrix}$

11. does not exist

12. $\begin{bmatrix} 3 & 2 \\ 4 & 3 \end{bmatrix} \begin{bmatrix} 10 \\ 8 \end{bmatrix} = \begin{bmatrix} 46 \\ 64 \end{bmatrix}$

   $46 in May and $64 in June

Lesson 4.2
Level B

1. $\begin{bmatrix} 16 & -2 \\ 9 & 13 \\ 5 & 5 \end{bmatrix}$

2. $\begin{bmatrix} -12 & 15 \end{bmatrix}$

3. does not exist

Answers

4. $\begin{bmatrix} 0 & 20 \\ -5 & 5 \end{bmatrix}$

5. $\begin{bmatrix} -\frac{5}{2} & 1 \\ -10 & \frac{11}{2} \\ 9 & \frac{1}{2} \end{bmatrix}$

6. $[-30]$

7. $\begin{bmatrix} 38 & -26 \\ 28 & 0 \\ 12 & 50 \end{bmatrix}$

8. does not exist

9. $\begin{bmatrix} 1 & 3 & -1 \\ 0 & 2 & 4 \\ 2 & -1 & 1 \end{bmatrix}$

10. $\begin{bmatrix} 2 & 6 \\ 3 & 2 \\ 1 & -4 \end{bmatrix}$

11. $\begin{bmatrix} 19 & 7 & 51 \\ 32 & 12 & 4 \\ -6 & 23 & 3 \end{bmatrix}$

12. $\begin{bmatrix} 29 & 99 & 63 \\ 2 & 110 & 70 \\ -4 & 19 & 15 \end{bmatrix}$

13. a. $\begin{matrix} A \\ B \\ C \\ D \end{matrix}\begin{bmatrix} 1 & 1 & 1 & 0 \\ 0 & 1 & 1 & 0 \\ 0 & 1 & 1 & 1 \\ 1 & 1 & 0 & 1 \end{bmatrix}$

b. $\begin{matrix} A \\ B \\ C \\ D \end{matrix}\begin{bmatrix} 1 & 3 & 3 & 1 \\ 0 & 2 & 2 & 1 \\ 1 & 3 & 2 & 2 \\ 2 & 3 & 2 & 1 \end{bmatrix}$

Lesson 4.2
Level C

1. a. $\begin{bmatrix} -3 & -2 & 4 & 5 \\ 0 & 3 & 1 & -4 \end{bmatrix}$

b. $\begin{bmatrix} -3 & -2 & 4 & 5 \\ 0 & -3 & -1 & 4 \end{bmatrix}$

c.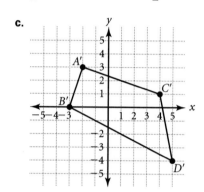

d.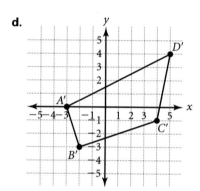

e. A rotation of ABCD 180 degrees.
f. It is reflected about the y-axis

2. a. $\begin{bmatrix} 38 \\ 35 \\ 35 \\ 29 \end{bmatrix}$

b. 137 points
c. PB is not defined because the dimensions are not appropriate for matrix multiplication.

Answers

Lesson 4.3
Level A

1. -2; yes
2. -7; yes
3. 0; no
4. -7; yes
5. 0; no
6. 20; yes
7. yes
8. no
9. $\begin{bmatrix} \frac{1}{8} & \frac{-3}{8} \\ \frac{1}{4} & \frac{1}{4} \end{bmatrix}$
10. $\begin{bmatrix} \frac{5}{3} & \frac{-4}{3} \\ \frac{2}{3} & \frac{-1}{3} \end{bmatrix}$
11. $\begin{bmatrix} \frac{-3}{2} & 3 \\ 3 & -2 \end{bmatrix}$
12. $\begin{bmatrix} \frac{1}{13} & \frac{-6}{13} \\ \frac{-3}{13} & \frac{5}{13} \end{bmatrix}$
13. $\begin{bmatrix} 1 & -1 \\ 0 & 1 \end{bmatrix}$
14. does not exist
15. $\begin{bmatrix} \frac{-1}{4} & 1 \\ \frac{3}{2} & 6 \end{bmatrix}$
16. $\begin{bmatrix} \frac{0.7}{0.019} & \frac{0.5}{0.019} \\ \frac{0.1}{0.019} & \frac{-0.02}{0.019} \end{bmatrix}$

Lesson 4.3
Level B

1. 7; yes
2. $\frac{-11}{180}$; yes
3. -143; yes
4. 0.066; yes
5. 0; no
6. 4; yes
7. $\begin{bmatrix} \frac{12}{11} & \frac{24}{11} \\ \frac{4}{11} & \frac{-3}{11} \end{bmatrix}$
8. $\begin{bmatrix} \frac{-15}{82} & \frac{48}{41} \\ \frac{45}{41} & \frac{40}{41} \end{bmatrix}$
9. $\begin{bmatrix} \frac{-3}{2} & 3 \\ 3 & -2 \end{bmatrix}$
10. no inverse
11. $\begin{bmatrix} 1.4 & 0.2 & 0.2 \\ -1.6 & -0.1\overline{3} & -0.4\overline{6} \\ 0.4 & 0.2 & 0.2 \end{bmatrix}$

Answers

12. $\begin{bmatrix} 0.\overline{3} & 0 & 0 \\ -0.0\overline{6} & 0.2 & 0.2 \\ -0.1\overline{6} & 0 & 0.5 \end{bmatrix}$

13. $\begin{bmatrix} 0.109 & 0.135 & -0.031 \\ -0.122 & 0.028 & 0.074 \\ 0.083 & -0.037 & 0.057 \end{bmatrix}$

14. $\begin{bmatrix} -0.005 & 0.012 & 0.005 \\ 0.008 & 0.002 & 0.014 \\ -0.009 & 0.004 & 0.013 \end{bmatrix}$

15. no inverse

16. no inverse

Lesson 4.3
Level C

1. $-0.26\overline{38}$; $\begin{bmatrix} 0.632 & 2.526 \\ 1.263 & -0.947 \end{bmatrix}$

2. -0.001994; $\begin{bmatrix} -1.505 & 20.060 \\ 25.075 & -1.003 \end{bmatrix}$

3. -3075; $\begin{bmatrix} 0.042 & 0.050 & -0.003 \\ 0.062 & -0.003 & 0.071 \\ 0.011 & 0.052 & 0.060 \end{bmatrix}$

4. 0; no inverse exists

5. -0.34375; $\begin{bmatrix} 0 & 1.\overline{09} & 1.\overline{09} \\ -0.\overline{8} & -2.\overline{42} & -0.\overline{42} \\ -1.\overline{3} & -1.\overline{45} & -1.\overline{45} \end{bmatrix}$

6. -32.32; $\begin{bmatrix} 0.192 & 0.125 & -0.292 \\ -0.093 & -0.060 & -0.230 \\ 0.209 & -0.299 & -0.317 \end{bmatrix}$

7. 77

8. -423

Lesson 4.4
Level A

1. $\begin{bmatrix} 2 & 1 & -2 \\ 3 & -2 & 4 \\ 1 & 3 & 2 \end{bmatrix} \begin{bmatrix} x \\ y \\ z \end{bmatrix} = \begin{bmatrix} 1 \\ -2 \\ 4 \end{bmatrix}$

2. $\begin{bmatrix} 6 & 2 & -1 \\ -2 & 3 & 2 \\ 3 & -5 & 0 \end{bmatrix} \begin{bmatrix} x \\ y \\ z \end{bmatrix} = \begin{bmatrix} 1 \\ -2 \\ 2 \end{bmatrix}$

3. $\begin{cases} 2x + y - 3z = 1 \\ -x - y + 2z = -2 \\ -2x + 4y + 5z = 3 \end{cases}$

4. $\begin{cases} 3x - y + z = -4 \\ 2x + 3y - 2z = 2 \\ 4x + 2y - 3z = 10 \end{cases}$

5. $x = -1, y = -2, z = 1$

6. $x = 7, y = 55, z = -29$

7. $x = 3, y = 1, z = 2$

8. $x = 1, y = 0, z = -1$

9. $x = 1, y = 0, z = 3$

10. $x = \dfrac{10}{3}, y = \dfrac{7}{6}, z = \dfrac{14}{3}$

Lesson 4.4
Level B

1. $\begin{bmatrix} 0.2 & 0.04 & -0.2 \\ 1.1 & -0.6 & 0.7 \\ 0.25 & 0.3 & -0.35 \end{bmatrix} \begin{bmatrix} x \\ y \\ z \end{bmatrix} = \begin{bmatrix} 0.03 \\ 0.5 \\ 0.1 \end{bmatrix}$

2. $\begin{bmatrix} 2 & -3 & 0 \\ 0 & 4 & 2 \\ -1 & 0 & 1 \end{bmatrix} \begin{bmatrix} x \\ y \\ z \end{bmatrix} = \begin{bmatrix} 1 \\ -3 \\ 0 \end{bmatrix}$

Answers

3. $\begin{cases} 4x - \dfrac{1}{2}z = 1 \\ -3x + \dfrac{2}{3}y = \dfrac{2}{3} \\ \dfrac{-1}{2}y - \dfrac{1}{2}z = \dfrac{-3}{4} \end{cases}$

4. $\begin{cases} -x + y - z = 0 \\ -x - y - z = 0 \\ x - y + z = 0 \end{cases}$

5. $x = 0, y = \dfrac{5}{4}, z = \dfrac{1}{8}$

6. $x = 3, y = -2, z = -2$

7. no solution

8. $x = -3.41, y = 72.12, z = -11.76$

9. $x = 4, y = -2, z = 0$

10. infinitely many solutions

Lesson 4.4
Level C

1. $x = -18.42, y = 7.5$

2. $x = -0.1, y = -0.26$

3. $x = 17.5, y = -71.5, z = -52$

4. $w = 3, x = -4, y = -2, z = 2$

5. a. $\begin{cases} 0.3x + 0.2y + 0.2z = 6500 \\ 0.5x + 0.4y + 0.2z = 10{,}000 \\ 0.2x + 0.4y + 0.6z = 11{,}500 \end{cases}$

b. $\begin{bmatrix} 0.3 & 0.2 & 0.2 \\ 0.5 & 0.4 & 0.2 \\ 0.2 & 0.4 & 0.6 \end{bmatrix} \begin{bmatrix} x \\ y \\ z \end{bmatrix} = \begin{bmatrix} 6500 \\ 10{,}000 \\ 11{,}500 \end{bmatrix}$

c. $9000 in low-risk stocks, $8500 in medium-risk-stocks and $10,500 in high-risk stocks

Lesson 4.5
Level A

1. $\begin{bmatrix} 3 & 4 & -5 & \vdots & 2 \\ -1 & 2 & 2 & \vdots & 10 \\ 2 & -1 & 3 & \vdots & 1 \end{bmatrix}$

2. $\begin{bmatrix} 5 & -4 & 0 & \vdots & 8 \\ 2 & 0 & -3 & \vdots & 1 \\ 0 & -3 & 1 & \vdots & -2 \end{bmatrix}$

3. $\begin{bmatrix} 1 & 0 & 0 & \vdots & -13 \\ 0 & 1 & 0 & \vdots & 3 \\ 0 & 0 & 1 & \vdots & 3 \end{bmatrix}$

4. $\begin{bmatrix} 1 & 0 & 0 & \vdots & -1.75 \\ 0 & 1 & 0 & \vdots & -3.5 \\ 0 & 0 & 1 & \vdots & 3.75 \end{bmatrix}$

5. $\begin{bmatrix} 1 & 0 & 0 & \vdots & 2 \\ 0 & 1 & 0 & \vdots & 5 \\ 0 & 0 & 1 & \vdots & 1 \end{bmatrix}$

6. $\begin{bmatrix} 1 & 0 & 0 & \vdots & 4 \\ 0 & 1 & 0 & \vdots & -5 \\ 0 & 0 & 1 & \vdots & 2 \end{bmatrix}$

7. $x = -4, y = -3$

8. $x = 1, y = 2$

9. $x = -2, y = 3, z = 2$

10. $x = -1, y = \dfrac{1}{2}, z = 2$

11. $x = 2, y = -1, z = 3$

12. $x = -1, y = 1, z = 4$

Lesson 4.5
Level B

1. $\begin{bmatrix} 1 & 0 & 0 & \vdots & 2 \\ 0 & 1 & 0 & \vdots & 2 \\ 0 & 0 & 1 & \vdots & -3 \end{bmatrix}$

Algebra 2   Practice Masters Levels A, B, and C

Answers

2. $\begin{bmatrix} 1 & 0 & 0 & \vdots & 2 \\ 0 & 1 & 0 & \vdots & -4 \\ 0 & 0 & 1 & \vdots & -2 \end{bmatrix}$

3. $\begin{bmatrix} 1 & 0 & 0 & \vdots & 4 \\ 0 & 1 & 0 & \vdots & 5 \\ 0 & 0 & 1 & \vdots & -3 \end{bmatrix}$

4. $x = 1, y = -2, z = 0$

5. $x = \dfrac{1}{2}, y = -1, z = \dfrac{1}{2}$

6. $x = -6, y = -8, z = 9$

7. $x = 1.7, y = -2.4, z = 1.2$

8. $x = 100, y = 110, z = 80$

9. $x = 0.2, y = 0.1, z = -0.4$

5. $x = \dfrac{1}{2}, y = \dfrac{-1}{2}, z = 3$

6. $x = -12, y = 10, z = \dfrac{1}{3}$

7. The solution can be described as $w = 4, x = -2, y = -3, z = 2$; It has infinitely more solutions.

8. $w = 0.2, x = -0.3, y = 0.2, z = 0.2$

Lesson 4.5
Level C

1. $\begin{bmatrix} 1 & 0 & 0 & \vdots & 2 \\ 0 & 1 & 0 & \vdots & -3 \\ 0 & 0 & 1 & \vdots & -2 \end{bmatrix}$

2. $\begin{bmatrix} 1 & 0 & 0 & \vdots & -6 \\ 0 & 1 & 0 & \vdots & 5.5 \\ 0 & 0 & 1 & \vdots & 3.1 \end{bmatrix}$

3. $\begin{bmatrix} 1 & 0 & 0 & 0 & \vdots & 14 \\ 0 & 1 & 0 & 1 & \vdots & -8 \\ 0 & 0 & 1 & 0 & \vdots & -6 \\ 0 & 0 & 0 & 1 & \vdots & 2 \end{bmatrix}$

4. $\begin{bmatrix} 1 & 0 & 0 & 0 & \vdots & 12 \\ 0 & 1 & 0 & 1 & \vdots & \dfrac{1}{2} \\ 0 & 0 & 1 & 0 & \vdots & \dfrac{1}{3} \\ 0 & 0 & 0 & 1 & \vdots & 4 \end{bmatrix}$

Answers

**Lesson 5.1
Level A**

1. $f(x) = x^2 - x - 6$;
   $a = 1, b = -1, c = -6$

2. $f(x) = x^2 + 3x - 2$;
   $a = 1, b = 3, c = -2$

3. $f(x) = -x^2 + x + 6$;
   $a = -1, b = 1, c = 6$

4. $f(x) = x^2 + 6x + 9$;
   $a = 1, b = 6, c = 9$

5. $f(x) = 6x^2 - 7x - 3$;
   $a = 6, b = -7, c = -3$

6. yes

7. no

8. yes

9. no

10. no

11. yes

12. up, minimum

13. down, maximum

14. down, maximum

15. up, minimum

16. V(0, 1)

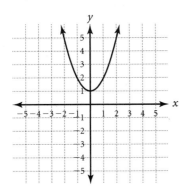

17. $V\left(\frac{1}{2}, 3\frac{1}{4}\right)$

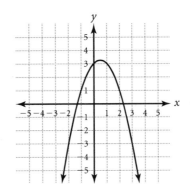

18. $V(-1, -4)$

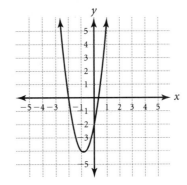

**Lesson 5.1
Level B**

1. no

2. yes; $a = -1, b = 5, c = -7$

3. yes; $a = -1, b = 1, c = 6$

4. no

5. yes; $a = 1, b = 12, c = 36$

6. yes; $a = 1, b = -3, c = 9$

Answers

7. opens up; minimum; $V\left(1\dfrac{1}{4}, \dfrac{-1}{8}\right)$

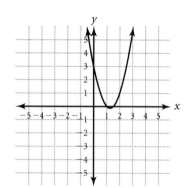

8. opens down; maximum; $V\left(\dfrac{1}{3}, \dfrac{4}{3}\right)$

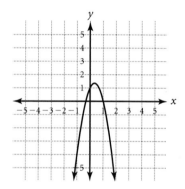

9. opens down; maximum; $V(0, -10)$

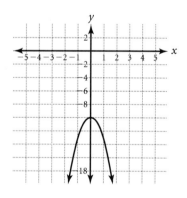

10. opens down; maximum; $V(-2.5, 0.25)$

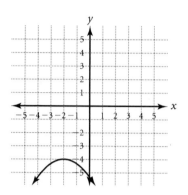

Lesson 5.1
Level C

1. yes; $a = 1, b = -6, c = 9$
2. yes; $a = 2, b = 10, c = -2$
3. no
4. yes; $a = \dfrac{3}{2}, b = \dfrac{5}{2}, c = -3$
5. no
6. yes; $a = 1, b = 1, c = 4$
7. opens up; minimum $V\left(1\dfrac{1}{2}, \dfrac{7}{8}\right)$

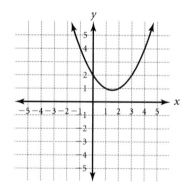

Answers

8. opens up; minimum; $V\left(\dfrac{1}{24}, \dfrac{-145}{144}\right)$

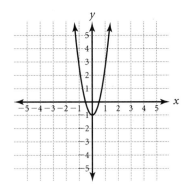

9. opens down; maximum; $V\left(\dfrac{1}{2}, \dfrac{1}{2}\right)$,

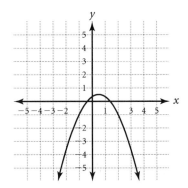

10. opens up; minimum; $V\left(-1\dfrac{1}{2}, \dfrac{-29}{400}\right)$,

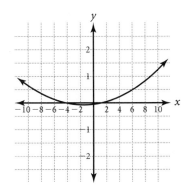

11. opens down; maximum; $V(-2, -4)$,

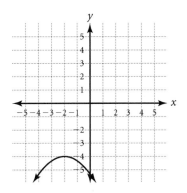

12. opens up; minimum; $V(-2, -3)$,

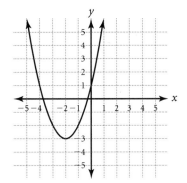

Lesson 5.2
Level A

1. $x = \pm 7$
2. $x = \pm 4$
3. $x = 12, -8$
4. $x = \pm 9$
5. $x = 2, -8$
6. $x = \pm 5$
7. $x = \pm \dfrac{\sqrt{34}}{2}$; $x = \pm 2.92$
8. no solution
9. $x = \pm\sqrt{2}$; $x = \pm 1.41$
10. $x = \pm \dfrac{5\sqrt{3}}{3}$; $x = \pm 2.89$

Answers

11. $x = 13$

12. $y = 3.6$

13. $z = 12.7$

14. $x = 24$

15. $y = 15.6$

16. $z = 6.9$

17. $y = 0.5$

Lesson 5.2
Level B

1. $x = \pm\dfrac{2}{3}$

2. $x = \pm\dfrac{1}{4}$

3. $x = \pm\dfrac{1}{2}$

4. $t = \pm 4.2$

5. $t = \pm 6.3$

6. $t = 0$

7. $x = \pm 2$

8. $x = \pm\sqrt{21}; x = \pm 4.58$

9. $z = 5.9$

10. $y = 10.2$

11. $x = 0.6$

12. $x = 4.4$

13. $y = 2.0$

14. $z = 3.2$

15. 13.4 feet

16. 5 cm

17. 9.2 by 9.2 inches

Lesson 5.2
Level C

1. $t = -0.1, 14.1$

2. $x = \dfrac{3 \pm \sqrt{72}}{5}; x = 2.30, -1.10$

3. $x = \dfrac{2}{3}, \dfrac{-4}{3}$

4. $x = \dfrac{3 \pm \sqrt{72}}{5}; x = 2.30, -1.10$

5. Sample answer: $x^2 - 16 = 0$

6. Sample answer: $x^2 - 15 = 0$

7. Sample answer: $x^2 - 42 = 0$

8. 11.28 cm

9. 11.48 feet

10. 7.94 feet from the building

11. 10.77 miles from home

12. 0.78 seconds

13. 106 yards

Lesson 5.3
Level A

1. $-3(4 + x)$

2. $-5x(x + 2)$

3. $6x^2(x - 3)$

4. $(3x + 4)(2x - 3)$

5. $-x(2x + 5)$

6. $(x + 7)(-2x - 1)$

7. $(x + 1)(x + 4)$

8. $(x - 2)(x + 3)$

9. $(x + 6)(x - 3)$

10. $(x - 7)(x - 5)$

11. $(x - 2)(x + 1)$

Answers

12. $(x + 10)(x + 1)$
13. $(x - 5)(x - 20)$
14. $(x + 12)(x - 2)$
15. $(x - 4)(x - 8)$
16. $(x + 18)(x + 3)$
17. $(2x + 5)(x - 3)$
18. $(3x - 1)(x + 4)$
19. $(4x - 1)(2x + 3)$
20. $(5x + 3)(5x - 1)$
21. $(3x - 7)(3x + 2)$
22. $(4x - 1)(3x + 5)$
23. $x = \pm 15$
24. $x = 3, 2$
25. $x = -2, -5$
26. $x = 5, -1$
27. $x = -1, 4$
28. $x = 4, 4$
29. $x = \pm 4$
30. $x = -5, -2$
31. $x = \pm 8$
32. $x = \pm 2$

Lesson 5.3
Level B

1. $3(x + 3)(x + 3)$
2. $(x - 2)(x + 10)$
3. $(4x + 3)(5x - 2)$
4. $(x - 13)(x - 3)$
5. $(x + 47)(x + 1)$
6. $-2x^3(4x + 5)$
7. $\dfrac{-1}{2} x(3x + 5)$
8. prime
9. $(2x + 3)(3x - 1)$
10. $(x + 0.1)(x + 0.2)$
11. $(4x - 5)(3x - 8)$
12. $4(x + 5)(x - 6)$
13. $3(2x + 9)(3x - 8)$
14. $-5x(x + 7)(x - 4)$
15. $(x^2 + 1)(x + 1)(x - 1)$
16. $-(x + 2)(3x + 8)$
17. $x = 0, -4$
18. $x = \pm 20$
19. $x = \dfrac{1}{2}, 3$
20. $x = \dfrac{2}{3}, \dfrac{-3}{4}$
21. $x = \dfrac{-1}{3}, \dfrac{-2}{3}$
22. $x = \pm 25$
23. $x = 3, -7$
24. $x = \dfrac{-3}{10}, \dfrac{7}{12}$
25. $x = -8$
26. $x = \dfrac{-3}{2}$

Answers

Lesson 5.3
Level C

1. $(x^2 + 1)^2$
2. $(x^2 + 4)(x + 2)(x - 2)$
3. $(x^4 + 16)(x^2 + 4)(x + 2)(x - 2)$
4. $x = -1, \dfrac{1}{3}$
5. $x = \dfrac{3}{4}, \dfrac{1}{2}$
6. $x = \pm 0.2$
7. $x = -0.1, 0.4$
8. $x = 4\dfrac{1}{2}, 3\dfrac{2}{3}$
9. $x = \dfrac{2}{5}, \dfrac{1}{10}$
10. $x = 0, \pm 7$
11. $x = 0, -1, -2$
12. $x = -1.27, 2.77$
13. $x = -0.43, 1.18$
14. no real roots
15. $x = -0.54, 1.87$
16. 5 seconds
17. a. 44.9 seconds
    b. 22.45 seconds
    c. 2469.4 feet
18. 2.61 inches wide

Lesson 5.4
Level A

1. $x^2 + 12x + 36; (x + 6)^2$
2. $x^2 - 14x + 49; (x - 7)^2$
3. $x^2 + 26x + 169; (x + 13)^2$
4. $x^2 + 5x + \dfrac{25}{4}; \left(x + \dfrac{5}{2}\right)^2$
5. $x^2 + 3x + \dfrac{9}{4}; \left(x - \dfrac{3}{2}\right)^2$
6. $x^2 + x + \dfrac{1}{4}; \left(x + \dfrac{1}{2}\right)^2$
7. $x = -4 \pm \sqrt{14}; x = -0.26, -7.74$
8. $x = 5 \pm \sqrt{10}; x = 8.16, 1.84$
9. $x = 0, -6$
10. $x = 0, 5$
11. no real solution
12. $x = \dfrac{-5}{2} \pm \dfrac{\sqrt{17}}{2}; x = -0.44, -4.56$
13. $f(x) = (x + 2)^2 - 1$
    Vertex $= (-2, -1)$;
    Axis of symmetry: $x = -2$
14. $f(x) = (x - 33)^2 - 25$
    Vertex $= (3, -25)$;
    Axis of symmetry: $x = 3$
15. $f(x) = (x + 1)^2 + 4$
    Vertex $= (-1, 4)$;
    Axis of symmetry: $x = -1$
16. $f(x) = (x + 5)^2 + (-24)$
    Vertex $= (-5, -24)$;
    Axis of symmetry: $x = -5$

Lesson 5.4
Level B

1. $x = -3 \pm \sqrt{19}; x = 1.36, -7.35$
2. $x = \dfrac{5 \pm 3\sqrt{5}}{2}; x = 5.85, -0.85$
3. $x = \dfrac{-3 \pm \sqrt{23}}{2}; x = 0.90, -3.89$

Answers

4. $x = \dfrac{3}{2} \pm \dfrac{\sqrt{21}}{6}$; $x = 2.26, 0.74$

5. $x = \dfrac{5 \pm \sqrt{17}}{4}$; $x = 2.28, 0.22$

6. $x = \dfrac{-7 \pm \sqrt{209}}{8}$; $x = 0.93, -2.68$

7. $\dfrac{7}{2} \pm \dfrac{\sqrt{129}}{2}$; $x = 9.18, -2.17$

8. no real solution

9. $x = 3, \dfrac{-5}{3}$

10. $\dfrac{5}{4} \pm \dfrac{\sqrt{193}}{4}$; $x = -2.22, 4.72$

11. $\dfrac{1}{10} \pm \dfrac{\sqrt{21}}{10}$; $x = 0.56, -0.36$

12. $-0.01 \pm \sqrt{0.0701}$; $x = 0.25, -0.27$

13. $h(x) = \left(x - \dfrac{3}{2}\right)^2 - 17\dfrac{1}{4}$

    Vertex: $\left(\dfrac{3}{2}, -17\dfrac{1}{4}\right)$;

    Axis of symmetry: $x = \dfrac{3}{2}$

    horizontal translation $\dfrac{3}{2}$ units to the right, vertical translation $17\dfrac{1}{4}$ units down

14. $h(x) = 2\left(x + \dfrac{1}{2}\right)^2 + \dfrac{5}{2}$

    Vertex $\left(\dfrac{-1}{2}, \dfrac{5}{2}\right)$;

    Axis of symmetry: $x = \dfrac{-1}{2}$

    horizontal translation $\dfrac{1}{2}$ unit to the left, vertical translation $\dfrac{5}{2}$ units up, vertical stretch by factor of 2

15. $h(x) = -3(x + 1)^2 + 13$

    Vertex $(-1, 13)$;
    Axis of symmetry: $x = -1$

    reflection across x-axis, horizontal translation 1 unit left, vertical translation 13 units up, vertical stretch by factor of 3

16. $h(x) = -5\left(x - \dfrac{1}{10}\right)^2 + 0.05$

    Vertex $\left(\dfrac{1}{10}, 0.05\right)$;

    Axis of symmetry: $x = \dfrac{1}{10}$

    reflected across the x-axis, horizontal translation $\dfrac{1}{10}$ unit right, vertical translation 0.05 units up, vertical stretch by factor of 5

Answers

17. $h(x) = \left(x + \dfrac{9}{2}\right)^2 - 20\dfrac{1}{4}$

    Vertex: $\left(\dfrac{-9}{2}, -20\dfrac{1}{4}\right)$;

    Axis of symmetry: $x = \dfrac{-9}{2}$

    horizontal translation $\dfrac{9}{2}$ units left,

    vertical translation $20\dfrac{1}{4}$ units down

18. $h(x) = -1(x + 5)^2 + 25$
    Vertex $(-5, 25)$;
    Axis of symmetry: $x = -5$

    reflected across the x-axis, horizontal translation 5 units left, vertical translation 25 units up

Lesson 5.4
Level C

1. $x = \dfrac{1 \pm \sqrt{21}}{4}$; $x = 1.39, -0.90$

2. $x = \dfrac{2 \pm \sqrt{22}}{6}$; $x = 1.12, -0.45$

3. $x = \dfrac{-5 \pm \sqrt{33}}{4}$; $x = 0.19, -2.69$

4. $x = \dfrac{-9 \pm \sqrt{185}}{4}$; $x = 1.15, -5.65$

5. no real solution

6. $x = 4 \pm \sqrt{22}$; $x = 8.69, -0.69$

7. $x = -1 \pm \sqrt{13}$; $x = 2.61, -4.61$

8. $x = \dfrac{15 \pm 3\sqrt{17}}{2}$; $x = 13.68, 1.32$

9. $x = \dfrac{3}{2} \pm \dfrac{\sqrt{11}}{2}$; $x = -0.16, 3.16$

10. no real solution

11. $x = -2 \pm \sqrt{14}$; $x = 1.74, -5.74$

12. $x = -3 \pm \sqrt{13}$; $x = 0.61, -6.61$

13. $h(x) = \dfrac{-1}{2}\left(x - \dfrac{1}{2}\right)^2 - \dfrac{7}{8}$

    Vertex $\left(\dfrac{1}{2}, -\dfrac{7}{8}\right)$

    reflected across the x-axis, horizontal translation of $\dfrac{1}{2}$ units to the right, vertical translation $\dfrac{7}{8}$ units down, and vertical compression by factor of $\dfrac{1}{2}$.

14. $h(x) = \dfrac{1}{4}(x - 4)^2 - 3$

    Vertex $(4, -3)$

    horizontal translation of 4 units to the right, vertical translation of 3 units down, and vertical compression by a factor of $\dfrac{1}{4}$.

15. $y = 5(x - 3)^2 + 1$

16. $y = \dfrac{-6}{25}(x + 2)^2 + 5$

Lesson 5.5
Level A

1. $x = -2, -5$

2. $x = \dfrac{1}{4}, 1$

3. $x = \dfrac{3}{4}, \dfrac{1}{2}$

Answers

4. $x = \dfrac{2}{3}, \dfrac{-3}{4}$

5. $x = 0, -4$

6. no real solutions

7. $x = \dfrac{-3}{2}, \dfrac{1}{3}$

8. $x = \dfrac{5 \pm \sqrt{17}}{2}$

9. $x = \dfrac{-3 \pm \sqrt{65}}{4}$

10. $x = \dfrac{-5 \pm \sqrt{13}}{6}$

11. $x = \dfrac{9 \pm \sqrt{21}}{2}$

12. no real solutions

13. $x = -2, V(-2, -1)$

14. $x = -1, V(-1, 13)$

15. $x = -5, V(-5, -24)$

16. $x = -1, V(-1, -1)$

17. $x = 1, V\left(1, \dfrac{7}{2}\right)$

18. $x = \dfrac{5}{4}, V\left(\dfrac{5}{4}, \dfrac{41}{8}\right)$

Lesson 5.5
Level B

1. $x = 4\dfrac{1}{2}, 3\dfrac{2}{3}$

2. $x = -0.1, 0.4$

3. $x = \dfrac{-9 \pm \sqrt{57}}{2}; x = -0.72, -8.27$

4. $x = 3 \pm \sqrt{5}; x = 5.24, 0.76$

5. $x = \dfrac{-3 \pm \sqrt{133}}{2}; x = 4.27, -7.27$

6. no real solutions

7. $x = \pm\sqrt{\dfrac{19}{6}}; x = \pm 1.78$

8. $x = \dfrac{-11 \pm \sqrt{433}}{12}; x = -2.65, 0.82$

9. $x = \dfrac{3 \pm \sqrt{43}}{2}; x = 4.78, -1.78$

10. $x = \dfrac{2}{3}, \dfrac{7}{9}$

11. $x = -4; V(-4, 16)$

12. $x = 2; V(2, -12)$

13. $x = -0.2; V(-0.2, -12.2)$

14. $x = 0.08; V(0.08, -13.04)$

15. $x = 0.25; V(0.25, 0.2875)$

16. $x = 2.5; V(2.5, 0.04)$

Lesson 5.5
Level C

1. $x = -4 \pm \dfrac{\sqrt{70}}{2}; x = 0.183, -8.183$

2. $x = \dfrac{1}{3} \pm \dfrac{\sqrt{5}}{3}; x = 1.08, -0.41$

3. $x = -20, 150$

4. $x = \dfrac{-3 \pm \sqrt{273}}{4}; x = 3.38, -4.88$

5. $\left(12, 71\dfrac{9}{10}\right)$

6. $\left(\dfrac{-9}{8}, \dfrac{-97}{192}\right)$

Algebra 2   Practice Masters Levels A, B, and C

Answers

7. $\left(\dfrac{23}{6}, \dfrac{-625}{144}\right)$

8. $\left(-\dfrac{7}{30}, \dfrac{169}{300}\right)$

9. a. 3.75 seconds
   b. 4.4 seconds

10. a. $A = x(200 - 2x)$
    b. 5000 sq. ft

Lesson 5.6
Level A

1. real = 4, imaginary = $2i$
2. real = -6, imaginary = $-3i$
3. real = 0, imaginary = $4i$
4. real = -5, imaginary = 0
5. $5i$
6. $i\sqrt{17}$
7. -16
8. -25
9. 1; 2 solutions, $x = -1, \dfrac{-3}{4}$
10. 41; 2 solutions, $x = \dfrac{7 \pm \sqrt{41}}{2}$
11. 29; 2 solutions, $x = \dfrac{-3 \pm \sqrt{29}}{2}$
12. -55; no real solutions
13. $2 - i$
14. $-3 + 2i$
15. $13 + 9i$

16. 5; no conjugate; 5

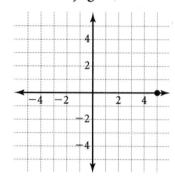

17. $-7 + i; 7 - i; \sqrt{50}$

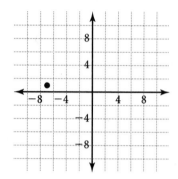

Lesson 5.6
Level B

1. 900; 2 solutions, $x = \pm 5$
2. $3\dfrac{1}{9}$; 2 solutions, $x = \dfrac{-1 \pm \sqrt{28}}{18}$
3. 0; 1 solutions, $x = \dfrac{-1}{3}$
4. -56; no real solutions
5. 32.09; 2 solutions, $x = \dfrac{-0.3 \pm \sqrt{32.09}}{4}$
6. 6.64; 2 solutions, $x = 4.48, 0.19$
7. $\dfrac{-11}{2} - \dfrac{1}{2}i$

Answers

8. $3 - 18i$

9. $3 - 0.2i$

10. $-27 + 5i$

11. $2 - 3i$

12. $\dfrac{-16i}{3}$

13. $\dfrac{25 + 2i}{37}$

14. $\dfrac{23 - 2i}{41}$

15. absolute value = 4

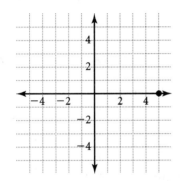

16. absolute value = $\sqrt{1000}$

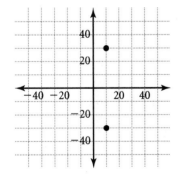

17. absolute value = $\sqrt{37}$

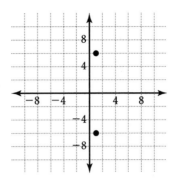

Lesson 5.6
Level C

1. 73; 2 solutions, $x = 4.77, -3.77$

2. 160; 2 solutions, $x = -1.08, 2.08$

3. 145; 2 solutions, $x = 0.54, -0.46$

4. -7; no real solution

5. $-\dfrac{1}{2} - \dfrac{8}{3}i; -\dfrac{1}{2} + \dfrac{8}{3}i$

6. $-7 + i\sqrt{5}; -7 - i\sqrt{5}$

7. $\dfrac{6 + 7i}{17}; \dfrac{6 - 7i}{17}$

8. $25i; -25i$

9. $4 - 4i; 4 + 4i$

10. $\dfrac{-71 - 43i}{30}; \dfrac{-71 + 43i}{30}$

11. $-8 + 6i\sqrt{17}; -8 - 6i\sqrt{17}$

12. $175 + 28i\sqrt{21}; 175 - 28i\sqrt{21}$

Answers

13. $\sqrt{16.25}$

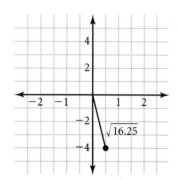

14. $\sqrt{0.53}$

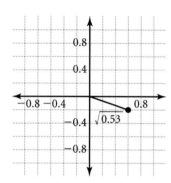

Lesson 5.7
Level A

1. $y = x^2 + 2x - 3$
2. $y = x^2 - 4x + 6$
3. $y = x^2 + x + 4$
4. $y = -x^2 - 2x + 1$
5. $y = 3x^2 + 2x + 6$
6. $y = 5x^2 + 2x + 3$
7. a. $d = n^2$
   b. 81
   c. 21
8. a. $y \approx -1.02x^2 + 16.65x - 1.20$
   b. about 66.7 feet
   c. about 8.15 seconds

Lesson 5.7
Level B

1. $y = -6x^2 - 5x - 10$
2. $y = 5x^2 - 4x + 12$
3. $y = 4x^2 + 10x - 9$
4. $y = -x^2 - 100x + 150$
5. $y = \frac{1}{2}x^2 + x + \frac{3}{2}$
6. $y = \frac{-1}{3}x^2 + \frac{2}{3}x + 4$
7. $y = \frac{1}{4}x^2 + \frac{3}{4}x - 5$
8. $y = 0.2x^2 + 0.5x + 1.2$
9. $y = -0.4x^2 + 0.1x - 0.6$
10. $y = 2.7x^2 - 1.4x - 2.5$
11. $h(t) \approx -1.84t^2 + 8.62t + 14.56$
12. about 24.66 feet
13. about 2.34 seconds
14. about 23.35 feet
15. about 6 seconds

Lesson 5.7
Level C

1. $y = 0.035x^2 - 0.172x + 1$
2. 0.8 million in 1991
3. 2.3 million
4. 14.2 million
5. 1995
6. $y \approx -74.18x^2 + 10,367.76x + 28.46$
7. approximately 527,500

Answers

8. 1912

9. $y \approx 675{,}843.17x^2 - 4{,}035{,}981.65x + 7{,}091{,}198.71$

10. 1,070,000

Lesson 5.8
Level A

1. $x < 2$ and $x > -1$

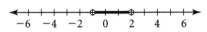

2. $x \geq 5$ or $x \leq 2$

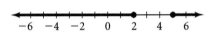

3. $x \geq 1$ and $x \leq 7$

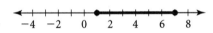

4. $x \geq -3$ and $x \leq 2$

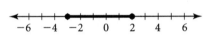

5. no solution

6. all real numbers

7.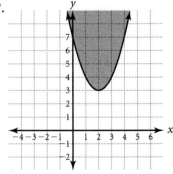

8.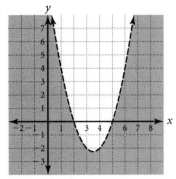

9. a. $P(p) = -10p^2 + 100p - 210$

   b. between \$3 and \$7

Lesson 5.8
Level B

1. $x < \dfrac{-3}{2}$ or $x > 1$

2. $x > \dfrac{5}{2}$ and $x < \dfrac{3}{2}$; no solution

3. $x \leq \dfrac{1}{3}$ and $x \geq \dfrac{-3}{4}$

4. $x \leq \dfrac{-4}{5}$ or $x \geq \dfrac{4}{9}$

5. $x > \dfrac{-1}{10}$ or $x < \dfrac{-2}{5}$

6. $x \geq \dfrac{1}{3}$ and $x \leq \dfrac{2}{3}$

Answers

7.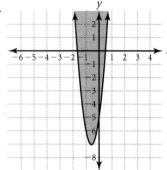

8.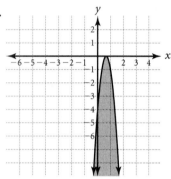

9. a. $y = -2x^2 + 114x$
   b. $1624.50
   c. 28 or 29

Lesson 5.8
Level C

1. $x > -2.21$ and $x < 1.21$

2. $x \geq 1.30$ or $x \leq -2.30$

3. $x \geq -1.25$ and $x \leq 0.68$

4. all real numbers

5.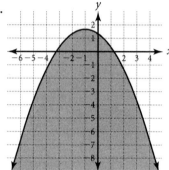

6.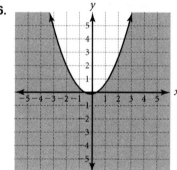

7. a. $R = -0.01x^2 + 39.91x + 7.70$
   b. $39,827.90
   c. 1995.5
   d. $C = 12x + 150$
   e. $P = -0.1x^2 + 27.9x - 142.3$
   f. 6
   g. $1803.73

Answers

Lesson 6.1
Level A

1. 0.93
2. 1.03
3. 0.95
4. 1.10
5. 1.13
6. 0.85
7. 1.052
8. 0.959
9. 81
10. 1.59
11. 743,614.82
12. 1250
13. 10,000
14. 158.49
15. 8.62
16. 0.00347
17. 7.6×10^{-5}
18. 2.08
19. exponential
20. linear
21. quadratic
22. $100(2)^n$
    a. 200
    b. 1600
    c. 50
23. $75(3)^n$
    a. 675
    b. 18,225
    c. $8\frac{1}{3}$

Lesson 6.1
Level B

1. 1.071
2. 0.929
3. 1.063
4. 0.937
5. 0.9986
6. 1.0007
7. -729
8. 729
9. 6.498
10. 274.37
11. 362,797,056
12. 417.96
13. quadratic
14. exponential
15. linear
16. $25(3)^{\frac{n}{2}}$
    a. 43.3
    b. 225
    c. 14.4
17. $50(2)^{2n}$
    a. 400
    b. 3200
    c. 3.125
18. $75(3)^{2n}$
    a. 657
    b. 17,739
    c. 1404
19. $550(2)^{\frac{n}{7}}$
    a. $777.82
    b. $996.30
    c. 35,200

Algebra 2   Practice Masters Levels A, B, and C

Answers

**Lesson 6.1
Level C**

1. 4.472
2. −0.047
3. 0.125
4. 3162.278
5. −1.102
6. −4.164
7. 0.015625
8. 5
9. a. 2,358,731 b. 2,394,201
   c. 2,244,252 d. 2,316,093
10. a. 6367 b. 6060
    c. 5882 d. 5988
11. 258 decades
12. a. 8100 b. 3487

**Lesson 6.2
Level A**

1. exponential
2. linear
3. quadratic
4. exponential
5. quadratic
6. linear
7. growth
8. decay
9. growth
10. growth
11. decay
12. growth

13.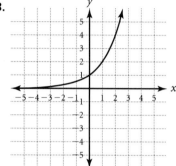

14.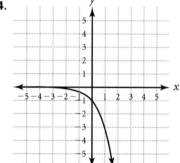

15.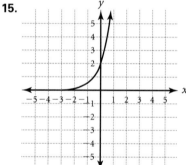

16.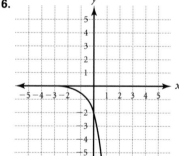

17. $115.76
18. $535.29

Answers

**Lesson 6.2
Level B**

1. decay

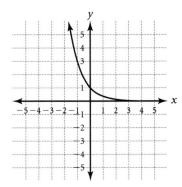

2. growth

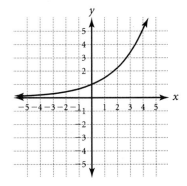

3. decay

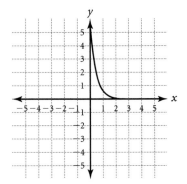

4. growth

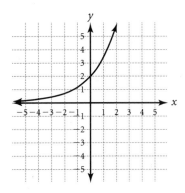

5. $1165.91
6. $1624.76
7. $1665.22
8. $3367.14
9. $3420.44
10. $2370.45
11. $3272.31

**Lesson 6.2
Level C**

1. growth

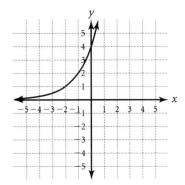

Answers

2. growth

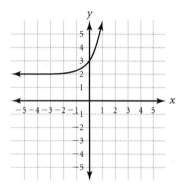

3. decay

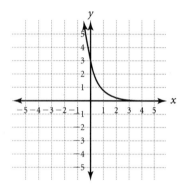

4. decay

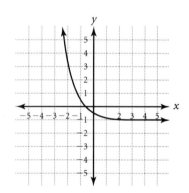

5. about $5000

6. about $1250

7. about -6.6%

8. a. $y = 6^x$
   b. $y = 3^x$
   c. $(0, 1)$

Lesson 6.3
Level A

1. $4 = \log_3 81$

2. $6 = \log_2 64$

3. $-2 = \log_5 \dfrac{1}{25}$

4. $\dfrac{1}{3} = \log_8 2$

5. $4 = \log_{\frac{1}{5}} \dfrac{1}{625}$

6. $2 = \log_{\frac{1}{11}} \dfrac{1}{121}$

7. $7^2 = 49$

8. $10^4 = 10{,}000$

9. $2^5 = 32$

10. $25^{\frac{1}{2}} = 5$

11. $125^{\frac{1}{3}} = 5$

12. $9^1 = 9$

13. $x = 2$

14. $x = 0$

15. $x = -1$

16. $x = 3.02$

17. $x = -3$

18. $x = 3.31$

19. $n = 4$

20. $n = 2$

21. $n = 2$

22. $n = 25$

Answers

23. $n = 81$

24. $n = 1024$

**Lesson 6.3
Level B**

1. $-2 = \log_{15} \dfrac{1}{225}$

2. $\dfrac{-1}{4} = \log_{16} \dfrac{1}{2}$

3. $-4 = \log_{\frac{1}{3}} 81$

4. $3 = \log_{\frac{1}{4}} \dfrac{1}{64}$

5. $-3 = \log_{\frac{1}{5}} 125$

6. $\dfrac{-1}{3} = \log_{1000} \dfrac{1}{10}$

7. $216^{\frac{1}{3}} = 6$

8. $10{,}000^{\frac{1}{4}} = 10$

9. $4^{-3} = \dfrac{1}{64}$

10. $4^{-4} = \dfrac{1}{256}$

11. $10^{-2} = 0.01$

12. $7^0 = 1$

13. $x = 1.30$

14. $x = 1.71$

15. $x = -0.40$

16. $x = 0.18$

17. $x = -2.12$

18. $x = 3.71$

19. $n = \dfrac{1}{8}$

20. $n = \dfrac{1}{625}$

21. $n = 2$

22. $n = 5$

23. $n = 10$

24. $n = 10$

**Lesson 6.3
Level C**

1. $81^{\frac{1}{2}} = 9$

2. $\log_{\frac{1}{2}} 8 = -3$

3. $2^{-6} = \dfrac{1}{64}$

4. $\log_{\frac{1}{4}} 256 = -4$

5. 3.2×10^{-6}

6. 5.0×10^{-10}

7. 3.2×10^{-4}

8. 1

9. vertical stretch by factor of 3

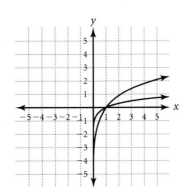

Answers

10. translation of 5 units up

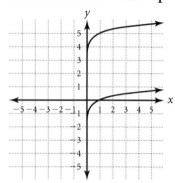

11. translation of 2 units down

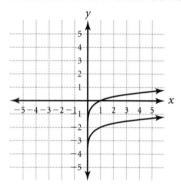

12. vertical compression by factor of $\frac{1}{2}$

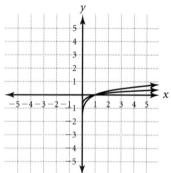

13. reflected across x-axis and translation of 4 units up

14. reflected about x-axis, vertical compression by factor of $\frac{1}{4}$ and translation of 5 units down

Lesson 6.4
Level A

1. $2 + \log_3 25$
2. -1
3. $\log_2 15 + \log_2 m$
4. $\log_5 512$
5. $1 - \log_4 x$
6. $\log_7 x - 2$
7. $\log_2 18$
8. $\log_{10} 6$
9. $\log_5 432$
10. 1
11. $\log_9 \frac{8}{3}$
12. $\log_3 \frac{xy}{z}$
13. 3
14. 20
15. 4.5
16. 100
17. $x = 3$
18. $x = 5$
19. $x = 12$
20. $x = -1$
21. $x = 15$
22. $x = 6$
23. $x = 4$
24. $x = 24$

Answers

Lesson 6.4
Level B

1. 3.907
2. 2.822
3. 0.661
4. 1.737
5. 2.322
6. 3.17
7. 7.229
8. 3.322
9. $\log_4 \dfrac{4x}{y}$
10. $\log_3 \dfrac{m^7}{n}$
11. $\log_3 5\sqrt{y}$
12. $\log_6 \dfrac{x^2 + 16x + 64}{x^4}$
13. $\log_4 \dfrac{x\sqrt{z}}{y}$
14. $\log_5 n^{\frac{4}{3}}$
15. true
16. false
17. $x = 24.3$
18. $x = -\dfrac{11}{3}$
19. $x = \pm 4$
20. $x = \pm 13$
21. $x = 8, -2$
22. $x = 27, -3$

Lesson 6.4
Level C

1. a. 0.564
   b. 0.986
   c. 0.357
2. a. $B = \log_{10} I + 16$
   b. 4 decibels
   c. 10^{59} watts per square meter
3. $x = -8, 6$
4. $x = \dfrac{-3}{4}, \dfrac{2}{3}$
5. $x = 20.628$
6. $x = 15.288$

Lesson 6.5
Level A

1. $x = 2.10$
2. $x = 1.95$
3. $x = 6.64$
4. $x = 1.43$
5. $x = 1.83$
6. $x = 1.32$
7. $x = -0.59$
8. $x = 1.32$
9. $x = -2.29$
10. $x = -2.03$
11. $x = -0.62$
12. $x = 2.22$
13. 4.19
14. 2.01
15. 4.01

Answers

16. 2.39
17. 2.52
18. 1.71
19. 0.29
20. -0.42
21. -2.96
22. -5
23. -1
24. -2

Lesson 6.5
Level B

1. $x = -2.40$
2. $x = 1.16$
3. $x = 1.21$
4. $x = 0.16$
5. $x = 4.29$
6. $x = 1.06$
7. $x = -0.75$
8. $x = 0.41$
9. $x = -1.92$
10. $x = 1$
11. $x = -2$
12. $x = \dfrac{2}{3}$
13. -0.68
14. -1.26
15. -0.12
16. -0.5
17. -2.70

18. -1.5
19. 1.42
20. -1.66
21. -2.17
22. 1.56
23. -0.10
24. 0.46

Lesson 6.5
Level C

1. a. ≈ 30
   b. ≈ 56
   c. ≈ 90
   d. ≈ 108
2. a. 2.4
   b. 1.5
   c. 8.3
   d. 9.8

3. $\log_b = \dfrac{n^{\frac{2}{3}}}{m^{\frac{26}{5}}}$

4. $\log_a = \dfrac{z^2}{y^{\frac{1}{2}} x^{\frac{1}{3}}}$

5. $x = 9.4$
6. $x = 32, -30.5$

Lesson 6.6
Level A

1. 7.389
2. 54.598
3. 40.447
4. 1.284
5. 1210.286

Answers

6. 5.050
7. 0.050
8. 0.961
9. −0.812
10. 4.605
11. 0.693
12. −1.609
13. 1.459
14. −1.609
15. undefined
16. 0.347
17. 4
18. 729
19. 3
20. 1
21. $x = 1.26$
22. $x = 3.11$
23. $x = 0.60$
24. $x = 2.51$
25. $x = 1.36$
26. $x = 4.19$
27. 19.88 mg

**Lesson 6.6
Level B**

1. 0.002
2. 0.779
3. 1.339
4. 9.356
5. −9.510
6. 0.729

7. 0.306
8. −0.513
9. 0.920
10. 1.875
11. 0.561
12. 0.770
13. −16
14. $\dfrac{1}{3}$
15. $e^{\frac{1}{3}}$
16. −3
17. $x = 2.71$
18. $x = -2.32$
19. $x = -1.18$
20. $x = 1.72$
21. no solution
22. $x = 0.26$
23. $2201.15; growth

**Lesson 6.6
Level C**

1. stretched vertically by factor of 4

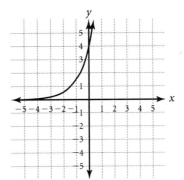

Answers

2. reflected about *y*-axis, horizontal compression by factor of $\frac{1}{2}$

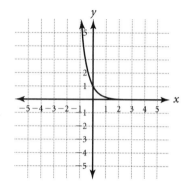

3. horizontal stretch by factor of 3

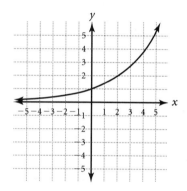

4. horizontal translation 2 units left, vertical compression by factor of $\frac{1}{3}$

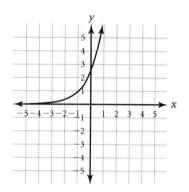

5. vertical stretch by factor of 2

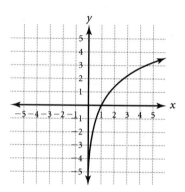

6. horizontal translation 3 units to the right

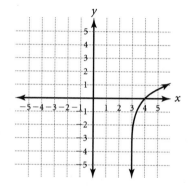

7. 0.135

8. 1.434

9. a. $P(t) = 26{,}959{,}000 e^{0.0145t}$
   b. 36,028,748
   c. 14,038,691

Lesson 6.7
Level A

1. $x = 4$

2. $x = 5$

3. $x = -4$

4. $b = 5$

5. $y = 0.0123;\ 3^{-4}$

6. $x = 1.151;\ \frac{1}{2}\ln 10$

Answers

7. $x = -0.828; \dfrac{-1}{3} \ln 12$

8. $x = 8103.084; e^9$

9. $x = 1.242; \dfrac{1}{2} \ln 12$

10. $y = 7.2 \times 10^{10}; e^{25}$

11. $y = 1.037 \times 10^{21}$

12. $x = 31,622.777$

13. $5809.17

14. $7476.46

15. a. 1,099,262
    b. 420,607

Lesson 6.7
Level B

1. $x = 4$
2. $x = -4.5$
3. $x = 2$
4. $x = \dfrac{7}{3}$
5. $b = 3$
6. $y = \dfrac{1}{4}^{\frac{-2}{3}} \approx 2.52$

7. $x = 12.169$
8. $x = 0.356$
9. $x \approx -0.505$
10. $x = 5$
11. $2396
12. 1,102,215
13. 5134.9 years old
14. 38,376 years old

Lesson 6.7
Level C

1. 2.82×10^{22}
2. 7.94×10^{22}
3. $75.5°C$
4. 31.9 minutes
5. 880,476
6. 658,503
7. a. 458,003
   b. 539,183
8. $x = \dfrac{25 \pm 5\sqrt{31}}{3}; x \approx 17.6$

Answers

Lesson 7.1
Level A

1. yes; degree 2, quadratic

2. yes; degree 3, cubic

3. no

4. yes; degree 4, quartic

5. 4

6. 19

7. 28

8. -20

9. $9x^3 + 2x + 7$

10. $3x^3 + 3x^2 - 6x - 8$

11. $-2x^2 - 8x + 8$

12. $6x^4 + 6x^3 - 3x^2 + 12x - 10$

13. $-2x^2 - 2x - 7$

14. $8.3x^3 + x^2 - 3.1x$

15. a cubic function with S-shape, 1 peak, 1 valley

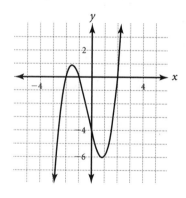

16. a quartic function with W-shape, 2 valleys and 1 peak.

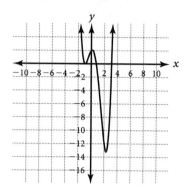

Lesson 7.1
Level B

1. yes, degree 2, quartic

2. no

3. yes, degree 4, quartic

4. yes, degree 3, cubic

5. -32

6. 100

7. 449.96

8. $4.7x^4 - 6.2x^3 + 5x^2 + 3x$

9. $\frac{3}{4}x^3 - \frac{3}{2}x^2 + x + \frac{5}{9}$

10. $-10x^4 - 3x^3 + 5x + 7$

11. $-4x^3 + 8.1x^2 - 0.3x - 4.6$

Answers

12. a quartic function with upside down W-shape with 2 peaks and 1 valley

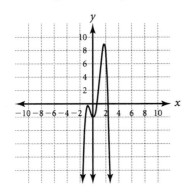

5. a quartic with upside down W-shape, 2 peaks and 1 valley

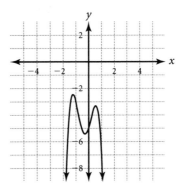

13. a cubic function with S-shape, 1 peak and 1 valley

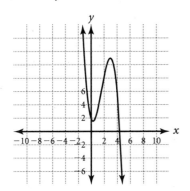

6. a cubic with S-shape, 1 peak and 1 valley

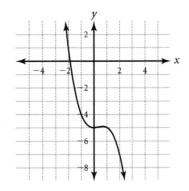

7. a. yes; cubic polynomial
b. $167.35

8. a. $2.90
b. $4.46
c. $16.57
d. $24.80

**Lesson 7.1
Level C**

1. $366\frac{10}{11}$

2. $-2\frac{5}{32}$

3. -0.0011247

4. $a = 2, b = 3, c = 6,$ and $d = 12$

**Lesson 7.2
Level A**

1. maxima of 5.8

2. maxima of 7.0, minima of -0.01

3. minima of -2

4. no maxima or minima

5. maxima of 5.1
increase on interval $-\infty < x < -0.75$
decrease on interval $-0.75 < x < \infty$

Algebra 2   Practice Masters Levels A, B, and C

Answers

6. maxima of 1.6, minima of 0.4
   increase on interval
   $-\infty < x < -0.47$, $0.47 < x < \infty$
   decrease on interval $-0.47 < x < 0.47$

7. maxima of 2
   increase on interval $-\infty < x < 0$
   decrease on interval $0 < x < \infty$

8. maxima of 0.1, minima of 0
   increase on interval $0 < x < 0.\overline{6}$
   decrease on interval
   $-\infty < x < 0$, $0.\overline{6} < x < \infty$

9. rises on left and right

10. rises on left and right

11. falls on left and rises on right

12. rises on left and falls on right

13. $y = 0.00008x^3 - 0.009x^2 + 0.12x + 13.78$

Lesson 7.2
Level B

1. none

2. maxima of 6.1, minima of -8.1

3. minima of 5.5

4. maxima of 8.5

5. maxima of 26.4, minima of 0.8
   increase on interval
   $-\infty < x < -3.5$ and $0.2 < x < \infty$
   decrease on interval $-3.5 < x < 0.2$

6. maxima of 6.3, minima of -2.3
   increase on interval $-0.3 < x < 2.3$
   decrease on interval
   $-\infty < x < -0.3$ and $2.3 < x < \infty$

7. maxima of -2.1
   increase on interval $0.5 < x < \infty$
   decrease on interval $-\infty < x < 0.5$

8. rises on left and right

9. falls on left, rises on right

10. falls on left and right

11. rises on left and falls on right

12. $y = 0.243x^4 - 4.547x^3 + 27.170x^2 - 29.490x + 812.801$

Lesson 7.2
Level C

1. a. maxima of 4.79
   b. increase on interval $-\infty < x < 0.78$
      decrease on $0.78 < x < \infty$
   c. falls on left and right

2. a. no maxima or minima
   b. increase on interval all values of x
   c. left end falls, right end rises

3. a. minima at 0.01
   b. increase on interval $0.34 < x < \infty$
      decrease on interval $-\infty < x < 0.34$
   c. left and right end rises

4. a. maxima of 0.64, minima of 0.49
   b. increase on interval $0.14 < x < 1.17$
      decrease on interval $-\infty < x < 0.14$
      and $1.17 < x < \infty$
   c. left end rises and right end falls

5. $y = x^4 + x^3 - 5x^2 + x - 1$

6. $y = -4x^3 - 6x^2 + 9x - 4$

7. a. $y = -0.00002x^4 + 0.002x^3 - 0.054x^2 + 0.409x + 7.874$
   b. 15.2 thousand
   c. the year 2025

Answers

Lesson 7.3
Level A

1. $8x^5 + 24x^4 - 12x^3$
2. $3x^3 + 14x^2 + 10x + 8$
3. $x^4 - 4x^3 + x^2 - 6x + 8$
4. $2x^3 - 5x^2 - 21x + 10$
5. yes
6. yes
7. no
8. yes
9. $x - 7$
10. $x - 4$
11. $x^2 - 2x + 5$
12. $x^2 - 5x - 4$
13. $x + 10$
14. $x^2 + 2x + 3$
15. $x^2 - 6x - 4 + \dfrac{7}{x - 3}$
16. $x^2 - 10x - 8 - \dfrac{11}{x + 4}$
17. 27
18. -2
19. 13
20. -4

Lesson 7.3
Level B

1. $6x^4 - 4x^3 - x^2 + 26x + 8$
2. $8x^4 - 20x^3 + 4x^2 - 22x + 30$
3. $8x^3 - 60x^2 + 150x - 125$
4. $18x^3 - 3x^2 - 28x - 12$

5. yes
6. no
7. yes
8. no
9. $-7x - 18 - \dfrac{44}{x - 3}$
10. $-4x^2 + 10x - 22 - \dfrac{41}{-x - 2}$
11. $-6x^2 + 20x - 66 + \dfrac{212x - 65}{x^2 + 3x - 1}$
12. $-x^4 - 5x^3 - 25x^2 - 125x - 627 + \dfrac{3135}{5 - x}$
13. $-6x + 25 - \dfrac{71}{x + 3}$
14. $-5x^2 - 22x - 106 - \dfrac{532}{x - 5}$
15. $x^2 - 4x + 16 - \dfrac{68}{x + 4}$
16. $\dfrac{1}{2}x^2 - \dfrac{7}{4}x - \dfrac{3}{8} - \dfrac{\frac{19}{16}}{x - \frac{1}{2}}$
17. 12
18. -5
19. 8.83247

Lesson 7.3
Level C

1. $3x^4 + 16x^3 + 17x^2 - 6x + 18$
2. $\dfrac{1}{2}x^3 - \dfrac{1}{3}x^2 - \dfrac{7}{9}x - \dfrac{2}{9}$
3. $x^4 - 16x^3 + 96x^2 - 256x + 256$

Answers

4. $9x^5 - 69x^4 + 139x^3 + 45x^2 - 216x - 108$
5. $x(x-6)^2$
6. $(x+4)(x^2-4x+16)$
7. $(x-5)(x^2+5x+25)$
8. $(x^2+1)(x-2)$
9. $\dfrac{1}{4}x - \dfrac{3}{8} + \dfrac{\frac{29}{16}}{x - \frac{1}{2}}$
10. $\dfrac{1}{3}x^3 - \dfrac{1}{9}x^2 + \dfrac{1}{27}x - \dfrac{1}{81} + \dfrac{\frac{1}{243}}{x + \frac{1}{3}}$
11. $0.05x^2 - 0.01x + 0.072 + \dfrac{0.0856}{x+0.2}$
12. $0.2x^3 + 0.24x^2 + 0.288x - 0.1544 - \dfrac{0.18528}{x-1.2}$
13. no
14. no
15. -1
16. 2.54112

Lesson 7.4
Level A

1. $x = 0, x = \pm 6$
2. $x = 0, -4, 7$
3. $x = 0, -3, 5$
4. $x = 0, -2, -2$
5. $x = 0, 3, 6$
6. $x = 0, 4, -5$
7. $x = 0, -4, 6$
8. $x = 1, 2, 4$
9. $x = -1, 2, -3$
10. $x = -2, 3, 5$
11. $x = \pm 3, \pm 2$
12. $x = \pm 5, \pm 1$
13. $x = \pm 4, \pm\sqrt{5}$
14. $x = \pm\sqrt{7}, \pm\sqrt{11}$
15. zeros at $x = 3, -1, -2$
16. zeros at $x = -5, 0, 2$
17. zeros at $x = -1.20, -3.49$
18. zeros at $x = -1.16, 1.77, 3.39$

Lesson 7.4
Level B

1. $x = 0, 2, -3, 4$
2. $x = \dfrac{3}{5}, \dfrac{1}{3}, 1$
3. $x = \dfrac{1}{3}, \dfrac{1}{4}, \dfrac{1}{5}$
4. $x = \dfrac{1}{3}, \dfrac{-1}{2}, \dfrac{-2}{3}$
5. $x = \dfrac{3}{2}, \dfrac{1}{3}, 5$
6. $x = \dfrac{-5}{3}, \dfrac{-7}{2}, 6$
7. $x = \dfrac{1}{4}, \dfrac{-1}{2}, -1$
8. $x = \dfrac{1}{2}, \dfrac{1}{2}, \dfrac{-1}{3}$

Answers

9. $x = \dfrac{-1}{2}, \dfrac{-1}{2}, \dfrac{-1}{2}$

10. $x = \dfrac{1}{2}, \dfrac{-1}{2}, 2$

11. $x = \pm\sqrt{5}, \pm\sqrt{13}$

12. $x = \dfrac{\pm 1}{2}, \dfrac{\pm 2}{3}$

13. $x = \dfrac{\pm 1}{4}, \pm 2$

14. $x = \dfrac{\pm 1}{10}, \dfrac{\pm 1}{5}$

15. zeros at $x = \dfrac{-3}{2}, \dfrac{-1}{2}, \dfrac{1}{2}$

16. zeros at $x = \dfrac{5}{2}, \dfrac{3}{2}, -1$

17. zeros at $x = 0, \dfrac{1}{3}, \dfrac{-2}{3}, -1$

18. zeros at $x = \pm 3, \pm 5$

Lesson 7.4
Level C

1. $x = 3, -3, 3$
2. $x = \pm 4, 2$
3. $x = 0, 2, \pm 5$
4. $x = 0, -10, \pm 2$
5. $x = \pm 3, \pm\sqrt{10}$
6. $x = \pm\sqrt{15}, \pm\sqrt{2}$
7. $x = \dfrac{1}{2}, \dfrac{1}{4}, \dfrac{1}{8}$
8. $x = \dfrac{\pm 2}{5}, 0, \dfrac{-4}{5}$

9. $x = \pm\sqrt{7}, \pm\sqrt{11}$

10. $x = \dfrac{1}{6}, \dfrac{1}{2}, \dfrac{5}{12}$

11. $x = \dfrac{2}{3}, \dfrac{-1}{9}, \dfrac{4}{9}$

12. $x = \pm\sqrt{3}, \pm 5$

13. $x = -3, \dfrac{-1}{7}, \dfrac{2}{9}$

14. $x = 0, -4, \pm\sqrt{10}$

15. zero at -2

16. zeros at $0.26, 0.95$

17. zero at -0.08

18. zeros at $-1.15, 0.94$

Lesson 7.5
Level A

1. $x = 3, -2, \dfrac{1}{2}$
2. $x = 4, -1, \dfrac{1}{3}$
3. $x = 1, -2, \dfrac{2}{3}$
4. $x = -3, 2, \dfrac{-2}{3}$
5. $x = 3, -1, \dfrac{-1}{2}$
6. $x = 2, \dfrac{1}{4}, \dfrac{1}{3}$
7. $x = 3, \pm\sqrt{7}$
8. $x = \pm 2, -4$

Answers

9. $x = -4, \pm\sqrt{5}$

10. $x = 11, \pm\sqrt{11}$

11. $x = 1, \dfrac{-3 \pm \sqrt{21}}{2}$

12. $x = -2, \dfrac{-1 \pm \sqrt{38}}{2}$

13. $x = 5, -2.73, 0.73$

14. $x = -3, 2.70, -3.70$

15. $x = 2, 1.59, 4.41$

16. $P(x) = (x+2)(x-3)(x-1)$

17. $P(x) = -2(x-2)(x+3i)(x-3i)$

Lesson 7.5
Level B

1. $x = \dfrac{1}{2}, \dfrac{1}{3}, -1$

2. $x = \dfrac{-1}{3}, \dfrac{1}{4}, 2$

3. $x = \dfrac{2}{3}, \dfrac{2}{3}, -3$

4. $x = \dfrac{-2}{2}, \dfrac{2}{3}, 4$

5. $x = \dfrac{-2}{5}, \dfrac{-2}{5}, -3$

6. $x = \dfrac{-1}{3}, \dfrac{1}{4}, \dfrac{1}{2}$

7. $x = \pm 2, \pm\sqrt{5}$

8. $x = \dfrac{1}{2}, \dfrac{-3 \pm \sqrt{37}}{2}$

9. $x = \dfrac{-1}{3}, \dfrac{2 \pm \sqrt{10}}{2}$

10. $x = \dfrac{1}{4}, \dfrac{-1 \pm \sqrt{61}}{6}$

11. $x = \dfrac{1}{2}$

12. $x = \dfrac{-2}{3}$

13. $x = \dfrac{-3}{2}, \dfrac{1}{2}, \dfrac{1}{2}$

14. $x = 3, -2, -4.37, 1.37$

15. $x = -2, \dfrac{1 \pm \sqrt{17}}{4}$

16. $P(x) = \dfrac{-1}{2}x^3 + \dfrac{1}{4}x^2 - 8x + 4$

17. $P(x) = x^4 + 2x^3 - 3x^2 - 8x - 4$

Lesson 7.5
Level C

1. $x = \dfrac{1}{2}, \dfrac{-1}{2}, \dfrac{1}{3}$

2. $x = \dfrac{2}{3}, \dfrac{3}{4}, \dfrac{-1}{2}$

3. $x = \dfrac{-2}{3}, \dfrac{1}{3}, \dfrac{1}{4}$

4. $x = \dfrac{-1}{4}, \dfrac{3}{4}, \dfrac{-1}{2}$

5. $x = \dfrac{2}{3}, \dfrac{1}{3}, \dfrac{1}{3}, \dfrac{-1}{3}$

6. $x = \dfrac{1}{2}, \dfrac{-1}{3}, \dfrac{1}{4}, 1$

7. $x = \dfrac{1}{2}, 3 \pm \sqrt{5}$

Answers

8. $x = \dfrac{-1}{3}, -1 \pm 2\sqrt{2}$

9. $x = \dfrac{2}{3}, \dfrac{3 \pm \sqrt{5}}{2}, 2$

10. $x = \dfrac{-3}{2}, \dfrac{-1}{2}, \dfrac{5 \pm \sqrt{5}}{2}$

11. a. about 4.7 seconds
    b. about 107 seconds

12. a. 101 items
    b. 362 items

13. $P(x) = 12x^4 - 10x^3 + 302x^2 - 250x + 50$

14. $P(x) = 2x^4 - 2x^3 + 200.5x^2 - 200x + 50$

Answers

Lesson 8.1
Level A

1. inverse
2. combined
3. inverse
4. joint
5. $y = 9x$
6. $y = -10x$
7. $y = -27x$
8. $y = \dfrac{1}{2}x$
9. $y = 12xz$
10. $y = -9xz$
11. $y = 8xz$
12. $y = \dfrac{1}{3}xz$
13. $z = \dfrac{5xy}{w}$
14. $z = \dfrac{-4xy}{w}$
15. $z = \dfrac{2xy}{w}$
16. $z = \dfrac{xy}{2w}$
17. 672

Lesson 8.1
Level B

1. joint
2. inverse
3. joint
4. combined
5. $y = 4x; y = 52$
6. $y = 25x; y = 150$
7. $y = -4.5x; y = 18$
8. $y = \dfrac{1}{2}x; y = 4$
9. $y = -6xz; y = -108$
10. $y = 12xz; y = -252$
11. $y = -2.5xz; y = -0.75$
12. $y = \dfrac{-1}{4}xz; y = -0.03125$
13. $z = \dfrac{-3xy}{w}; z = 120$
14. $z = \dfrac{xy}{3w}; z = -1.2$
15. $z = \dfrac{xy}{10w}; z = 5$
16. $z = \dfrac{-7xy}{10w}; z = -0.21$

Lesson 8.1
Level C

1. $y = \dfrac{364}{x}; y = 30.\overline{3}$
2. $y = \dfrac{48}{x}; y = 96$
3. $y = \dfrac{636}{x}; y = -181.7$
4. $y = \dfrac{21.75}{x}; y = -9.\overline{6}$
5. $y = \dfrac{-50}{21}xz; y = -150$

Answers

6. $y = \dfrac{-19}{12} xz;\ y = 12.\overline{6}$

7. $y = \dfrac{-22}{5} xz;\ y = 198$

8. $y = -12xz;\ y = -3.84$

9. $y = \dfrac{2}{3}$

10. $x = 600$

11. $x = -2.14$

12. $y = -3.75$

13. $x = -4.96$

14. $y = -0.47$

15. a. 140.6 revolutions per minute
    b. 11.3 miles per hour

Lesson 8.2
Level A

1. yes, all real numbers except 2

2. No, $|x|$ is not a polynomial.

3. No, e^x is not a polynomial.

4. yes, all real numbers except ± 1

5. vertical asymptote at $x = -5$; horizontal asymptote at $y = 3$

6. vertical asymptote at $x = 2$; horizontal asymptote at $y = 1$

7. vertical asymptote at $x = 5$; horizontal asymptote at $y = 0$; hole at $x = -5$

8. $x \neq -2$; vertical asymptote at $x = -2$; horizontal asymptote at $y = 4$

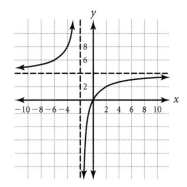

9. $x \neq 0$; vertical asymptote at $x = 0$; horizontal asymptote at $y = 1$

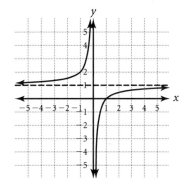

10. $x \neq -1, -2$; vertical asymptote at $x = -2$; horizontal asymptote at $y = 0$; hole at $x = -1$

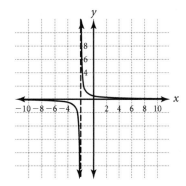

Answers

11. $x \neq \pm 4$; vertical asymptote at $x = -4$; horizontal asymptote at $y = 0$; hole at $x = 4$

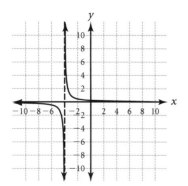

Lesson 8.2
Level B

1. yes, all real numbers except 0

2. yes, all real numbers

3. No, $|x| - 2$ is not a polynomial.

4. yes, all real numbers except $1, -6$

5. vertical asymptote at $x = \pm\sqrt{10}$; horizontal asymptote at $y = 0$

6. vertical asymptote at $x = 3$; horizontal asymptote at $y = 0$; hole at $x = -7$

7. vertical asymptote at $x = -2, -4$

8. vertical asymptote at $x = 4$; horizontal asymptote at $y = 1$; hole at $x = -1$

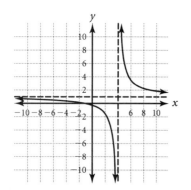

9. vertical asymptote at $x = -\frac{1}{2}$ and $\frac{1}{2}$; horizontal asymptote at $y = \frac{1}{4}$

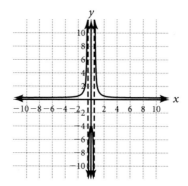

10. vertical asymptote at $x = 5$; horizontal asymptote at $y = 0$; hole at $x = -2$

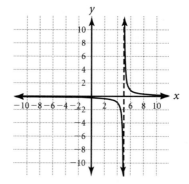

11. horizontal asymptote at $y = 0$; hole at $x = -1$

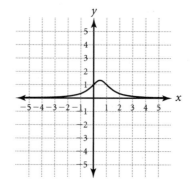

358 Practice Masters Levels A, B, and C Algebra 2

Answers

Lesson 8.2
Level C

1. yes, all real numbers except 1

2. yes, all real numbers except ± 2

3. yes, all real numbers

4. No, e^x is not a polynomial.

5. vertical asymptote at $x = -0.70, -4.30$; horizontal asymptote at $y = 3$

6. vertical asymptote at $x = 2, -3, -4$; horizontal asymptote at $y = 1$

7. vertical asymptote at $x = 1$; horizontal asymptote at $y = 0$; hole at -1

8. vertical asymptote at $x = 4.65, -0.65$

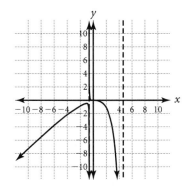

9. vertical asymptote at $x = \pm 1$

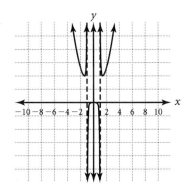

10. vertical asymptote at $x = -3$; horizontal asymptote at $y = 0$; holes at $x = -2$ and 3

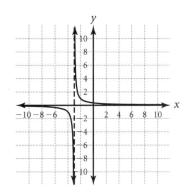

11. vertical asymptote at $x = 4$ and 5; horizontal asymptote at $y = 0$; holes at $x = -4, -5$

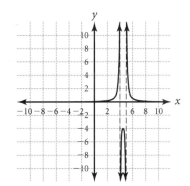

Lesson 8.3
Level A

1. $\dfrac{x+2}{x-3}$

2. $\dfrac{x+5}{x-1}$

3. $\dfrac{x-4}{x+2}$

4. $\dfrac{x+6}{x-7}$

5. $\dfrac{x^{11}}{3}$

Answers

6. $\dfrac{x^7}{4}$

7. $\dfrac{3(x+4)}{5(x-2)}$

8. $\dfrac{x-3}{4(x-4)}$

9. $\dfrac{x+5}{3x}$

10. $\dfrac{x^2}{x+6}$

11. $\dfrac{x^3}{5}$

12. $\dfrac{(x+1)}{7}$

13. $\dfrac{x+3}{x+2}$

14. $\dfrac{(x-5)}{2(x-1)}$

15. $\dfrac{x-7}{3x}$

16. $\dfrac{(x+2)}{(x+4)(x-1)}$

Lesson 8.3
Level B

1. $\dfrac{x+4}{x-4}$

2. $\dfrac{x^2-4}{x+3}$

3. $4x(x+5)$

4. $\dfrac{x(x-1)}{x+1}$

5. $\dfrac{14x}{3}$

6. $\dfrac{1}{4x^5}$

7. $\dfrac{(x+3)(x+5)}{x(x-2)}$

8. $\dfrac{(x+4)^2}{x+2}$

9. $\dfrac{6x(x-3)}{(x+1)(x+2)}$

10. $\dfrac{2x^3(x-3)}{(x+4)(x-1)}$

11. $\dfrac{(x+4)(x-6)}{x-2}$

12. $\dfrac{4(x+2)}{5(x+3)(x-6)}$

13. $\dfrac{(x+3)(x-3)}{x+4}$

14. $\dfrac{2x(x+3)}{5(x-5)}$

15. $\dfrac{3x^2(x+10)}{x-9}$

16. $\dfrac{4x(x-3)}{(x+2)(x+3)}$

Lesson 8.3
Level C

1. $\dfrac{(x+2)(x-3)}{(x+1)(x+4)}$

2. $\dfrac{6x(x-4)}{(x-5)^2}$

3. $\dfrac{1}{x^4}$

Answers

4. $\dfrac{x^2}{8}$

5. $\dfrac{3x + 2}{4x - 1}$

6. $\dfrac{2x - 3}{3x + 5}$

7. $\dfrac{2x(2x - 3)(x - 4)}{(5x + 2)(x - 2)}$

8. $x^2 - 3x + 9$

9. $\dfrac{3x}{5(7x - 2)}$

10. 1

11. $\dfrac{(2x - 5)^2}{2}$

12. $\dfrac{x + 3}{x(2x + 1)}$

13. $\dfrac{8x^2 + 18x + 7}{9x^2 + 12x - 5}$

14. $\dfrac{3x^3 - x^2}{25x^2 + 40x + 16}$

Lesson 8.4
Level A

1. $\dfrac{10}{x}$

2. $\dfrac{3}{x + 2}$

3. $\dfrac{-x}{6}$

4. $\dfrac{21 - 5x}{18}$

5. $\dfrac{-3x + 4}{30}$

6. $\dfrac{5x + 30}{(x + 2)(x - 3)}$

7. $\dfrac{-3x^2 - 75x}{(x + 5)(x - 7)}$

8. $\dfrac{2x^2 - 2x - 13}{(x + 4)(x - 5)}$

9. $\dfrac{x^3 + 14x^2 + 35x + 6}{x(x + 8)(x - 3)}$

10. $\dfrac{x^4 - 3x^3 - x^2 - 11x - 28}{x^2(x - 3)(x + 7)}$

11. $\dfrac{10x + 24}{(x + 4)(x - 4)}$

12. $\dfrac{x - 35}{(x + 5)(x + 2)(x - 3)}$

13. $\dfrac{2x^2 + 12x}{(x + 3)(x - 3)(x + 9)}$

14. $\dfrac{x - 35}{(x + 5)(x - 3)(x + 2)}$

15. $\dfrac{x^2 + 3x + 10}{(x + 2)^2(x - 2)}$

16. $\dfrac{-9}{(x + 1)(x + 10)}$

Answers

Lesson 8.4
Level B

1. $\dfrac{3x^2 + 7}{x^4}$

2. $\dfrac{x - 2}{x}$

3. $\dfrac{-3x^2 - 19x}{(x + 3)(x - 2)}$

4. $\dfrac{-4x + 92}{(x + 7)(x - 5)}$

5. $\dfrac{-x^3 - 6x^2 + 12x - 24}{x(x - 2)(x + 6)}$

6. $\dfrac{7x + 2}{(x + 2)(x - 2)}$

7. $\dfrac{2x^2 + 2x - 36}{(x + 2)(x + 7)(x - 5)}$

8. $\dfrac{9x + 14}{(x + 1)(x - 7)(x + 6)}$

9. $\dfrac{2x^2 + 5x + 6}{(x + 4)(x - 4)(x + 2)}$

10. $\dfrac{22x + 80}{(x + 5)(x + 10)(x - 10)}$

11. $\dfrac{-5x - 22}{x(x + 2)(x + 3)(x - 4)}$

12. $\dfrac{2x^2 + 2x - 10}{x(x + 1)^2(x + 11)}$

13. $\dfrac{-14}{(x - 3)(x + 2)}$

14. $\dfrac{2x^3 - 3x^2 - 39x - 12}{(x + 3)(x + 1)(x - 5)}$

15. $\dfrac{5x + 2}{x^2}$

16. $\dfrac{4x^2 + 16x + 13}{3(x + 2)(x - 1)}$

Lesson 8.4
Level C

1. $\dfrac{3x^3 - 17x^2 - 56x - 32}{2(x + 4)(x + 1)(x - 4)}$

2. $\dfrac{x^3 + 7x^2 + 36x + 90}{2(x + 3)(x - 3)(x + 4)}$

3. $\dfrac{x^2 + 4y}{xy}$

4. $2x^3y^2 + xy^3$

5. $\dfrac{8x^2 + 44x - 20}{(x - 5)(x + 5)}$

6. $\dfrac{x - y}{x + y}$

7. $\dfrac{y + x}{xy}$

8. $\dfrac{-3}{(x + 1)(x - 2)}$

9. $\dfrac{2x^2 - 14x + 25}{(x - 3)^2(x - 4)^2}$

10. $\dfrac{2x^3 + 3x^2 + 3x + 1}{x^3(x + 1)^3}$

11. $A = x + 4;\ B = (x + 1)(x - 2)(x + 2)$

12. $A = 2a^2 - 4ab;\ B = (a - 3b)^2$

13. $A = x + 3;\ B = (x + 1)(x - 2)(x + 4)$

14. $A = 6x^3 + 16x^2 - 26x + 6$

Answers

Lesson 8.5
Level A

1. $x = 8$
2. $x = 2$
3. $x = \dfrac{-1}{4}$
4. $x = \dfrac{8}{11}$
5. $x = 5 \pm \sqrt{41}$
6. $x = \dfrac{11 \pm \sqrt{129}}{4}$
7. $x = 5$
8. $x = 5$
9. $x = 1, \dfrac{-2}{3}$
10. $x = 5.3$
11. $x > 1.2$
12. $x < -0.25$ or $x > 0$
13. $1.6 < x < 2$
14. $-4 < x < 0$
15. $x \leq 0$ or $x \geq \dfrac{3}{2}$
16. $x < 2$ or $x > 2.8$
17. $-1.52 < x < 0$
18. all real numbers
19. $x > \dfrac{-1}{4}$
20. $-0.4 < x < 0.5$ or $1.9 < x < 2$

Lesson 8.5
Level B

1. $x = 1$
2. no solution
3. $x = -2.6$
4. $x = -2.9$
5. all real numbers
6. $x = 0.9, -6.4$
7. $x = 0$
8. $x = -0.4, -2, 2.4$
9. $x = -5.9, -1.4, 1.2$
10. $x = -1$
11. $-3.3 < x < 0$ or $x > 5$
12. $x > -2.\overline{3}$
13. $x < 3.2$
14. all real numbers
15. no solution
16. $-1.5 < x < 1.5$
17. $x < -1.9$
18. $x < 0.79$
19. $x < -0.9$ or $x > 2.2$
20. $-3.6 < x < 2.2$

Lesson 8.5
Level C

1. sometimes
2. sometimes
3. never
4. always
5. $x < -0.8, 4 < x < 4.1$

Answers

6. $x < 0, 0.6 < x < 5.5$

7. $x < -0.8$

8. a. 4051 km
   b. 5285 km

9. a. $T(s) = \dfrac{15}{s} + \dfrac{40}{s} + \dfrac{6.2}{s-25}$
   b. 33.8 mph

Lesson 8.6
Level A

1. $x \geq \dfrac{1}{3}$

2. $x \geq 2$

3. $x \geq 3$ or $x \leq -3$

4. all real numbers

5.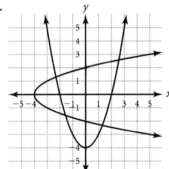

6.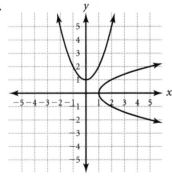

7.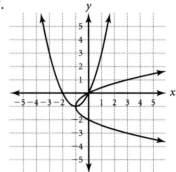

8. $\dfrac{5}{3}$

9. $\dfrac{-1}{2}$

10. 9

11. -6

12. $\dfrac{3}{5}$

13. $\dfrac{1}{5}$

14. vertical translation 2 units down
    horizontal translation 3 units left

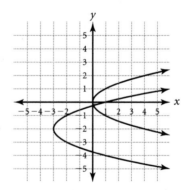

Lesson 8.6
Level B

1. $x \geq -2$ or $x \leq -8$

2. $x \geq 3$ or $x \leq -1$

Answers

3. $x \geq 1$ or $x \leq -5$

4. all real numbers

5.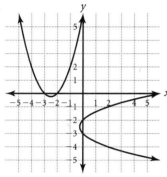

6.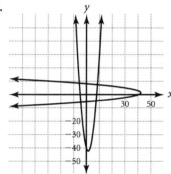

7.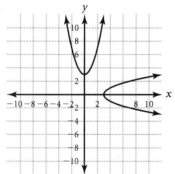

8. -5

9. $\dfrac{-1}{5}$

10. $\dfrac{-1}{2}$

11. 33

12. -32

13. 144

14. horizontal translation 2 units right
horizontal compression by a factor of $\dfrac{1}{3}$

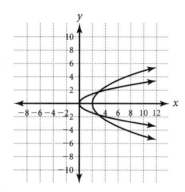

15. vertical translation 2 units down
horizontal translation 2.5 units left
horizontal compression by a factor of $\dfrac{1}{4}$

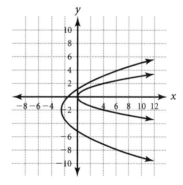

Answers

Lesson 8.6
Level C

1. a. $x \geq 3$ or $x \leq 1$
   b. $y = 2 \pm \sqrt{1 + x^2}$
   c.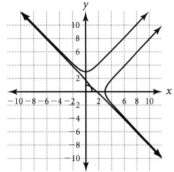

2. a. all real numbers
   b. $y = -5 \pm x$
   c.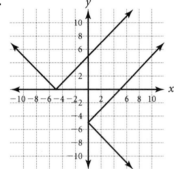

3. a. $x > 10$ or $x < -3$
   b. $y = \dfrac{7 \pm \sqrt{169 + 4x^2}}{2}$
   c.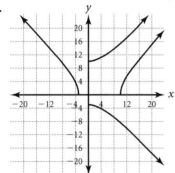

4. a. $r = 5$
   b. $r = 12.41$

5. $-2\sqrt[3]{15} + 2\sqrt[3]{50}$

6. $2\sqrt[4]{9}$

7. 518,400

8. $3\sqrt[3]{75} + 4\sqrt[3]{45} + 4$

9. vertical transformation 3 units down
   horizontal translation 2.5 units left
   vertical compression by factor of $\dfrac{1}{4}$
   horizontal compression by factor of $\dfrac{1}{2}$

10. vertical stretch by factor of 6 reflected across x-axis
    horizontal compression by factor of $\dfrac{1}{2}$ reflected across y-axis
    vertical translation 5 units up
    horizontal translation 2 units right

Lesson 8.7
Level A

1. $2\sqrt{10}$

2. $4\sqrt{2}$

3. $2\sqrt[3]{7}$

4. $3x\sqrt[3]{2x^2}$

5. $2|x||y|\sqrt[4]{2xy^2}$

6. $-4x^3y^2$

7. $2x^6\sqrt{3}$

8. $5x^3|y|$

9. $3x^3y^2\sqrt[3]{y}$

10. $5x^2y^2\sqrt[3]{2x}$

11. $3\sqrt{2} - 3\sqrt{3} + 2\sqrt{6} - 6$

12. $-33 - 18\sqrt{2}$

Answers

13. $38 - 3\sqrt{3}$
14. $2 - 10\sqrt{2}$
15. $15\sqrt{2} - 40\sqrt{6}$
16. $-4 + 11\sqrt{5}$
17. $\dfrac{5\sqrt{3}}{3}$
18. $\dfrac{9\sqrt{2}}{2}$
19. $\dfrac{4 - \sqrt{2}}{7}$
20. $\dfrac{-9 - 3\sqrt{5}}{2}$

Lesson 8.7
Level B

1. $4\sqrt{3}$
2. $6\sqrt{2}$
3. $-2\sqrt[3]{6}$
4. $3|x|\sqrt[4]{2x^3}$
5. $4xy\sqrt[3]{2x^2y}$
6. $2xy^2\sqrt[5]{2x}$
7. $2x^3\sqrt[3]{3}$
8. $-343x$
9. $5x^3y^4\sqrt[3]{x}$
10. $2|x|y^2\sqrt[4]{2xy}$
11. $7 + \sqrt{5}$
12. $1 - 8\sqrt{3}$
13. $-48 - 10\sqrt{5}$
14. $-26 - 75\sqrt{3}$
15. $-156 - 21\sqrt{35}$

16. $12\sqrt{6} + 72\sqrt{2} - 324$
17. $\dfrac{2\sqrt{5}}{5}$
18. $\dfrac{5\sqrt{5} + 30}{-31}$
19. $\dfrac{3\sqrt{11} - 3\sqrt{5}}{2}$
20. $\dfrac{-8\sqrt{6} - 8\sqrt{13}}{7}$

Lesson 8.7
Level C

1. $2\sqrt[5]{2}$
2. $10|y|\sqrt[6]{x^5y^2}$
3. $10x^2y^3\sqrt[5]{4x^2}$
4. $-xy^2z\sqrt[5]{y^3z^3}$
5. $10x^{30}\sqrt[4]{50}$
6. $-2x^6y^{12}\sqrt[5]{4x^2y^4}$
7. $5x^2y^3$
8. $243x^2y^6\sqrt[3]{3x^2y^2}$
9. $5x^2y^4\sqrt[3]{2y}$
10. $x^3y\sqrt[4]{2}$
11. $17\sqrt{5} - 2\sqrt{10} - 68 + 8\sqrt{2}$
12. $-2 + 4\sqrt{5}$
13. $x^2y^3\sqrt[3]{9y}$
14. $96\sqrt{2}$
15. $24\sqrt{3} + 24\sqrt{15}$
16. $50\sqrt[3]{2} - 5\sqrt[3]{4}$
17. $\dfrac{30}{11} - \dfrac{68\sqrt{5}}{11}$

Answers

18. $\dfrac{-7}{2} + \dfrac{21\sqrt{6}}{4}$

19. $\dfrac{36}{13} - \dfrac{9\sqrt{3}}{13}$

20. $\dfrac{-67}{34} - \dfrac{91\sqrt{2}}{68}$

Lesson 8.8
Level A

1. $x = 19$
2. $x = 139$
3. $x = 5$
4. $x = 21$
5. $x = 32$
6. no solution
7. $x \geq 23$
8. $3 \leq x \leq 67$
9. $\dfrac{-5}{2} \leq x < 142$
10. $x > 57$
11. no solution
12. $6 \leq x < 22$
13. $x = 2.3$
14. $x = 3.4$
15. $x = \pm 1.4$
16. no solution
17. $x \leq -2.8$ and $0 \leq x \leq 2.8$
18. $-2.6 \leq x \leq 2.6$

Lesson 8.8
Level B

1. $x = 2$
2. $x = \dfrac{54}{7}$
3. $x = 1$
4. no solution
5. $x = \dfrac{13 + \sqrt{33}}{2}$
6. $x = \dfrac{13 + \sqrt{93}}{2}$
7. $x > 2$
8. $\dfrac{-1}{4} \leq x \leq 1$
9. $x \geq 3$
10. $-5 \leq x \leq 5$
11. $x \geq \dfrac{1}{2}$
12. $\dfrac{1}{3} \leq x \leq \dfrac{19}{54}$
13. $x = 4.7$
14. $x = 4.5$
15. $x = 7.1$
16. $\dfrac{1}{2} \leq x < 3$
17. $0 \leq x < 4.5$
18. $\dfrac{-11}{2} \leq x \leq -1$ and $x \geq 1.3$

Answers

Lesson 8.8
Level C

1. $x = \dfrac{1 \pm \sqrt{37}}{9}$

2. $x = \dfrac{13}{14}$

3. $x = \dfrac{1 + 3\sqrt{19}}{4}$

4. $x = \dfrac{-8}{3}$

5. no solution

6. $x = \dfrac{-1}{2}$

7. $0 \le x \le 1$ and $4 \le x \le 5$

8. $x < 3 - 2\sqrt{6}$ and $x > 3 + 2\sqrt{6}$

9. $x \ge \dfrac{-3 + \sqrt{353}}{2}$ and $x \le \dfrac{-3 - \sqrt{353}}{2}$

10. $-7.1 < x \le -5.2$ and $1.2 \le x < 3.1$

11. no solution

12. $-4 \le x \le 0$ and $x \ge 5$

13. $-2.3 < x < 1.3$

14. $x \ge -0.6$

15. no solution

16. $x > 8.4$

17. all real numbers

18. $x \ge -0.9$

Answers

Lesson 9.1
Level A

1. $y = \pm\sqrt{16 - x^2}$; circle

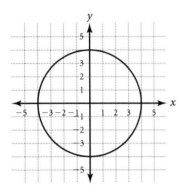

2. $y = 4x^2$; parabola

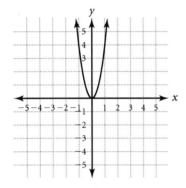

3. $y = \pm\sqrt{\dfrac{12 - 3x^2}{6}}$; ellipse

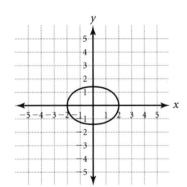

4. $5; \left(3\dfrac{1}{2}, 6\right)$

5. $13; \left(7\dfrac{1}{2}, 9\right)$

6. $7.07; \left(-2\dfrac{1}{2}, -2\dfrac{1}{2}\right)$

7. $7.62; \left(\dfrac{-1}{2}, \dfrac{1}{2}\right)$

8. $6.08; \left(-5\dfrac{1}{2}, -7\right)$

9. $13.89; \left(\dfrac{1}{2}, -1\right)$

10. $(7, 9); C = 10\pi; A = 25\pi$

11. $\left(4\dfrac{1}{2}, 9\right); C = 15\pi; A = \dfrac{225\pi}{4}$

12. $\left(\dfrac{-3}{2}, 0\right); C = 13\pi; A = \dfrac{169\pi}{4}$

13. $\left(9, 9\dfrac{1}{2}\right); C = 17\pi; A = \dfrac{289\pi}{4}$

Lesson 9.1
Level B

1. $y = \dfrac{1}{4}x^2$; parabola

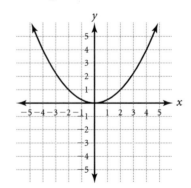

Answers

2. $y = \pm\sqrt{\dfrac{10 - 4x^2}{5}}$; ellipse

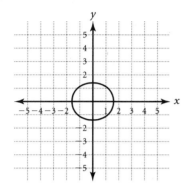

3. $y = \pm\sqrt{\dfrac{3x^2 - 6}{2}}$; hyperbola

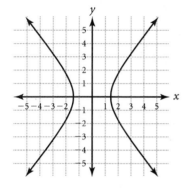

4. $17; \left(-8, -19\dfrac{1}{2}\right)$

5. $4.24; \left(\dfrac{-1}{2}, \dfrac{11}{2}\right)$

6. $1.41; \left(0, 5\dfrac{1}{2}\right)$

7. $7.16; \left(4\dfrac{1}{4}, 1\dfrac{1}{2}\right)$

8. $5.70; \left(\dfrac{3}{4}, \dfrac{-3}{4}\right)$

9. $5.39; \left(\dfrac{3}{2}, 1\right)$

10. $\left(-2\dfrac{1}{2}, \dfrac{-1}{2}\right); C = \pi\sqrt{82}; A = \dfrac{41\pi}{2}$

11. $\left(\dfrac{-1}{2}, \dfrac{1}{2}\right); C = \pi\sqrt{146}; A = \dfrac{73\pi}{2}$

12. $(3, -4); C = \pi\sqrt{13}; A = \dfrac{13\pi}{4}$

13. $\left(\dfrac{-5}{4}, \dfrac{7}{4}\right); C = \dfrac{5\pi\sqrt{2}}{2}; A = \dfrac{25\pi}{8}$

Lesson 9.1
Level C

1. $y = \pm\sqrt{\dfrac{8 - x^2}{6}}$; ellipse

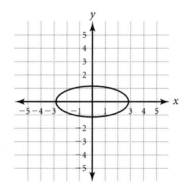

2. $y = \pm\sqrt{\dfrac{2x^2 - 10}{4}}$; hyperbola

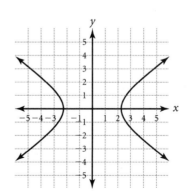

Answers

3. $y = \dfrac{-4}{3}x^2$; parabola

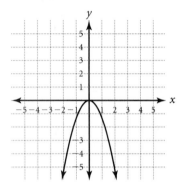

4. yes; $\left(3\dfrac{1}{2}, -1\right)$

5. no

6. yes; $\left(2, 3\dfrac{1}{2}\right)$

7. yes; $\left(-2\dfrac{1}{2}, 5\dfrac{1}{2}\right)$

8. yes

9. no

10. yes

11. yes

12. 26 square units

Lesson 9.2
Level A

1. $y - 3 = \dfrac{1}{4}x^2$

2. $y + 2 = \dfrac{-1}{4}x^2$

3. $y + 1 = \dfrac{1}{8}(x+3)^2$

4.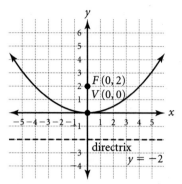

5.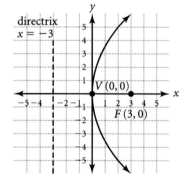

6.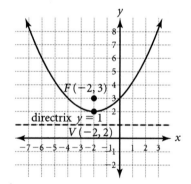

7. $y = \dfrac{1}{20}x^2$

8. $x = \dfrac{-1}{16}y^2$

9. $x + 3 = \dfrac{1}{12}y^2$

10. $y - 4 = -\dfrac{1}{20}x^2$

Answers

11. $y - 7 = \dfrac{1}{24}(x - 2)^2$

12. $x - 5 = \dfrac{1}{10}(y - 1)^2$

Lesson 9.2
Level B

1. $y - 4 = \dfrac{1}{4}(x + 2)^2$

2. $x + 5 = \dfrac{-1}{8}(y - 3)^2$

3. $y - 2 = \dfrac{-1}{16}(x + 1)^2$

4.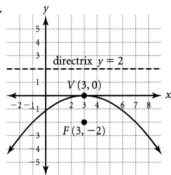

5.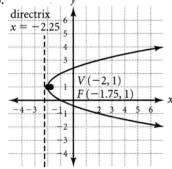

6.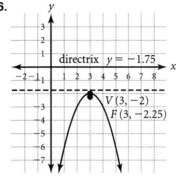

7. $y - 2 = \dfrac{-1}{20}(x + 3)^2$

8. $x - 5 = \dfrac{1}{24}(y - 1)^2$

9. $y + 1 = \dfrac{1}{12}(x - 4)^2$

10. $y + 1 = \dfrac{1}{8}(x + 5)^2$

11. $x - 4 = \dfrac{-1}{4}(y - 5)^2$

12. $y + 2 = \dfrac{1}{32}(x - 1)^2$

Lesson 9.2
Level C

1. $y - 3 = \dfrac{1}{2}(x - 1)^2$

2. $x + 1 = (y + 2)^2$

3. $y - 3 = -(x - 3)^2$

Answers

4.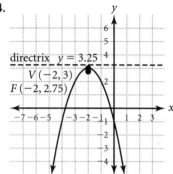

5.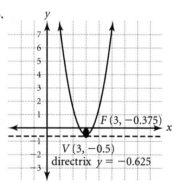

6.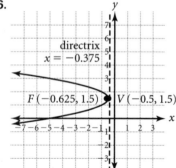

7. $y + 1 = \left(x - \dfrac{5}{2}\right)^2$

8. $y - \dfrac{3}{2} = \dfrac{1}{8}\left(x - \dfrac{5}{2}\right)^2$

9. $x - \dfrac{9}{4} = \dfrac{1}{3}\left(y - \dfrac{1}{2}\right)^2$

10. $x - \dfrac{1}{2} = \dfrac{-1}{8}\left(y - \dfrac{7}{2}\right)^2$

Lesson 9.3
Level A

1. $x^2 + y^2 = 16$
2. $(x + 2)^2 + (y - 1)^2 = 4$
3. $(x + 1)^2 + (y + 2)^2 = 1$
4. $x^2 + y^2 = 9$
5. $(x - 2)^2 + (y - 1)^2 = 25$
6. $(x + 3)^2 + (y - 4)^2 = 16$
7. $C = (0, 0)$

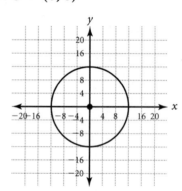

8. $C = (3, 1)$

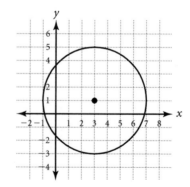

Answers

8. $C = (-2, -2)$

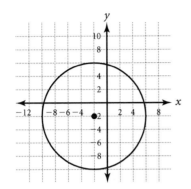

10. $(x + 3)^2 + y^2 = 34$;
    $C(-3, 0); r = \sqrt{34}$

11. $x^2 + (y + 2)^2 = 16$;
    $C(0, -2); r = 4$

Lesson 9.3
Level B

1. $(x - 4)^2 + y^2 = 12.25$

2. $(x + 2)^2 + (y - 3)^2 = \dfrac{9}{4}$

3. $(x - 3)^2 + (y - 2)^2 = 6\dfrac{1}{4}$

4. $\left(x - \dfrac{1}{2}\right)^2 + (y - 4)^2 = 4$

5. $(x + 2)^2 + \left(y - \dfrac{3}{2}\right)^2 = \dfrac{9}{16}$

6. $(x - 0.6)^2 + (y + 0.3)^2 = 0.49$

7. $C(2, -1)$

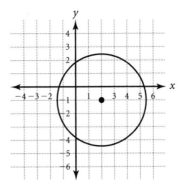

8. $C = \left(\dfrac{-1}{2}, \dfrac{1}{2}\right)$

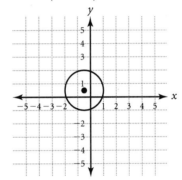

9. $C = \left(2\dfrac{1}{2}, 3\dfrac{1}{2}\right)$

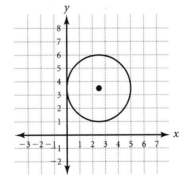

10. $(x + 2)^2 + (y - 4)^2 = 36$;
    $C(-2, 4); r = 6$

11. $\left(x - \dfrac{3}{2}\right)^2 + \left(y - \dfrac{5}{2}\right)^2 = 9$;
    $C\left(\dfrac{3}{2}, \dfrac{5}{2}\right); r = 3$

Answers

Lesson 9.3
Level C

1. $\left(x+\dfrac{3}{2}\right)^2 + \left(y-\dfrac{1}{2}\right)^2 = \dfrac{9}{4}$

2. $\left(x-\dfrac{3}{2}\right)^2 + \left(y+\dfrac{5}{2}\right)^2 = \dfrac{1}{4}$

3. $\left(x+\dfrac{5}{2}\right)^2 + \left(y-\dfrac{3}{2}\right)^2 = \dfrac{49}{4}$

4. circle
5. parabola
6. circle
7. on
8. inside
9. outside

10.
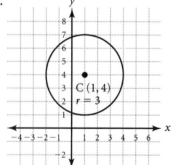

11.
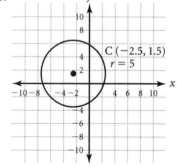

Lesson 9.4
Level A

1. $\dfrac{x^2}{9} + \dfrac{y^2}{4} = 1$

2. $\dfrac{(x-2)^2}{16} + \dfrac{y^2}{25} = 1$

3. $(x+1)^2 + \dfrac{(y-1)^2}{16} = 1$

4. Center: $(0, 0)$; Vertices: $(\pm 4, 0)$; Co-Vertices: $(0, \pm 2)$; Foci: $\left(\pm 2\sqrt{3}, 0\right)$

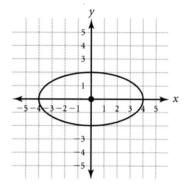

5. Center: $(0, 0)$; Vertices: $(0, \pm 3)$; Co-Vertices: $(\pm 1, 0)$; Foci: $\left(0, 2\sqrt{2}\right)$

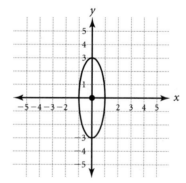

Answers

6. Center: (2, 1); Vertices: (5, 1) and (−1, 1); Co-Vertices: (2, 3) and (2, −1); Foci: $(2 \pm \sqrt{5}, 1)$

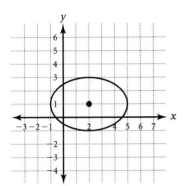

7. $\dfrac{x^2}{81} + \dfrac{y^2}{49} = 1$

8. $\dfrac{x^2}{25} + \dfrac{y^2}{144} = 1$

9. $\dfrac{x^2}{144} + \dfrac{y^2}{144} = 1$

10. $\dfrac{x^2}{16} + \dfrac{y^2}{52} = 1$

Lesson 9.4
Level B

1. $\dfrac{(x-2)^2}{25} + \dfrac{(y+1)^2}{9} = 1$

2. $\dfrac{(x+3)^2}{9} + \dfrac{(y-2)^2}{16} = 1$

3. $\dfrac{(x-4)^2}{36} + \dfrac{(y-2)^2}{4} = 1$

4. Center: (−3, −4); Vertices: (1, −4) and (−7, −4); Co-Vertices: $(-3, -4 \pm \sqrt{10})$; Foci: $(-3 \pm \sqrt{6}, -4)$

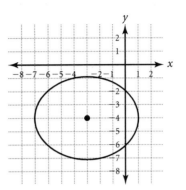

5. Center: (2, 4); Vertices: (2, 7) and (2, 1); Co-Vertices: $(2 \pm \sqrt{6}, 4)$; Foci: $(2, 4 \pm \sqrt{3})$

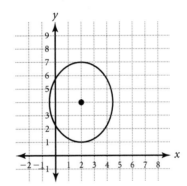

6. Center: (0, 2); Vertices: $(\pm 2\sqrt{3}, 2)$; Co-Vertices: (0, 4) and (0, 0); Foci: $(\pm 2\sqrt{2}, 2)$

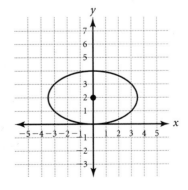

Algebra 2   Practice Masters Levels A, B, and C   **377**

Answers

7. $\dfrac{(x+2)^2}{12} + \dfrac{(y-1)^2}{4} = 1$

8. $\dfrac{(x+2)^2}{100} + \dfrac{(y-3)^2}{25} = 1$

9. $\dfrac{(x-1)^2}{4} + \dfrac{(y+1)^2}{9} = 1$

10. $\dfrac{(x-1)^2}{24} + \dfrac{(y+2)^2}{6} = 1$

Lesson 9.4
Level C

1. $\dfrac{x^2}{0.25} + \dfrac{y^2}{6.25} = 1$

2. $\dfrac{x^2}{2.25} + \dfrac{y^2}{4} = 1$

3. $\dfrac{x^2}{12.25} + \dfrac{y^2}{6.25} = 1$

4. Center: $(0, 0)$; Vertices: $(\pm 2, 0)$; Co-Vertices: $(0, \pm 1.5)$; Foci: $\left(\dfrac{\pm\sqrt{7}}{2}, 0\right)$

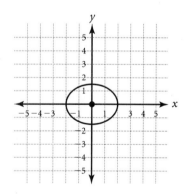

5. Center: $(0, 0)$; Vertices: $(\pm 2.5, 0)$; Co-Vertices: $(0, \pm 0.5)$; Foci: $(\pm\sqrt{6}, 0)$

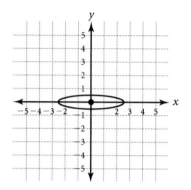

6. Center: $(0, 0)$; Vertices: $(0, \pm 3.5)$; Co-Vertices: $(\pm 1.5, 0)$; Foci: $(0, \pm\sqrt{10})$

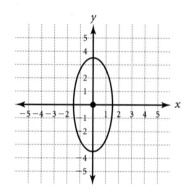

7. $\dfrac{\left(x+\dfrac{3}{2}\right)^2}{20} + \dfrac{(y-1)^2}{5} = 1$

8. $\dfrac{\left(x-\dfrac{3}{2}\right)^2}{4} + \dfrac{\left(y-\dfrac{7}{2}\right)^2}{8} = 1$

9. $\dfrac{(x+3)^2}{15} + \dfrac{(y-2)^2}{9} = 1$

10. $\dfrac{\left(x-\dfrac{3}{2}\right)^2}{0.4} + \dfrac{\left(y+\dfrac{3}{2}\right)^2}{0.\overline{6}} = 1$

Answers

Lesson 9.5
Level A

1. $\dfrac{x^2}{4} - \dfrac{y^2}{9} = 1$

2. $\dfrac{y^2}{9} - \dfrac{x^2}{16} = 1$

3. Center: $(0, 0)$; Vertices: $(\pm 2, 0)$; Co-Vertices $(0, \pm 4)$; Foci $(\pm 2\sqrt{5}, 0)$

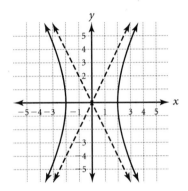

4. Center: $(0, 0)$; Vertices: $(0, \pm 3)$; Co-Vertices $(\pm 5, 0)$; Foci $(0, \pm\sqrt{34})$

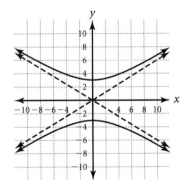

5. $\dfrac{x^2}{25} - \dfrac{y^2}{36} = 1$

6. $\dfrac{y^2}{16} - \dfrac{x^2}{20} = 1$

7. $\dfrac{y^2}{15} - \dfrac{x^2}{49} = 1$

8. $\dfrac{x^2}{100} - \dfrac{y^2}{44} = 1$

Lesson 9.5
Level B

1. $\dfrac{(x+1)^2}{25} - \dfrac{(y-2)^2}{16} = 1$

2. $\dfrac{(y+3)^2}{9} - \dfrac{(x-1)^2}{4} = 1$

3. Center: $(-2, 1)$;
   Vertices $(-3, 1), (-1, 1)$;
   Co-vertices $(-2, 3), (-2, -1)$;
   Foci $(-2 \pm \sqrt{5}, 1)$

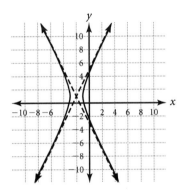

4. Center: $(-2, -1)$;
   Vertices $(-2, -5), (-2, 3)$;
   Co-vertices $(-5, -1), (1, -1)$;
   Foci: $(-2, 4), (-2, -6)$

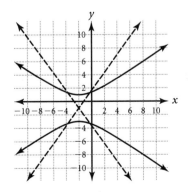

Answers

5. $\dfrac{(x+3)^2}{4} - (y-2)^2 = 1$

6. $\dfrac{(x-1)^2}{3} - \dfrac{(y-5)^2}{12} = 1$

7. $\dfrac{(y-4)^2}{10} - \dfrac{(x-3)^2}{5} = 1$

8. $\dfrac{(y-2)^2}{3} - \dfrac{(x+2)^2}{8} = 1$

Lesson 9.5
Level C

1. $\dfrac{(x+4)^2}{0.25} - \dfrac{(y-3)^2}{2.25} = 1$

2. $\dfrac{(y+3)^2}{6.25} - \dfrac{(x-1)^2}{2.25} = 1$

3. Center: (3, 1); Vertices (1.5, 1), (4.5, 1); Co-vertices (3, −1.5), (3, 3.5); Foci: $\left(3 \pm \sqrt{\dfrac{17}{2}}, 1\right)$

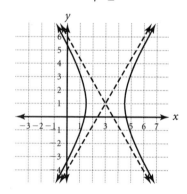

4. Center: (2, −1); Vertices (2, 0.5), (2, −2.5); Co-vertices (2.5, −1), (1.5, −1); Foci: $\left(2, -1 \pm \sqrt{2.5}\right)$

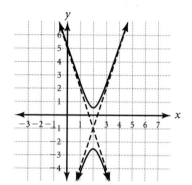

5. $\dfrac{(x+3)^2}{4} - \dfrac{(y-3)^2}{2} = 1$

6. $\dfrac{(y-6)^2}{9} - \dfrac{(x+5)^2}{15} = 1$

7. $\dfrac{\left(x+\dfrac{3}{2}\right)^2}{5} - \dfrac{\left(y-\dfrac{5}{2}\right)^2}{6} = 1$

8. $\dfrac{\left(y-\dfrac{7}{2}\right)^2}{18} - \dfrac{\left(x+\dfrac{3}{2}\right)^2}{8} = 1$

Lesson 9.6
Level A

1. (0, 0), (4, 16)
2. (1, 1), (4, −2)
3. (2, 4), (−2, 4)
4. (2, 1), (−2, 1), (2, −1), (−2, −1)
5. (1, 3), (1, −3), (−1, 3), (−1, −3)
6. (3, 2), (3, −2), (−3, 2), (−3, −2)
7. (2, 0), (−2, 0)
8. (5, 3), (5, −3), (−5, 3), (−5, −3)

Answers

9. (0, 0), (2, 4)
10. circle
11. parabola
12. hyperbola
13. ellipse
14. parabola
15. ellipse

Lesson 9.6
Level B

1. (3, 4), (−3, 4), ($\sqrt{4.75}$, −4.5), (−$\sqrt{4.75}$, −4.5)
2. (1, 4), (1, −4)
3. (−2, 3), (1.19, 2.42)
4. (3, 5), (3, −5), (−3, 5), (−3, −5)
5. (1, 5), (1, −5), (−1, 5), (−1, −5)
6. (±5, 0)
7. (2, 0), (−2, 0)
8. (0, 3)
9. none
10. parabola
11. hyperbola
12. ellipse

13. circle
14. parabola
15. hyperbola

Lesson 9.6
Level C

1. (2, 3), (2, −3), (−2.3, 3.2), (−2.3, −3.2)
2. (2, 4), (2, −4), (−2, 4), (−2, −4)
3. $\left(\frac{1}{2}, \frac{11}{4}\right), \left(\frac{-1}{2}, \frac{11}{4}\right)$
4. $\left(\frac{1}{4}, \frac{3}{2}\right), \left(\frac{1}{4}, \frac{-3}{2}\right)$
5. $\left(\frac{1}{2}, \frac{1}{2}\right), \left(\frac{1}{2}, \frac{-1}{2}\right), \left(\frac{-1}{2}, \frac{1}{2}\right), \left(\frac{-1}{2}, \frac{-1}{2}\right)$
6. (± 3, 0)
7. (0.96, 2.84), (−2.59, 1.52)
8. none
9. (−2.91, −0.95), (2.96, 0.64)
10. hyperbola
11. circle
12. parabola
13. ellipse
14. hyperbola
15. parabola

Algebra 2   Practice Masters Levels A, B, and C

Answers

Lesson 10.1
Level A

1. $\dfrac{2}{5}$
2. $\dfrac{7}{20}$
3. $\dfrac{1}{4}$
4. $\dfrac{1}{12}$
5. $\dfrac{1}{12}$
6. $\dfrac{1}{2}$
7. $\dfrac{1}{2}$
8. $\dfrac{5}{12}$
9. $\dfrac{3}{4}$
10. $\dfrac{7}{12}$
11. $\dfrac{11}{12}$
12. 1
13. $\dfrac{2}{3}$
14. $\dfrac{33}{60}$
15. $\dfrac{1}{60}$
16. 960
17. 96
18. 450

Lesson 10.1
Level B

1. $\dfrac{1}{4}$
2. $\dfrac{1}{2}$
3. $\dfrac{3}{13}$
4. $\dfrac{1}{13}$
5. $\dfrac{1}{52}$
6. $\dfrac{1}{26}$
7. $\dfrac{12}{13}$
8. $\dfrac{2}{13}$
9. $\dfrac{11}{13}$
10. $\dfrac{11}{21}$
11. $\dfrac{3}{35}$
12. $\dfrac{1}{5}$
13. $\dfrac{\pi}{4} \approx 78.5\%$

Answers

14. 540

15. 5,184,000

**Lesson 10.1
Level C**

1. $\dfrac{21}{26}$

2. $\dfrac{5}{13}$

3. 0

4. 1

5. $\dfrac{3}{26}$

6. $\dfrac{3}{13}$

7. $\dfrac{67}{380}$

8. $\dfrac{313}{380}$

9. $\dfrac{1}{2}$

10. $\dfrac{1}{4}$

11. $\dfrac{93}{95}$

12. $\dfrac{97}{380}$

13. about 21.5%

14. 21,233,664

15. 7,077,888

**Lesson 10.2
Level A**

1. 3,628,800

2. 30,240

3. 3,628,800

4. 720

5. 5040

6. 604,800

7. 1,663,200

8. 1.0×10^{10}

9. 24

10. 120

11. 720

12. 5040

13. 40,320

14. 10,080

15. 1000

16. 39,916,800

17. 479,001,600

18. 5040

**Lesson 10.2
Level B**

1. 100

2. 10,000

3. 1,000,000

4. 10,000,000,000

5. 3,628,800

6. 0

7. 39,916,800

Answers

8. 6.7×10^{11}

9. 181,440

10. 4,989,600

11. 19,958,400

12. 129,729,600

13. 59,875,200

14. 34,650

15. 657,720

16. 1.0×10^{10}

17. 4.1×10^{16}

18. 2.2×10^{30}

19. 30

20. 24,360

**Lesson 10.2
Level C**

1. 358,800

2. 165,765,600

3. 6.5×10^{16}

4. 4.0×10^{26}

5. 6840

6. 310,080

7. 465,120

8. 3,488,400

9. a. 24
   b. $\frac{1}{4}$
   c. $\frac{1}{2}$

10. a. 518,400
    b. 43,200

11. 900,000,000

12. 362,880

13. 540,540

14. 2.9×10^{10}

**Lesson 10.3
Level A**

1. 20

2. 56

3. 21

4. 1

5. 10

6. 2970

7. 120

8. 56

9. 1

10. 924

11. 90

12. 15

13. 30

14. 1

15. 660

16. 660

17. 5

18. 924

19. permutation

20. combination

Answers

**Lesson 10.3
Level B**

1. 11,760
2. 3.75
3. $73.\overline{3}$
4. 1323
5. 18,370.625
6. 9,018,009
7. 5.4×10^{20}
8. 4.9×10^{12}
9. 495
10. 1.2×10^{37}
11. 60
12. 60
13. 20
14. 120
15. 0.03
16. 0.00043
17. 0.00034
18. 0.13
19. combination
20. permutation

**Lesson 10.3
Level C**

1. 1326
2. 0.005
3. 2,598,960
4. 0.000002
5. 0.000009
6. 133,784,560
7. 78,960,960
8. 658,008
9. 0.000002
10. a. 161,700
    b. 161,700
    c. The number of ways taking 100 items three at a time (part a) would be the same as not taking 100 items three at a time (part b).
11. permutation
12. combination
13. permutation
14. combination
15. combination

**Lesson 10.4
Level A**

1. mutually exclusive; $\frac{2}{9}$
2. mutually exclusive; $\frac{2}{9}$
3. inclusive; $\frac{13}{36}$
4. mutually exclusive; $\frac{13}{36}$
5. inclusive; 1
6. inclusive; $\frac{1}{12}$
7. mutually exclusive; $\frac{7}{12}$
8. inclusive; $\frac{3}{4}$

Answers

9. inclusive; $\dfrac{7}{9}$

10. mutually exclusive; $\dfrac{13}{36}$

11. $\dfrac{8}{25}$

12. $\dfrac{4}{25}$

13. $\dfrac{11}{25}$

14. $\dfrac{4}{5}$

15. $\dfrac{3}{25}$

16. $\dfrac{11}{25}$

Lesson 10.4
Level B

1. inclusive; $\dfrac{7}{13}$

2. inclusive; $\dfrac{11}{26}$

3. mutually exclusive; 1

4. inclusive; 1

5. mutually exclusive; $\dfrac{21}{26}$

6. inclusive; $\dfrac{10}{13}$

7. mutually exclusive; $\dfrac{27}{52}$

8. mutually exclusive; 1

9. $\dfrac{5}{8}$

10. $\dfrac{1}{2}$

11. $\dfrac{1}{2}$

12. $\dfrac{5}{8}$

13. $\dfrac{3}{4}$

14. $\dfrac{7}{8}$

15. $\dfrac{5}{8}$

16. $\dfrac{1}{2}$

17. $\dfrac{1}{2}$

18. $\dfrac{7}{8}$

Lesson 10.4
Level C

1. mutually exclusive; $\dfrac{1}{2}$

2. inclusive; 1

3. mutually exclusive; $\dfrac{5}{8}$

4. inclusive; $\dfrac{11}{16}$

5. $\dfrac{5}{16}$

Answers

6. $\frac{1}{2}$

7. 1

8. $\frac{25}{32}$

9. $\frac{18}{125}$

10. $\frac{3}{25}$

11. $\frac{27}{125}$

12. $\frac{11}{25}$

13. $\frac{12}{25}$

14. $\frac{2}{25}$

15. $\frac{1}{2}$

16. $\frac{4}{5}$

17. $\frac{1}{2}$

18. $\frac{1}{5}$

**Lesson 10.5
Level A**

1. 0.15
2. 0.2
3. 0.03
4. 0.05

5. 0.12
6. 0.04
7. independent
8. independent
9. dependent
10. independent
11. $\frac{81}{400}$
12. $\frac{9}{16}$
13. $\frac{27}{200}$
14. $\frac{3}{16}$
15. $\frac{1}{16}$
16. $\frac{77}{200}$
17. 0
18. 1
19. 0
20. $\frac{45}{400}$

**Lesson 10.5
Level B**

1. 0.1875
2. 0.15
3. 0.225
4. 0.05
5. 0.075

Algebra 2   Practice Masters Levels A, B, and C   387

Answers

6. 0.06
7. dependent
8. dependent
9. dependent
10. dependent
11. $\dfrac{4}{25}$
12. $\dfrac{9}{25}$
13. $\dfrac{6}{25}$
14. $\dfrac{6}{25}$
15. $\dfrac{12}{125}$
16. $\dfrac{27}{125}$
17. $\dfrac{108}{3125}$
18. $\dfrac{1944}{390,625}$

**Lesson 10.5
Level C**

1. $\dfrac{1}{8}$
2. $\dfrac{1}{64}$
3. $\dfrac{1}{2197}$
4. $\dfrac{125}{2197}$
5. $\dfrac{9}{2197}$
6. $\dfrac{1}{2704}$
7. $\dfrac{243}{2197}$
8. $\dfrac{1}{64}$
9. 0
10. $\dfrac{300}{2197}$
11. $\dfrac{1}{16}$
12. $\dfrac{27}{128}$
13. $\dfrac{1}{8000}$
14. $\dfrac{2907}{4000}$
15. $\dfrac{171}{4000}$
16. $\dfrac{81}{10,000}$
17. $\dfrac{9}{4000}$
18. $\dfrac{3}{256}$
19. $\dfrac{16}{625}$
20. $\dfrac{16}{625}$

Answers

Lesson 10.6
Level A

1. $\dfrac{1}{9}$

2. $\dfrac{1}{6}$

3. $\dfrac{1}{15}$

4. $\dfrac{1}{15}$

5. $\dfrac{1}{9}$

6. $\dfrac{1}{45}$

7. $\dfrac{2}{9}$

8. 0

9. $\dfrac{3}{4}$

10. 0.6

11. $\dfrac{1}{16}$

12. 0.1554

13. $\dfrac{3}{10}$

14. 0.43

15. $\dfrac{1}{6}$

16. $\dfrac{1}{6}$

17. $\dfrac{1}{2}$

18. 0

19. $\dfrac{1}{2}$

20. $\dfrac{1}{2}$

21. 1

22. $\dfrac{3}{5}$

Lesson 10.6
Level B

1. $\dfrac{1}{57}$

2. $\dfrac{7}{95}$

3. $\dfrac{1}{19}$

4. $\dfrac{1}{57}$

5. $\dfrac{3}{76}$

6. $\dfrac{6}{95}$

7. $0.\overline{3}$

8. 0.1728

9. $\dfrac{9}{28}$

10. $\dfrac{4}{9}$

11. $\dfrac{24}{91}$

Answers

12. $\dfrac{2}{55}$

13. $\dfrac{1}{55}$

14. $\dfrac{36}{55}$

15. $\dfrac{18}{55}$

16. $\dfrac{21}{55}$

17. 0

18. $\dfrac{1}{55}$

19. $\dfrac{6}{55}$

**Lesson 10.6
Level C**

1. 0.03
2. 0.002
3. 0.0004
4. 0.04

5. 0.002
6. 0.00002
7. 0.000009
8. 0.004
9. 0.000002
10. 0.00000154
11. 7,059,052
12. 0.00000014
13. 0.000032

**Lesson 10.7
Level A**

Answers will vary.

**Lesson 10.7
Level B**

Answers will vary.

**Lesson 10.7
Level C**

Answers will vary.

Answers

Lesson 11.1
Level A

1. 4, 7, 10, 13, 16
2. 3, 8, 13, 18, 23
3. $-5, -3, -1, 1, 3$
4. $1, -1, -3, -5, -7$
5. $2, -2, -6, -10, -14$
6. $-10, -17, -24, -31, -38$
7. 1, 4, 9, 16, 25
8. 3, 12, 27, 48, 75
9. 5, 8, 13, 20, 29
10. 0, 3, 8, 15, 24
11. 2, 5, 8, 11, 14
12. $0, -2, -4, -6, -8$
13. 1, 4, 16, 64, 256
14. $2, -6, 18, -54, 162$
15. 0, 3, 0, 3, 0
16. 2, 4, 16, 256, 65,536
17. $2 + 4 + 6 + 8 = 20$
18. $4 + 5 + 6 + 7 = 22$
19. $1 + 4 + 9 + 16 + 25 = 55$
20. $1 + 3 + 5 + 7 + 9 = 25$

Lesson 11.1
Level B

1. 8, 11, 16, 23, 32
2. 4, 16, 36, 64, 100
3. $-3, -12, -27, -48, -75$
4. 1, 7, 17, 31, 49
5. 1, 8, 27, 64, 125
6. 3, 24, 81, 192, 375
7. $-2, -16, -54, -128, -250$
8. 3, 10, 29, 66, 127
9. 1, 7, 31, 127, 511
10. $1, 0, 2, -2, 6$
11. $1, -4, 11, -34, 101$
12. 1, 4, 64, 16384, 1,073,741,824
13. $0, 4, -8, 28, -80$
14. $0, -1, -8, -729, -389,017,000$
15. $6 + 9 + 14 + 21 = 50$
16. $4 + 16 + 36 + 64 = 120$
17. $1 + (-5) + (-15) + (-29) = -48$
18. $1 + 8 + 27 + 64 = 100$
19. 20
20. 30

Lesson 11.1
Level C

1. $\dfrac{16}{5}, \dfrac{17}{5}, \dfrac{18}{5}, \dfrac{19}{5}$
2. $\dfrac{-5}{4}, \dfrac{-3}{2}, \dfrac{-7}{4}, -2$
3. $\dfrac{1}{4}, 1, \dfrac{9}{4}, 4$
4. 2, 4, 8, 16
5. $\dfrac{1}{3}, \dfrac{1}{9}, \dfrac{1}{27}, \dfrac{1}{81}$
6. 4, 10, 28, 82
7. $t_n = 2t_{n-1} + 1$; 31, 63
8. $t_n = n^2 + 2$; 2,090,918, 4.4×10^{12}

Answers

9. 1, 4, 28, 868

10. 0, 2, 3, $\frac{7}{2}$

11. 1, 2, 4, -16

12. 2, 16, 65536, 1.8×10^{19}

13. 6

14. 240

15. 8

16. 288

Lesson 11.2
Level A

1. yes, $d = 3$

2. yes, $d = 5$

3. no

4. yes, $d = -2$

5. yes, $d = -5$

6. no

7. 0, 3, 6, 9

8. $-1, -11, -21, -31$

9. 1, 4, 7, 10

10. $-2, \frac{-5}{2}, -3, \frac{-7}{2}$

11. $t_5 = 12$

12. $t_5 = 3$

13. $t_1 = -3$

14. $t_8 = -1$

15. 23, 28

16. 6, 2

17. 1, -1, -3

18. $-9, 2, 13$

19. 27, 17, 7

20. $-38, -16, 6$

Lesson 11.2
Level B

1. yes; $d = -7$

2. no

3. yes; $d = 1$

4. no

5. yes; $d = \frac{2}{3}$

6. yes; $d = \frac{-1}{4}$

7. $-11, -7, -3, 1$

8. $\frac{2}{3}, 1, \frac{4}{3}, \frac{5}{3}$

9. 4, 5, 6, 7

10. $-2, -7, -12, -17$

11. $t_4 = 6$

12. $t_3 = 0$

13. $t_6 = \frac{8}{5}$

14. $t_1 = \frac{-10}{3}$

15. $\frac{1}{2}, \frac{1}{4}, 0$

16. $\frac{-3}{5}, \frac{1}{5}, 1$

17. $-2, \frac{-3}{2}, -1$

18. $\frac{-2}{5}, \frac{1}{5}, \frac{4}{5}$

Answers

**Lesson 11.2
Level C**

1. $\dfrac{-1}{4}, \dfrac{1}{4}, \dfrac{3}{4}, \dfrac{5}{4}$

2. $\dfrac{3}{2}, \dfrac{7}{6}, \dfrac{5}{6}, \dfrac{1}{2}$

3. $\dfrac{-1}{2}, \dfrac{1}{6}, \dfrac{5}{6}, \dfrac{3}{2}$

4. $\dfrac{1}{3}, \dfrac{-5}{12}, \dfrac{-7}{6}, \dfrac{-23}{12}$

5. $-31, -10.5, -17.9, -25.3$

6. $-5.7, 5.6, 16.9, 28.2$

7. $t_n = 3(n-1) - 3$

8. $t_n = 7 + (n-1)5$

9. $t_n = 47 - (n-1)13$

10. $t_n = -7 + (n-1)8$

11. $t_n = -500 + (n-1)12$

12. $t_n = \dfrac{3}{5} - (n-1)\dfrac{2}{5}$

13. $t_n = -1 + (n-1)\dfrac{4}{3}$

14. $t_n = -7 + (n-1)\dfrac{3}{2}$

15. $\dfrac{7}{3}, \dfrac{11}{3}, 5$

16. $\dfrac{-5}{2}, -2, \dfrac{-3}{2}, -1$

17. $\dfrac{4}{3}, \dfrac{5}{3}, 2, \dfrac{7}{3}$

18. $\dfrac{-1}{3}, -1, \dfrac{-5}{3}, \dfrac{-7}{3}$

19. $63, 76, 89, 102$

20. $-58.5, -70, -81.5, -93$

**Lesson 11.3
Level A**

1. 35
2. -35
3. 140
4. -65
5. 105
6. -65
7. 5050
8. 2550
9. 780
10. 798
11. 340
12. 285
13. -65
14. 55
15. 60
16. 595
17. 55
18. 87
19. 128
20. -120

**Lesson 11.3
Level B**

1. -120
2. 986
3. 10

Answers

4. -140

5. 45.5

6. -102.5

7. $125,250$

8. $10,000$

9. 3150

10. 240

11. 1290

12. -2330

13. 355

14. -305

15. 650

16. 245

17. -5

18. 56

19. 472.5

20. -307

**Lesson 11.3
Level C**

1. $500,500$

2. $50,005,000$

3. $15,150$

4. 2525

5. 6700

6. 434

7. -1175

8. $-15,125$

9. $2158.\overline{3}$

10. -606.25

11. $1933.\overline{3}$

12. 7300

13. 52.5

14. 126

15. 85

16. $20,500$

17. $22,650$

18. $-405,000$

19. 2500

20. $125,251,000$

**Lesson 11.4
Level A**

1. yes; 3; 162, 486, 1458

2. yes; 4; 256, 1024, 4096

3. no

4. yes; -3; 81, -243, 729

5. no

6. yes; -2; 32, -64, 128

7. 2, 8, 32, 128

8. 3, 9, 27, 81

9. $-2, 2, -2, 2$

10. $1, -5, 25, -125$

11. $2, -20, 200, -2000$

12. 1, 0.5, 0.25, 0.125

13. 36, 108

14. 180, 1080

15. $-135, -405, -1215$

16. $36, -72$

Answers

Lesson 11.4
Level B

1. yes; -3; 324, -972, 2916
2. no
3. yes; $\frac{1}{2}$; $\frac{1}{16}, \frac{1}{32}, \frac{1}{64}$
4. yes; $\frac{1}{5}$; 0.0016, 0.00032, 0.000064
5. no
6. yes; 0.1; 0.00006, 0.000006, 0.0000006
7. 2, -60, 1800, $-54,000$
8. 3, 1, $\frac{1}{3}, \frac{1}{9}$
9. $-1, \frac{-2}{5}, \frac{-4}{25}, \frac{-8}{125}$
10. 2, $\frac{-3}{2}, \frac{9}{8}, \frac{-27}{32}$
11. $-2, -1.4, -0.98, -0.686$
12. 1, -0.2, 0.04, -0.008
13. -54, 324
14. 3.72, 11.532, 35.7492
15. $\frac{5}{16}, \frac{5}{64}, \frac{5}{256}$
16. $-1, \frac{1}{2}, \frac{-1}{4}, \frac{1}{8}$

Lesson 11.4
Level C

1. -2048
2. approximately $\frac{-3}{500,000}$
3. 18,750
4. $-960,000$
5. 0.00405
6. 35.15625
7. $\frac{1600}{81}$
8. 160
9. $t_n = -3t_{n-1}$
10. $t_n = \frac{2}{3} t_{n-1}$
11. $t_n = 3t_{n-1}$
12. $t_n = \frac{1}{2} t_{n-1}$
13. $t_n = \frac{2}{5} t_{n-1}$
14. $t_n = -0.3 t_{n-1}$
15. $-100, 1000, -10,000, 100,000$
16. 2, 0.5, 0.125
17. $-640, -2560, -10,240$
18. $\frac{1}{2}, \frac{1}{4}, \frac{1}{8}, \frac{1}{16}$

Lesson 11.5
Level A

1. -5
2. 11
3. -341
4. 10,923
5. 80
6. 728

Answers

7. 59,048

8. 531,440

9. 213

10. -62

11. -33

12. -605

13. 1023

14. $-19,680$

15. Basis step: $\begin{array}{c} 3 \leq 1 + 2 \\ 3 \leq 3 \end{array}$
Induction step: Statement is true for natural number k.
$k + 1 : 3 \leq (k + 1) + 2$
$3 \leq k + 3$
$0 \leq k$ true

Lesson 11.5
Level B

1. 160
2. 118,096
3. 6,973,568,800
4. 1.7×10^{12}
5. 19.375
6. 19.98047
7. 19.99998
8. 20
9. 55
10. 12.4
11. 1.9
12. 0.2
13. -170.5
14. -2.7

15. Basis step: $1 = \dfrac{1(2)(3)}{6}$
$1 = 1$

Induction step:
$$k^2 = \dfrac{k(k+1)(2k+1)}{6}$$
$$(k+1)^2 = k^2 + 2k + 1$$
$$k^2 + (k+1)^2 = \dfrac{2k^3 + 3k^2 + 1}{6} + k^2 + 2k + 1$$
$$= \dfrac{2k^3 + 9k^2 + 13k + 1}{6}$$
$$(k+1)^2 = \dfrac{(k+1)(k+2)(2k+3)}{6}$$
$$= \dfrac{2k^3 + 9k^2 + 13k + 1}{6}$$

Lesson 11.5
Level C

1. 7.407
2. 7.469
3. 7.500
4. 7.500
5. 12.8
6. $\dfrac{-2343}{128}$
7. 1318.75
8. 0.276672
9. 0.7
10. -63.75
11. -2.5
12. 0.4
13. 1428.6

Answers

14. 3.0×10^{56}

15. Basis step: $1^3 = \dfrac{1^2(1+1)^2}{4} = 1$

    Induction step:

    $1^3 + 2^3 + \ldots + n^3 = \dfrac{n^2(n+1)^2}{4}$

    $1^3 + 2^3 + \ldots + n^3 + (n+1)^3$

    $= \dfrac{n^2(n+1)^2}{4} + n^3 + 3n^2 + 3n + 1$

    $= \dfrac{n^4 + 6n^3 + 13n^2 + 12n + 4}{4}$

    $= \dfrac{(n+1)^2(n+2)^2}{4}$

Lesson 11.6
Level A

1. 20
2. $26\dfrac{2}{3}$
3. none
4. 37.5
5. none
6. $111.\overline{1}$
7. $\dfrac{1}{6}$
8. $\dfrac{1}{3}$
9. $\dfrac{-3}{13}$
10. none
11. $\dfrac{4}{9}$
12. $\dfrac{-1}{3}$
13. $\displaystyle\sum_{n=1}^{\infty} 4\left(\dfrac{1}{10}\right)^n$
14. $\displaystyle\sum_{n=1}^{\infty} 32\left(\dfrac{1}{100}\right)^n$
15. $\displaystyle\sum_{n=0}^{\infty} 65\left(\dfrac{1}{100}\right)^n$
16. $\displaystyle\sum_{n=1}^{\infty} 90\left(\dfrac{1}{100}\right)^n$
17. $\dfrac{1}{3}$
18. $\dfrac{2}{3}$
19. $\dfrac{8}{9}$
20. $\dfrac{29}{99}$

Lesson 11.6
Level B

1. -20
2. none
3. $-13.\overline{3}$
4. 128
5. none
6. 1142.9
7. $\dfrac{1}{2}$
8. $\dfrac{-1}{11}$

Algebra 2   Practice Masters Levels A, B, and C

Answers

9. 10

10. $\dfrac{-10}{3}$

11. $\dfrac{5}{4}$

12. none

13. $\sum_{n=1}^{\infty} 63\left(\dfrac{1}{100}\right)^n$

14. $\sum_{n=0}^{\infty} 72\left(\dfrac{1}{100}\right)^n$

15. $\sum_{n=1}^{\infty} 134\left(\dfrac{1}{1000}\right)^n$

16. $\sum_{n=1}^{\infty} 703\left(\dfrac{1}{1000}\right)^n$

17. $\dfrac{47}{99}$

18. $\dfrac{19}{99}$

19. $\dfrac{362}{999}$

20. $\dfrac{20}{37}$

Lesson 11.6
Level C

1. $\dfrac{1}{4}$

2. $\dfrac{-4}{9}$

3. $\dfrac{1}{2}$

4. none

5. $-2\dfrac{1}{4}$

6. 0.93

7. $0.\overline{18}$

8. none

9. $0.0\overline{47}$

10. $0.1\overline{26}$

11. none

12. $0.0\overline{001}$

13. $\sum_{n=1}^{\infty} 2430\left(\dfrac{1}{10{,}000}\right)^n$

14. $\sum_{n=0}^{\infty} 9299\left(\dfrac{1}{10{,}000}\right)^n$

15. $\dfrac{38}{333}$

16. $\dfrac{28}{111}$

17. $\dfrac{491}{3333}$

18. $\dfrac{911}{1111}$

Lesson 11.7
Level A

1. third entry, row 4; 6

2. fourth entry, row 5; 10

3. second entry, row 6; 6

4. sixth entry, row 7; 21

5. ninth entry, row 8; 1

6. sixth entry, row 10; 252

7. 3

Answers

8. 1

9. 36

10. 4

11. $\dfrac{3}{8}$

12. $\dfrac{5}{32}$

13. $\dfrac{5}{16}$

14. $\dfrac{7}{16}$

15. $\dfrac{21}{64}$

16. $\dfrac{21}{32}$

17. $\dfrac{1}{16}$

18. $\dfrac{1}{4}$

19. $\dfrac{11}{16}$

20. $\dfrac{15}{16}$

Lesson 11.7
Level B

1. fourth entry, row 5; 10

2. seventh entry, row 10; 210

3. eighth entry, row 10; 120

4. fifth entry, row 20; 4845

5. sixteenth entry, row 15; 1

6. fourth entry, row 9; 84

7. 20

8. 165

9. 495

10. 190

11. $\dfrac{5}{16}$

12. $\dfrac{21}{64}$

13. $\dfrac{57}{64}$

14. $\dfrac{29}{128}$

15. $\dfrac{63}{128}$

16. $\dfrac{15}{128}$

17. $\dfrac{63}{256}$

18. $\dfrac{165}{512}$

19. $\dfrac{193}{512}$

Lesson 11.7
Level C

1. 1, 10, 45, 120, 210, 252, 210, 120, 45, 10, 1

2. 1, 15, 105, 455, 1365, 3003, 5005, 6435, 6435, 5005, 3003, 1365, 455, 105, 15, 1

3. about 23%

4. about 99%

Answers

5. about 57%
6. about 17%
7. about 94%
8. about 59%
9. 0%
10. about 25%
11. about 4%
12. about 8%
13. about 99%
14. about 62%
15. about 58%

Lesson 11.8
Level A

1. $x^3 + 3x^2y + 3xy^2 + y^3$
2. $a^6 + 6a^5b + 15a^4b^2 + 20a^3b^3 + 15a^2b^4 + 6ab^5 + b^6$
3. $x^7 + 7x^6y + 21x^5y^2 + 35x^4y^3 + 35x^3y^4 + 21x^2y^5 + 7xy^6 + y^7$
4. $a^8 - 8a^7b + 28a^6b^2 - 56a^5b^3 + 70a^4b^4 - 56a^3b^5 + 28a^2b^6 - 8ab^7 + b^8$
5. $32x^5 + 240x^4y + 720x^3y^2 + 1080x^2y^3 + 810xy^4 + 243y^5$
6. $(x + y)^5 = x^5 + 5x^4y + 10x^3y^2 + 10x^2y^3 + 5xy^4 + y^5$
7. $(a + b)^8 = a^8 + 8a^7b + 28a^6b^2 + 56a^5b^3 + 70a^4b^4 + 56a^3b^5 + 28a^2b^6 + 8ab^7 + b^8$
8. $120a^7b^3$
9. $210a^4b^6$

10. 25%
11. 6%
12. 76%
13. 6%

Lesson 11.8
Level B

1. $x^5 - 5x^4y + 10x^3y^2 - 10x^2y^3 + 5xy^4 - y^5$
2. $a^{10} + 10a^9b + 45a^8b^2 + 120a^7b^3 + 210a^6b^4 + 252a^5b^5 + 210a^4b^6 + 120a^3b^7 + 45a^2b^8 + 10ab^9 + b^{10}$
3. $x^7 - 7x^6y + 21x^5y^2 - 35x^4y^3 + 35x^3y^4 - 21x^2y^5 + 7xy^6 - y^7$
4. $729x^6 + 5832x^5y + 19440x^4y^2 + 34560x^3y^3 + 34560x^2y^4 + 18432xy^5 + 4096y^6$
5. $32x^5 - 400x^4y + 2000x^3y^2 - 5000x^2y^3 + 6250xy^4 - 3125y^5$
6. $(p + q)^7 = p^7 + 7p^6q + 21p^5q^2 + 35p^4q^3 + 35p^3q^4 + 21p^2q^5 + 7pq^6 + q^7$
7. $(x + y)^8 = x^8 + 8x^7y + 28x^6y^2 + 56x^5y^3 + 70x^4y^4 + 56x^3y^5 + 28x^2y^6 + 8xy^7 + y^8$
8. $495x^4y^8$
9. $-220x^9y^3$
10. 1%
11. 20%
12. 73%
13. 0%

Answers

Lesson 11.8
Level C

1. $x^5 - 25x^4y + 250x^3y^2 - 1250x^2y^3 + 3125xy^4 - 3125y^5$

2. $4096x^6 - 24576x^5y + 61440x^4y^2 - 81920x^3y^3 + 61440x^2y^4 - 24576xy^5 + 4096y^6$

3. $243y^5 - 405y^4x + 270y^3x^2 - 90y^2x^3 + 15yx^4 - x^5$

4. $\dfrac{1}{32}x^5 + \dfrac{5}{64}x^4y + \dfrac{5}{64}x^3y^2 + \dfrac{5}{128}x^2y^3 + \dfrac{5}{512}xy^4 + \dfrac{1}{1024}y^5$

5. $\dfrac{64}{729}x^6 - \dfrac{64}{243}x^5y + \dfrac{80}{243}x^4y^2 - \dfrac{160}{729}x^3y^3 + \dfrac{20}{243}x^2y^4 - \dfrac{4}{243}xy^5 + \dfrac{1}{729}y^6$

6. $15,504a^{15}b^5$

7. $167,960a^{11}b^9$

8. 21

9. 39%

10. 6%

11. 24%

12. 63%

13. 65%

Answers

Lesson 12.1
Level A

1. mean = 7.111; median = 8; mode = 4

2. mean = 34.429; median = 35; mode = 32

3. mean = 15.6; median = 16; mode = 14, 20

4. mean = 14; median = 14; mode = none

5. mean = −6.333; median = −6; mode = −6

6. mean = 7.543, median = 7.575, mode = none. The mean and median are very close. Therefore, the data is evenly spread out.

7. mean = $11.\overline{3}$

Shoe Size	Tally	Frequency
9	/	1
10	///	3
11	//	2
12	////	4
13	/	1
14	/	1

8. mean = 6.1

Number of days	Class Mean	Frequency	Product
0–2	1	////	4
3–5	4	///// /	24
6–8	7	////	28
9–11	10	////	40
12–14	13	//	26

Lesson 12.1
Level B

1. mean = 25.818; median = 24; mode = 22

2. mean = −12.375; median = −11.5; mode = −11, −15

3. mean = 2.667; median = 2.65; mode = none

4. mean = 14.717; median = 14.75; mode = 14.2, 14.8

5. mean = $\dfrac{23}{48}$; median = $\dfrac{1}{2}$; mode = $\dfrac{2}{3}$

6. mean = 20.8; median = 20.8; mode = 20.6, 21; These values are close together because each of the data points is close to the others.

7. mean = 5.27

Number of members	Tally	Frequency
2	/	1
3	/	1
4	//	2
5	/////	5
6	///	3
7	/	1
8	//	2

8. mean = 11.56

Number of days	Class Mean	Frequency	Product
3–5	4	///	12
6–8	7	/////	35
9–11	10	////	40
12–14	13	/////	65
15–17	16	/////	80
18–20	19	///	57

Answers

**Lesson 12.1
Level C**

1. mean = 131.875
   median = 130
   mode = 130

Number of Home Runs	Class Mean	Frequency	Product
101–110	105.5	//	211
111–120	115.5	///	346.5
121–130	125.5	𝓗𝓗	627.5
131–140	135.5	//	
141–150	145.5		
151–160	155.5	//	311
161–170	165.5	/	165.5
171–180	175.5	/	175.5
	Total	16	2108

2. mean = 0.276
   median = 0.277
   mode = 0.289, 0.270

Number of Home Runs	Class Mean	Frequency	Product
0.261–0.265	0.263	/	0.263
0.266–0.270	0.268	𝓗𝓗	1.34
0.271–0.275	0.273	/	0.273
0.276–0.280	0.278	//	0.556
0.281–0.285	0.283	///	0.849
0.286–0.290	0.288	//	0.576
	Total	14	3.857

**Lesson 12.2
Level A**

1. 3 | 22238
   4 | 2266
   5 | 12

2. 0 | 69
   1 | 33777
   2 | 112669

3.
Number	Frequency
1	𝓗𝓗 /
2	//
3	////
4	
5	/
6	/
7	//
8	////

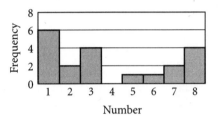

4.
Number	Frequency	Relative Frequency
1–2	8	40%
3–4	4	20%
5–6	2	10%
7–8	6	30%
9–10	0	0%

5. $\dfrac{7}{20} = 35\%$

**Lesson 12.2
Level B**

1. 5. | 1579
   6. | 00679
   7. | 3388
   8. | 11255

2. 0.1 | 6
   0.2 | 046
   0.3 | 023477
   0.4 | 11349
   0.5 | 04699

Algebra 2   Practice Masters Levels A, B, and C

Answers

3.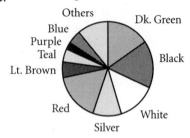
Most Popular Car Colors 1998

4.
Number	Frequency	Relative Frequency
0.9–1.0	2	10%
1.1–1.2	5	25%
1.3–1.4	4	20%
1.5–1.6	4	20%
1.7–1.8	4	20%
1.9–2.0	1	5%

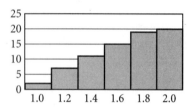

5. $\dfrac{7}{20} = 35\%$

Lesson 12.2
Level C

1. a.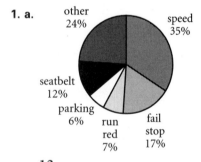

b. $\dfrac{13}{14}$

2. a.
Number	Frequency	Relative Frequency
0–0.9	6	25%
1–1.9	9	37.5%
2–2.9	6	25%
3–3.9	3	12.5%

b.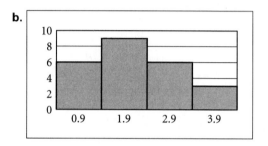

c. 1.7

d. $\dfrac{7}{8}$

Lesson 12.3
Level A

1. $Q_1 = 3, Q_2 = 5, Q_3 = 6.5$
   range $= 9; IQR = 3.5$

2. $Q_1 = 13, Q_2 = 14.5, Q_3 = 17.5$
   range $= 8; IQR = 4.5$

3. $Q_1 = 22, Q_2 = 41, Q_3 = 91$
   range $= 99; IQR = 69$

4. min $= 2$; max $= 20, Q_1 = 9, Q_2 = 14,$
   $Q_3 = 18,$ range $= 18; IRQ = 9$

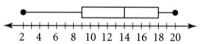

5. min $= 31$; max $= 40, Q_1 = 32, Q_2 = 35,$
   $Q_3 = 39,$ range $= 9; IRQ = 7$

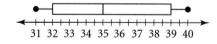

6. min $= 56$; max $= 82, Q_1 = 66, Q_2 = 69,$
   $Q_3 = 74,$ range $= 26; IRQ = 8$

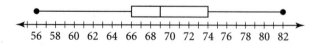

Answers

7. min = 0.99; max = 1.59, Q_1 = 1.09, Q_2 = 1.19, Q_3 = 1.33, range = 0.60; IRQ = 0.24

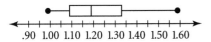

Lesson 12.3
Level B

1. Q_1 = 2.3, Q_2 = 3.95, Q_3 = 6.85 range = 8.6; IQR = 4.55

2. Q_1 = 374, Q_2 = 557, Q_3 = 886 range = 889; IQR = 512

3. min = 46; max = 90, Q_1 = 61.5, Q_2 = 68, Q_3 = 78, range = 44; IRQ = 16.5

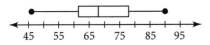

Lesson 12.3
Level C

1. min = 14; max = 55, Q_1 = 24, Q_2 = 31, Q_3 = 38, range = 41; IRQ = 14

2. min = 3; max = 31, Q_1 = 10, Q_2 = 14, Q_3 = 19, range = 28; IRQ = 9

3.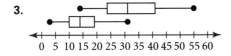

Lesson 12.4
Level A

1. range = 6, mean deviation = 1.5
2. range = 5, mean deviation = 1.5
3. range = 28, mean deviation = 9

4. range = 8, mean deviation = $2.\overline{6}$
5. range = 8, mean deviation = $2\frac{1}{3}$
6. range = 10, mean deviation = $3.\overline{3}$
7. variance = 3.58; st. dev. = 1.89
8. variance = 3.14; st. dev. = 1.77
9. variance = 106.92; st. dev. = 10.34
10. variance = 8.89; st. dev. = 2.98
11. variance = 7.22; st. dev. = 2.69
12. variance = 13.22; st. dev. = 3.64
13. mean = 11.04, median = 11
14. range = 1.06, mean deviation = 0.20
15. 0.28

Lesson 12.4
Level B

1. range = 29, mean deviation = $6.8\overline{4}$
2. range = 10, mean deviation = 2.73
3. range = 18, mean deviation = 6.31
4. range = 2.6, mean deviation = 0.69
5. range = $\frac{13}{20}$, mean deviation = 0.20
6. range = 1.25, mean deviation = 0.43
7. variance = 73.80; st. dev. = 8.59
8. variance = 10.79; st. dev. = 3.27
9. variance = 45.42; st. dev. = 6.74
10. variance = 0.72; st. dev. = 0.85
11. variance = 0.05; st. dev. = 0.23
12. variance = 0.21; st. dev. = 0.46

Answers

13. mean = 3:12.15, median = 3:12.00
14. range = 35.21, mean deviation = 12.26
15. st. deviation = 13.08

Lesson 12.4
Level C

1. mean deviation = 24.4, st. dev. = 28.90
2. mean deviation = 1.32, st. dev. = 1.47
3. mean deviation = 17.08, st. dev. = 20.55
4. mean deviation = 46.8, st. dev. = 49.73
5. mean deviation = 0.16, st. dev. = 0.18
6. mean deviation = 245.7, st. dev. = 259.89
7. mean deviation = 0.51, st. dev. = 0.56
8. mean deviation = 18.91, st. dev. = 20.70
9. mean deviation
10. mean deviation
11. 4, 12
12. −4, −5, −13
13. mean = 8.64, median = 4
14. st. dev. = 21.79

Lesson 12.5
Level A

1. 31.25%
2. 15.63%
3. 50%
4. 81.25%
5. 0.01%
6. 3%
7. 0.3%
8. 20%
9. 99.7%
10. 99.9%
11. 31%
12. 8.7%
13. 10%
14. 90%
15. 68%
16. 99.9%
17. 31%
18. 97%

Lesson 12.5
Level B

1. 0.004%
2. 3%
3. 0.08%
4. 99%
5. 76%
6. 41%
7. 25%
8. 28%
9. 94%
10. 24%
11. 94%
12. 98%
13. 20%
14. 25%

Answers

15. 24%

16. 15%

17. 17%

18. 21%

**Lesson 12.5
Level C**

1. 0.1%

2. 39%

3. 93%

4. 7%

5. approximately 0%

6. 5%

7. 8%

8. 5%

9. 6%

10. 21%

11. 11%

12. 21%

13. 99%

14. 79%

**Lesson 12.6
Level A**

1. 0.4207

2. 0.0359

3. 0.8849

4. 0.8413

5. 0.9192

6. 0.3446

7. 0.8413

8. 0.2743

9. 0.62%

10. 5%

11. 95%

12. 45%

13. 99%

14. 25%

15. 16%

16. 41%

**Lesson 12.6
Level B**

1. 0.2743

2. 0.2119

3. 0.6587

4. 0.9772

5. 0.0449

6. 0.1359

7. 0.7262

8. 0.9413

9. 44%

10. 4%

11. 99%

12. 86%

13. 33%

14. 49%

15. 97%

16. 82%

Answers

**Lesson 12.6
Level C**

1. 3%

2. 26%

3. 94%

4. 16%

5. 30%

6. 17 and 33 minutes

7. 8%

8. 8%

9. 2%

10. 42%

11. 18%

12. 7:25 and 7:45

13. 5%

14. 25%

15. 63%

16. 75%

17. 25%

18. 7.4 and 8.6 ounces

Answers

Lesson 13.1
Level A

1. $\dfrac{4}{5} = 0.8$

2. $\dfrac{3}{5} = 0.6$

3. $\dfrac{4}{3} = 1.\overline{3}$

4. $\dfrac{3}{5} = 0.6$

5. $\dfrac{4}{5} = 0.8$

6. $\dfrac{3}{4} = 0.75$

7. $\dfrac{5}{3} = 1.\overline{6}$

8. $\dfrac{5}{3} = 1.\overline{6}$

9. 33.7°
10. 30°
11. 41.8°
12. 26.6°
13. 41.4°
14. 78.5°
15. $AB = 10$, $m\angle A = 37°$, $m\angle B = 53°$
16. $AB = 20$, $AC = 17.3$, $m\angle B = 60°$
17. $AB = 10.4$, $AC = 6.7$, $m\angle A = 50°$

Lesson 13.1
Level B

1. $\dfrac{7}{25} = 0.28$

2. $\dfrac{24}{25} = 0.96$

3. $\dfrac{7}{24} = 0.2917$

4. $\dfrac{24}{25} = 0.96$

5. $\dfrac{7}{25} = 0.28$

6. $\dfrac{24}{7} = 3.4286$

7. $\dfrac{25}{24} = 1.0417$

8. $\dfrac{25}{24} = 1.0417$

9. 41.8°
10. 26.6°
11. 46.6°
12. 35.4°
13. 61.3°
14. 76.9°
15. $AB = 5.8$, $m\angle A = 41°$, $m\angle B = 49°$
16. $AC = 12.2$, $m\angle A = 41°$, $m\angle B = 49°$
17. $AB = 15.1$, $BC = 13.7$, $m\angle B = 65°$

Answers

Lesson 13.1
Level C

1. $AB = 6.3$, m$\angle A = 55°$, m$\angle B = 35°$
2. $AC = 2.1$, m$\angle A = 27°$, m$\angle B = 63°$
3. $AB = 3.9$, m$\angle A = 45°$, m$\angle B = 45°$
4. $AB = 2$, m$\angle A = 60°$, m$\angle B = 30°$
5. $AC = 5$, m$\angle A = 60°$, m$\angle B = 30°$
6. $AB = 6.2$, m$\angle A = 23°$, m$\angle B = 67°$
7. $AB = 8.7$, m$\angle A = 69°$, m$\angle B = 21°$
8. $CB = 9.9$, m$\angle A = 82°$, m$\angle B = 8°$
9. $AC = 24$, m$\angle A = 30°$, m$\angle B = 60°$
10. $70.5°$
11. legs = 14.1 feet each
    angles = $45°$

Lesson 13.2
Level A

1. $-340°$
2. $-305°$
3. $-260°$
4. $-181°$
5. $-120°$
6. $-90°$
7. $-30°$
8. $40, -320°$
9. $280°$
10. $50°$
11. $65°$
12. $20°$
13. $70°$
14. $5°$
15. $68°$
16. $40°$
17. $80°$
18. $40°$
19. $\sin = \dfrac{\sqrt{2}}{2}$, $\cos = \dfrac{\sqrt{2}}{2}$, $\tan = 1$
    $\csc = \sqrt{2}$, $\sec = \sqrt{2}$, $\cot = 1$
20. $\sin = \dfrac{5\sqrt{34}}{34}$, $\cos = \dfrac{3\sqrt{34}}{34}$, $\tan = \dfrac{5}{3}$
    $\csc = \dfrac{\sqrt{34}}{5}$, $\sec = \dfrac{\sqrt{34}}{3}$, $\cot = \dfrac{3}{5}$
21. $\sin = \dfrac{\sqrt{5}}{5}$, $\cos = \dfrac{-2\sqrt{5}}{5}$, $\tan = \dfrac{-1}{2}$
    $\csc = \sqrt{5}$, $\sec = \dfrac{-\sqrt{5}}{2}$, $\cot = -2$
22. $\dfrac{-\sqrt{3}}{2}$
23. $\dfrac{2\sqrt{29}}{29}$
24. $\dfrac{-\sqrt{5}}{3}$
25. $\dfrac{8\sqrt{89}}{89}$
26. $\dfrac{4\sqrt{65}}{65}$
27. $\dfrac{11\sqrt{146}}{146}$

Answers

Lesson 13.2
Level B

1. $-283°$
2. $-248°$
3. $300°$
4. $240°$
5. $140°, -220°$
6. $125°, -235°$
7. $60°$
8. $20°$
9. $90°$
10. $60°$
11. $65°$
12. $20°$
13. $\sin\theta = \dfrac{-4}{5}, \cos\theta = \dfrac{3}{5}, \tan\theta = \dfrac{-4}{3}$
    $\csc\theta = \dfrac{-5}{4}, \sec\theta = \dfrac{5}{3}, \cot\theta = \dfrac{-3}{4}$
14. $\sin\theta = \dfrac{2\sqrt{5}}{5}, \cos\theta = \dfrac{-\sqrt{5}}{5}, \tan\theta = -2$
    $\csc\theta = \dfrac{\sqrt{5}}{2}, \sec\theta = -\sqrt{5}, \cot\theta = \dfrac{-1}{2}$
15. $\sin\theta = \dfrac{-3\sqrt{13}}{13}, \cos\theta = \dfrac{-2\sqrt{13}}{13},$
    $\tan\theta = \dfrac{3}{2}, \csc\theta = \dfrac{-\sqrt{13}}{3},$
    $\sec\theta = \dfrac{-\sqrt{13}}{2}, \cot\theta = \dfrac{2}{3}$
16. $\dfrac{-\sqrt{33}}{7}$
17. $\sqrt{3}$
18. $\dfrac{-2\sqrt{29}}{29}$
19. $\dfrac{-7\sqrt{65}}{65}$
20. $\dfrac{-\sqrt{5}}{2}$
21. $\dfrac{5\sqrt{106}}{106}$
22. $\dfrac{1}{2}$
23. $1\dfrac{1}{4}$
24. $1\dfrac{3}{4}$

Lesson 13.2
Level C

1. $190°, -170°; \theta_{ref} = 10°$
2. $10°; \theta_{ref} = 10°$
3. $-20°, 340°; \theta_{ref} = 20°$
4. $2\dfrac{5}{6}$; counterclockwise
5. $1\dfrac{11}{24}$; clockwise
6. $1\dfrac{271}{360}$; clockwise
7. $\sin\theta = \dfrac{6\sqrt{61}}{61}, \cos\theta = \dfrac{-5\sqrt{61}}{61},$
   $\tan\theta = \dfrac{-6}{5}, \csc\theta = \dfrac{\sqrt{61}}{6},$
   $\sec\theta = \dfrac{-\sqrt{61}}{5}, \cot\theta = \dfrac{-5}{6}$

Algebra 2   Practice Masters Levels A, B, and C

Answers

8. $\sin \theta = \dfrac{-5\sqrt{29}}{29}, \cos \theta = \dfrac{2\sqrt{29}}{29},$
   $\tan \theta = \dfrac{-5}{2}, \csc \theta = \dfrac{-\sqrt{29}}{5},$
   $\sec \theta = \dfrac{\sqrt{29}}{2}, \cot \theta = \dfrac{-2}{5}$

9. $\sin \theta = \dfrac{4\sqrt{17}}{17}, \cos \theta = \dfrac{\sqrt{17}}{17}, \tan \theta = 4$
   $\csc \theta = \dfrac{\sqrt{17}}{4}, \sec \theta = \sqrt{17}, \cot \theta = \dfrac{1}{4}$

10. $\sin \theta = \dfrac{-2}{5}, \cos \theta = \dfrac{\sqrt{21}}{5}, \tan \theta = \dfrac{-\sqrt{21}}{21},$
    $\csc \theta = \dfrac{-5}{2}, \sec \theta = \dfrac{5\sqrt{21}}{21}, \cot \theta = \dfrac{-\sqrt{21}}{2}$

11. $\sin \theta = \dfrac{2}{9}, \cos \theta = \dfrac{-\sqrt{77}}{9}, \tan \theta = \dfrac{-2\sqrt{77}}{77},$
    $\csc \theta = \dfrac{9}{2}, \sec \theta = \dfrac{-9\sqrt{77}}{77}, \cot \theta = \dfrac{-\sqrt{77}}{2}$

12. $\sin \theta = \dfrac{-4\sqrt{65}}{65}, \cos \theta = \dfrac{-7\sqrt{65}}{65}, \tan \theta = \dfrac{4}{7},$
    $\csc \theta = \dfrac{-\sqrt{65}}{4}, \sec \theta = -\dfrac{\sqrt{65}}{7}, \cot \theta = \dfrac{7}{4}$

Lesson 13.3
Level A

1. $\sin = \dfrac{\sqrt{2}}{2}, \cos = \dfrac{\sqrt{2}}{2}, \tan = 1$

2. $\sin = \dfrac{\sqrt{3}}{2}, \cos = \dfrac{1}{2}, \tan = \sqrt{3}$

3. $\sin = 1, \cos = 0, \tan = \text{undefined}$

4. $\sin = \dfrac{\sqrt{3}}{2}, \cos = \dfrac{-1}{2}, \tan = -\sqrt{3}$

5. $\sin = -1, \cos = 0, \tan = \text{undefined}$

6. $\sin = \dfrac{-1}{2}, \cos = \dfrac{\sqrt{3}}{2}, \tan = \dfrac{-\sqrt{3}}{3}$

7. $(0.34, 0.94)$

8. $(-0.5, 0.87)$

9. $(-0.98, -0.17)$

10. $(-0.34, -0.94)$

11. $(0.98, -0.17)$

12. $(0.64, -0.77)$

13. $\sin = 0, \cos = -1, \tan = 0$

14. $\sin = \dfrac{1}{2}, \cos = \dfrac{\sqrt{3}}{2}, \tan = \dfrac{\sqrt{3}}{3}$

15. $\sin = \dfrac{\sqrt{2}}{2}, \cos = \dfrac{\sqrt{2}}{2}, \tan = 1$

16. -5.759

17. -2.924

18. 0.017

Lesson 13.3
Level B

1. $\sin = \dfrac{1}{2}, \cos = \dfrac{\sqrt{3}}{2}, \tan = \dfrac{\sqrt{3}}{3}$

2. $\sin = \dfrac{-\sqrt{2}}{2}, \cos = \dfrac{-\sqrt{2}}{2}, \tan = 1$

3. $\sin = \dfrac{\sqrt{3}}{2}, \cos = \dfrac{-1}{2}, \tan = -\sqrt{3}$

4. $\sin = \dfrac{-1}{2}, \cos = \dfrac{\sqrt{3}}{2}, \tan = \dfrac{-\sqrt{3}}{3}$

5. $\sin = \dfrac{-\sqrt{2}}{2}, \cos = \dfrac{-\sqrt{2}}{2}, \tan = 1$

6. $\sin = \dfrac{-\sqrt{3}}{2}, \cos = \dfrac{1}{2}, \tan = -\sqrt{3}$

7. $(-0.34, 0.94)$

8. $(-0.91, -0.42)$

Answers

9. (0.75, 0.66)
10. (0.80, −0.60)
11. (0.53, −0.85)
12. (−0.95, −0.33)
13. $\sin = \dfrac{\sqrt{3}}{2}, \cos = \dfrac{-1}{2}, \tan = -\sqrt{3}$
14. $\sin = \dfrac{-\sqrt{3}}{2}, \cos = \dfrac{1}{2}, \tan = -\sqrt{3}$
15. $\sin = 0, \cos = -1, \tan = 0$
16. 2
17. −1.556
18. 1.732

Lesson 13.3
Level C

1. $(2, -2\sqrt{3})$
2. $(-3\sqrt{2}, -3\sqrt{2})$
3. $(-5\sqrt{2}, 5\sqrt{2})$
4. $\left(\dfrac{-5}{2}, \dfrac{-5\sqrt{3}}{2}\right)$
5. $(4\sqrt{3}, 4)$
6. (−5, 0)
7. (−0.12, −0.99)
8. (0.85, −0.53)
9. (−0.91, 0.42)
10. $\sin = \dfrac{\sqrt{2}}{2}, \cos = \dfrac{\sqrt{2}}{2}, \tan = 1$
11. $\sin = \dfrac{\sqrt{3}}{2}, \cos = \dfrac{-1}{2}, \tan = -\sqrt{3}$
12. $\sin = \dfrac{-\sqrt{2}}{2}, \cos = \dfrac{-\sqrt{2}}{2}, \tan = 1$
13. $\sin = \dfrac{\sqrt{3}}{2}, \cos = \dfrac{1}{2}, \tan = \sqrt{3}$
14. $\sin = -1, \cos = 0, \tan = $ undefined
15. $\sin = \dfrac{-\sqrt{2}}{2}, \cos = \dfrac{\sqrt{2}}{2}, \tan = -1$
16. −2
17. 1.155
18. 0.577

Lesson 13.4
Level A

1. $\dfrac{\pi}{4}$
2. $\dfrac{\pi}{3}$
3. $\dfrac{4\pi}{3}$
4. $\dfrac{5\pi}{6}$
5. $\dfrac{-2\pi}{3}$
6. $\dfrac{-5\pi}{6}$
7. $\dfrac{5\pi}{9}$
8. $\dfrac{-35\pi}{36}$
9. $\dfrac{7\pi}{9}$
10. 540°
11. 45°
12. 60°

Algebra 2   Practice Masters Levels A, B, and C

Answers

13. $-60°$
14. $150°$
15. $120°$
16. $57.3°$
17. $-114.6°$
18. $177.6°$
19. 6π inches
20. π inches
21. $\dfrac{9\pi}{2}$ inches
22. 12 inches
23. 27 inches
24. 9π inches
25. 1
26. $\dfrac{\sqrt{2}}{2}$
27. $\sqrt{3}$

Lesson 13.4
Level B

1. $\dfrac{2\pi}{9}$
2. $\dfrac{-7\pi}{18}$
3. $\dfrac{13\pi}{18}$
4. $\dfrac{10\pi}{9}$
5. $\dfrac{-25\pi}{36}$
6. $\dfrac{29\pi}{36}$
7. $\dfrac{3\pi}{20}$
8. $\dfrac{3\pi}{10}$
9. $\dfrac{-\pi}{5}$
10. $135°$
11. $108°$
12. $270°$
13. $315°$
14. $-300°$
15. $-15°$
16. $200.5°$
17. $-154.7°$
18. $68.8°$
19. 7.9 inches
20. 27.5 inches
21. 41.2 inches
22. 7.5 inches
23. 17.3 inches
24. 23.9 inches
25. $\dfrac{-1}{2}$
26. 0
27. 1

Answers

Lesson 13.4
Level C

1. $\dfrac{7\pi}{3}$

2. $\dfrac{-98\pi}{45}$

3. $\dfrac{134\pi}{45}$

4. $\dfrac{-103\pi}{45}$

5. $\dfrac{25\pi}{24}$

6. $\dfrac{-29\pi}{180}$

7. 195°

8. 1035°

9. −1530°

10. 13.6 inches

11. 25.3 inches

12. 10.1 inches

13. $\dfrac{\sqrt{3}}{2}$

14. $\dfrac{\sqrt{2}}{2}$

15. $\dfrac{\sqrt{3}}{3}$

16. undefined

17. $-\sqrt{2}$

18. 1

19. about 4712.4 feet

20. $\dfrac{5\pi}{6}$

Lesson 13.5
Level A

1. amplitude = 3, period = 2π

2. amplitude = 2.5, period = 2π

3. amplitude = 6, period = $\dfrac{\pi}{2}$

4. amplitude = 2, period = π

5. amplitude = 2, period = π

6. amplitude = 1.5, period = 2π

7. phase shift = 90° to the left, vertical shift = 2 units up

8. phase shift = 30° to the right, vertical shift = 1 unit down

9. phase shift = 60° to the right, vertical shift = 3 units up

10. phase shift = 45° to the left, vertical shift = 3 units down

11. amplitude = 3, vertical shift = 1 unit down, period = $\dfrac{2\pi}{3}$

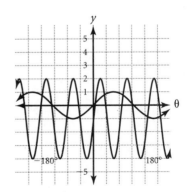

Answers

12. amplitude = −1, vertical shift = 2 units up, phase shift = 45° to the right

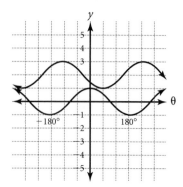

13. $y = 1.5 \sin 200\pi t$

Lesson 13.5
Level B

1. amplitude = $\frac{1}{2}$, period = 4π

2. amplitude = 3.5, period = 8π

3. amplitude = −1, period = 2π

4. amplitude = $\frac{1}{4}$, period = $\frac{\pi}{2}$

5. amplitude = 2.5, period = $\frac{\pi}{2}$

6. amplitude = $\frac{2}{5}$, period = 6π

7. phase shift = 180° to right

8. phase shift = 10° to left, vertical shift = 2 units down

9. phase shift = 60° to right, vertical shift = 5 units up

10. phase shift = 30° to left, vertical shift = $\frac{1}{2}$ unit up

11. phase shift = $\frac{\pi}{2}$ radians right, vertical shift = 1 unit up

12. phase shift = π radians left, vertical shift = 3 units down

13. amplitude = 2, vertical shift = 1 unit up, phase shift = $\frac{\pi}{2}$ units to right

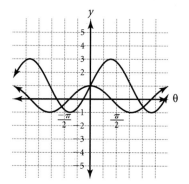

14. vertical shift = 2 units up, phase shift = $\frac{3\pi}{4}$ units left, reflected about x-axis

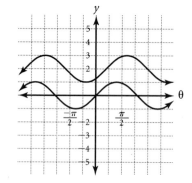

15. $2\sin 150\pi \left(t - \frac{1}{225}\right)$

Answers

**Lesson 13.5
Level C**

1. $3\sin 360\pi\left(t - \dfrac{1}{260}\right)$

2. period $= \dfrac{\pi}{2}$, amplitude $= 1$ (reflected)

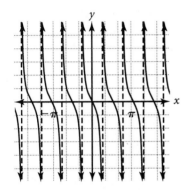

3. amplitude $= -2$, vertical shift $= 1$, phase shift $= \dfrac{\pi}{4}$ radians to left

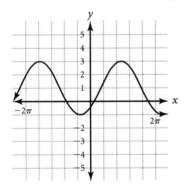

4. amplitude $= 3$, vertical shift $= 1$ unit down, period $= \pi$, phase shift $= \dfrac{\pi}{2}$ radians to right

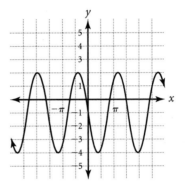

5. $y = 2\sin 3\left(x + \dfrac{\pi}{2}\right)$

**Lesson 13.6
Level A**

1. $30°$
2. $-90°$
3. $30°$
4. $135°$
5. $-30°$
6. $90°$
7. $30° + 360°n$, $150° + 360°n$
8. $360°n$
9. $45° + 180°n$
10. $240° + 360°n$, $300° + 360°n$
11. $180° + 360°n$
12. $120° + 180°n$
13. $\dfrac{\sqrt{3}}{2}$

Answers

14. $\dfrac{\sqrt{2}}{2}$

15. undefined

16. -1

17. $\dfrac{1}{2}$

18. 30°

19. 31°

20. 72.5°

21. 14°

Lesson 13.6
Level B

1. 0°
2. 180°
3. $-45°$
4. 45°
5. 60°
6. $-90°$
7. $270° + 360°n$
8. $90° + 180°n$
9. $180°n$
10. $180°n$
11. $120° + 360°n$; $240° + 360°n$
12. $30° + 180°n$
13. $\dfrac{1}{2}$
14. $\dfrac{\sqrt{2}}{2}$
15. 0

16. 45°
17. 40.9°
18. 54.7°
19. a. 11.3°
    b. 4.4 feet
20. a. 287.7 feet
    b. 54.1°

Lesson 13.6
Level C

1. 17.5°
2. 101.5°
3. 84.5°
4. $336.4° + 360°n$, $203.6° + 360°n$
5. $36.9° + 360°n$, $323.1° + 360°n$
6. $83° + 180°n$
7. 0.53
8. -0.97
9. 79.84°
10. a. 84 feet
    b. 16.7 feet
11. a. 31.6°
    b. 15.5°
    c. 42.9°
    d. 7.6
    e. 9.6

Answers

Lesson 14.1
Level A

1. m∠B = 88°, AB = 15.3,
   BC = 13.4, area = 102.4 sq. units

2. m∠B = 22°, AC = 3.8,
   BC = 9.7, area = 18.4 sq. units

3. m∠B = 115°, AB = 5.7,
   AC = 9.1, area = 13 sq. units

4. m∠C = 62°, AB = 10.6,
   AC = 5.4, area = 28.9 sq. units

5. m∠B = 74°, AC = 24.1,
   BC = 20, area = 192.3 sq. units

6. m∠C = 165°, AC = 16.2,
   BC = 24.2, area = 50.7 sq. units

7. m∠A = 53.1°, m∠B = 36.9°,
   AB = 10, area = 24 sq. units

8. m∠B = 45°, AC = 7.1,
   BC = 7.1, area = 25.2 sq. units

9. m∠A = 53.4°, m∠C = 41.6°,
   BC = 24.2, area = 241.1 ft^2

Lesson 14.1
Level B

1. 7.0 in.^{2}
2. 14.1 m^2
3. 52.0 cm^2
4. 19.1 in.^{2}
5. 11.7 ft^2
6. 28.3 m^2
7. 17.7
8. 14.9
9. 4.1
10. 5.1

11. a = 9.9, c = 16.2, C = 73°
12. b = 21.4, c = 27.6, A = 30°
13. a = 1.4, b = 6.9, B = 123°
14. a = 7.5, c = 9.4, A = 50°
15. b = 17.5, c = 14.4, C = 44°
16. a = 7.3, b = 5.6, B = 50°
17. 33.5 meters from x, 36.6 from y

Lesson 14.1
Level C

1. 15.3 m^2
2. 16.3 cm^2
3. 63.4 cm^2
4. 21.7 m^2
5. 12.7
6. 2.0
7. 75.2
8. 2.9
9. 3.1
10. 2.6
11. a = 2.8, c = 4.6, C = 81°
12. a = 10.4, b = 10.55, A = 71°
13. a = 4.5, b = 2.4, C = 100°
14. b = 6.4, c = 5.9, B = 98°
15. 1 triangle, a = 4.6, b = 7.0, C = 100°
16. 2 triangles, A = 59°, C = 81°, c = 9.2;
    A = 121°, C = 19°, c = 3.0
17. 447.6 feet

Answers

**Lesson 14.2
Level A**

1. SAS, 8.2
2. SAS, 15.6
3. SSS, 56.3°
4. SSS, 14.6°
5. $A = 31.6°, B = 38.9°, C = 109.5°$
6. $A = 21.3°, B = 93.6°, C = 65.1°$
7. $A = 115°, B = 42.8°, C = 22.2°$
8. $A = 111.8°, B = 27.7°, C = 40.5°$
9. $a = 15.8, B = 87.5°, C = 40.5°$
10. $A = 42.8°, b = 13.3, C = 72.2°$
11. $A = 52.9°, B = 32.1°, c = 11.2$
12. $b = 2.4, B = 28°, C = 52°$

**Lesson 14.2
Level B**

1. SAS, 8.0
2. SAS, 19.8
3. SAS, 7.9
4. SAS, 17.8
5. SAS, 5.5
6. SAS, 9.8
7. $A = 122.6°, B = 17.3°, C = 40.1°$
8. $A = 57.1°, B = 44.4°, C = 78.5°$
9. $a = 24.8, B = 63.7°, C = 48.3°$
10. $A = 40.6°, b = 3.7, C = 102.4°$
11. $A = 50.7°, B = 75.30°, c = 1.7$
12. $b = 7.2, c = 9.1, C = 87°$
13. SAS, $a = 3.7, B = 85.8°, C = 56.2°$
14. not possible, SSS
15. SAS, $A = 18.4°, B = 31.6°, c = 7.3$
16. SAS, $A = 39.5°, b = 5.8, C = 72.5°$
17. SAS, $A = 37.8°, B = 125.2°, c = 1.4$
18. AAA, infinitely many triangles possible

**Lesson 14.2
Level C**

1. SAS, 4.4
2. SAS, 4.3
3. SAS, 4.2
4. SAS, 0.5
5. SAS, 15.5
6. SAS, 22.3
7. $A = 59.1°, B = 50.9°, C = 70°$
8. $a = 10.2, B = 34.0°, C = 74.0°$
9. $A = 55.4°, b = 3.0, C = 88.4°$
10. $A = 42.6°, b = 14.9, C = 89.6°$
11. $A = 123.6°, B = 33.7°, C = 22.7°$
12. $A = 78.5°, B = 68.9°, c = 3.4$
13. SAS, $A = 92.8°, B = 40.2°, c = 11.8$
14. SAS, $a = 9.9, B = 65.4°, C = 28.4°$
15. AAA, infinitely many triangles possible
16. SAS, $A = 83.9°, b = 7.4, C = 59.7°$
17. not possible, SSS
18. AAS, $b = 2.9, c = 7.7, C = 55.5°$

Answers

Lesson 14.3
Level A

1. $\cos\theta \cdot \dfrac{\sin\theta}{\cos\theta} = \sin\theta$
$\sin\theta = \sin\theta$

2. $\sin^2\theta = \dfrac{y^2}{r^2} = \dfrac{r^2 - x^2}{r^2}$
$= 1 - \dfrac{x^2}{r^2} = 1 - \left(\dfrac{x}{r}\right)^2$
$= 1 - \cos^2\theta$

3. $\dfrac{1}{\sin\theta} \cdot \dfrac{\sin\theta}{\cos\theta} = \sec\theta$
$\dfrac{1}{\cos\theta} = \sec\theta$
$\sec\theta = \sec\theta$

4. $\tan^2\theta = \dfrac{1}{\cos^2\theta} - 1 = \dfrac{1}{\cos^2\theta} - \dfrac{\cos^2\theta}{\cos^2\theta}$
$= \dfrac{1 - \cos^2\theta}{\cos^2\theta} = \dfrac{\sin^2\theta}{\cos^2\theta} = \tan^2\theta$

5. $\sec\theta$
6. $\csc\theta$
7. $\csc\theta$
8. $\sec^2\theta$
9. $\sin\theta$
10. $\cot^2\theta$
11. $2\sin\theta$
12. $\sin^2\theta$
13. $\sin\theta$

Lesson 14.3
Level B

1. $\sin\theta \cdot \dfrac{\cos\theta}{\sin\theta} \cdot \dfrac{1}{\cos\theta} = 1$
$\dfrac{\sin\theta}{\sin\theta} \cdot \dfrac{\cos\theta}{\cos\theta} = 1$

2. $\csc\theta = \sin\theta\,(\csc^2\theta)$
$\csc\theta = \sin\theta\,\dfrac{1}{\sin^2\theta}$
$\csc\theta = \dfrac{1}{\sin\theta}$
$\csc\theta = \csc\theta$

3. $\dfrac{\sec^2\theta}{\csc^2\theta} = \tan^2\theta$
$\dfrac{1}{\cos^2\theta} \cdot \sin^2\theta = \tan^2\theta$
$\dfrac{\sin^2\theta}{\cos^2\theta} = \tan^2\theta$
$\tan^2\theta = \tan^2\theta$

4. $\csc^2\theta\,\sin^2\theta = 1$
$\dfrac{1}{\sin^2\theta} \cdot \sin^2\theta = 1$
$\dfrac{\sin^2\theta}{\sin^2\theta} = 1$
$1 = 1$

5. 1
6. $\sec\theta$
7. $\tan^2\theta$
8. 0
9. $\tan^2\theta$
10. $\sec\theta$
11. $1 - \cos^2\theta$
12. $\cos^2\theta$
13. $3\cos^2\theta$

Answers

Lesson 14.3
Level C

1. $1 + \tan^2\theta = \sec^2\theta$

$\dfrac{\cos^2\theta}{\cos^2\theta} + \dfrac{\sin^2\theta}{\cos^2\theta} = \sec^2\theta$

$\dfrac{\cos^2\theta + \sin^2\theta}{\cos^2\theta} = \sec^2\theta$

$\dfrac{1}{\cos^2\theta} = \sec^2\theta$

$\sec^2\theta = \sec^2\theta$

2. $\dfrac{\csc^2\theta}{\cot^2\theta} = \sec^2\theta = \dfrac{\frac{1}{\sin^2\theta}}{\frac{\cos^2\theta}{\sin^2\theta}} = \sec^2\theta$

$\dfrac{1}{\sin^2\theta} \cdot \dfrac{\sin^2\theta}{\cos^2\theta} = \sec^2\theta = \dfrac{1}{\cos^2\theta} = \sec^2\theta$

$\sec^2\theta = \sec^2\theta$

3. $\tan\theta = \dfrac{\cos\theta}{\frac{1}{\sin\theta} - \frac{\sin^2\theta}{\sin\theta}} = \dfrac{\cos\theta}{\frac{1-\sin^2\theta}{\sin\theta}}$

$\dfrac{\frac{\cos\theta}{\cos^2\theta}}{\sin\theta} = \tan\theta = \dfrac{\frac{\sin\theta}{\cos\theta}}$

4. $\dfrac{\frac{1}{\cos\theta} - \frac{1}{\cos\theta}}{\frac{1}{\sin\theta}} = -\cot\theta$

$\dfrac{\sin\theta}{\cos\theta} - \dfrac{1}{\sin\theta\cos\theta} = -\cot\theta$

$\dfrac{\sin^2\theta}{\sin\theta\cos\theta} - \dfrac{1}{\sin\theta\cos\theta} = \dfrac{\sin^2\theta - 1}{\sin\theta\cos\theta}$

$= \dfrac{-\cos^2\theta}{\sin\theta\cos\theta} = \dfrac{-\cos\theta}{\sin\theta}$

5. $\cot^2\theta$

6. $\csc^2\theta$

7. $2\sec^2\theta$

8. $\csc^2\theta$

9. $-2\tan^2\theta$

10. $\cos\theta$

11. $\dfrac{4\cos\theta}{\sin^2\theta}$

12. 1

13. $\sec^2\theta$

Lesson 14.4
Level A

1. $\dfrac{\sqrt{2}+\sqrt{6}}{4}$

2. $\dfrac{-1}{2}$

3. 0

4. $\dfrac{-\sqrt{2}}{2}$

5. $\dfrac{\sqrt{2}}{2}$

6. $\dfrac{-\sqrt{2}}{2}$

7. $\dfrac{\sqrt{6}+\sqrt{2}}{4}$

8. $\dfrac{\sqrt{3}}{2}$

9. $\dfrac{\sqrt{3}}{2}$

10. $\dfrac{-\sqrt{3}}{2}$

11. $\dfrac{-\sqrt{2}}{2}$

12. $\dfrac{\sqrt{3}}{2}$

13. $\dfrac{\sqrt{2}}{2}$

14. $\dfrac{-1}{2}$

Answers

15. $\begin{bmatrix} 0.87 & -0.5 \\ 0.5 & 0.87 \end{bmatrix}$

16. $\begin{bmatrix} 0.71 & 0.71 \\ -0.71 & 0.71 \end{bmatrix}$

17. $\begin{bmatrix} 0.5 & 0.87 \\ -0.87 & 0.5 \end{bmatrix}$

18. $\begin{bmatrix} 0.87 & 0.5 \\ -0.5 & 0.87 \end{bmatrix}$

19. $\begin{bmatrix} -1 & 0 \\ 0 & -1 \end{bmatrix}$

20. $\begin{bmatrix} -0.87 & -0.5 \\ 0.5 & -0.87 \end{bmatrix}$

21.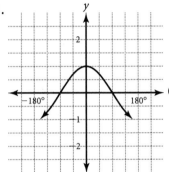

Lesson 14.4
Level B

1. $\dfrac{-\sqrt{2}}{2}$

2. $\dfrac{-\sqrt{6} + \sqrt{2}}{4}$

3. $\dfrac{\sqrt{6} + \sqrt{2}}{4}$

4. $\dfrac{\sqrt{6} + \sqrt{2}}{4}$

5. $\dfrac{-\sqrt{6} - \sqrt{2}}{4}$

6. 0

7. $\dfrac{-1}{2}$

8. $\dfrac{-\sqrt{3}}{2}$

9. $\dfrac{-1}{2}$

10. $\dfrac{\sqrt{2}}{2}$

11. $\begin{bmatrix} -0.5 & -0.87 \\ 0.87 & -0.5 \end{bmatrix}$

12. $\begin{bmatrix} -0.87 & 0.5 \\ -0.5 & -0.87 \end{bmatrix}$

13. $\begin{bmatrix} -0.87 & 0.5 \\ -0.5 & -0.87 \end{bmatrix}$

14. $\begin{bmatrix} 0.82 & -0.57 \\ 0.57 & 0.82 \end{bmatrix}$

15. $\begin{bmatrix} 0.64 & -0.77 \\ 0.77 & 0.64 \end{bmatrix}$

16. $\begin{bmatrix} -0.17 & 0.98 \\ -0.98 & -0.17 \end{bmatrix}$

17.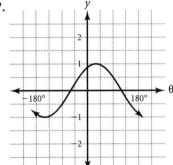

18.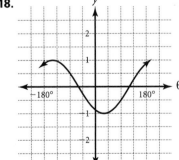

19. $(0.71, 4.95)$

20. $(2.13, -4.95)$

21. $(-5.66, -2.83)$

Algebra 2 — Practice Masters Levels A, B, and C

Answers

Lesson 14.4
Level C

1. $\dfrac{-\sqrt{6}+\sqrt{2}}{4}$
2. $\dfrac{-\sqrt{3}}{2}$
3. $\dfrac{-\sqrt{6}+\sqrt{2}}{4}$
4. $\dfrac{\sqrt{6}-\sqrt{2}}{4}$
5. $\dfrac{-\sqrt{2}}{2}$
6. $\dfrac{\sqrt{6}+\sqrt{2}}{4}$
7. $\dfrac{\sqrt{2}}{2}$
8. $\dfrac{-\sqrt{2}}{2}$
9. $\dfrac{\sqrt{2}}{2}$
10. $\dfrac{\sqrt{6}+\sqrt{2}}{4}$
11. $\begin{bmatrix} -0.34 & -0.94 \\ 0.94 & -0.34 \end{bmatrix}$
12. $\begin{bmatrix} 0.26 & 0.97 \\ -0.97 & 0.26 \end{bmatrix}$
13. $\begin{bmatrix} -0.91 & -0.42 \\ 0.42 & -0.91 \end{bmatrix}$
14. $\begin{bmatrix} -0.98 & 0.17 \\ -0.17 & -0.98 \end{bmatrix}$
15. $\begin{bmatrix} -0.94 & -0.34 \\ 0.34 & -0.94 \end{bmatrix}$
16. $\begin{bmatrix} -0.64 & -0.77 \\ 0.77 & -0.64 \end{bmatrix}$
17. $(-8.93, -0.54)$
18. $(-3.33, -4.23)$
19. $(10.66, 1.54)$

20.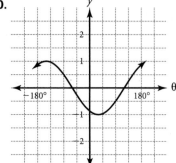

21.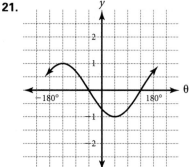

Lesson 14.5
Level A

1. $2\cos\theta$
2. $-\sin^2\theta$
3. $|\sin\theta|$
4. $\cos 2\theta$
5. $1 - \tan^2\theta$
6. $\cos^2\theta$
7. $\dfrac{\sqrt{2+\sqrt{2}}}{2}$
8. $\dfrac{\sqrt{2-\sqrt{2}}}{2}$
9. $\dfrac{\sqrt{2+\sqrt{2}}}{2}$
10. $\sin 2\theta = \dfrac{24}{25}$; $\cos 2\theta = \dfrac{7}{25}$
11. $\sin 2\theta = \dfrac{-24}{25}$; $\cos 2\theta = \dfrac{-7}{25}$
12. $\sin 2\theta = \dfrac{-240}{289}$; $\cos 2\theta = \dfrac{-161}{289}$

Answers

13. $\sin 2\theta = \dfrac{336}{625}$; $\cos 2\theta = \dfrac{-527}{625}$

14. $\sin \dfrac{\theta}{2} = \dfrac{\sqrt{5}}{5}$; $\cos \dfrac{\theta}{2} = \dfrac{2\sqrt{5}}{5}$

15. $\sin \dfrac{\theta}{2} = \dfrac{2\sqrt{5}}{5}$; $\cos \dfrac{\theta}{2} = \dfrac{-\sqrt{5}}{5}$

16. $\sin \dfrac{\theta}{2} = \dfrac{3\sqrt{10}}{10}$; $\cos \dfrac{\theta}{2} = \dfrac{-\sqrt{10}}{10}$

17. $\sin \dfrac{\theta}{2} = \dfrac{\sqrt{10}}{10}$; $\cos \dfrac{\theta}{2} = \dfrac{-3\sqrt{10}}{10}$

Lesson 14.5
Level B

1. $\cos 2\theta$
2. $\cos \theta$
3. $4\cos 8\theta$
4. $|\cos 2\theta|$
5. $\dfrac{1}{\cos \theta}$
6. $-\sin^2 \theta$
7. $\dfrac{\sqrt{2-\sqrt{2}}}{2}$
8. $\dfrac{\sqrt{2+\sqrt{3}}}{2}$
9. $\dfrac{-\sqrt{2-\sqrt{3}}}{2}$
10. $\sin 2\theta = \dfrac{240}{289}$; $\cos 2\theta = \dfrac{-161}{289}$
11. $\sin 2\theta = \dfrac{-240}{289}$; $\cos 2\theta = \dfrac{-161}{289}$
12. $\sin 2\theta = \dfrac{336}{625}$; $\cos 2\theta = \dfrac{-527}{625}$
13. $\sin 2\theta = \dfrac{-336}{625}$; $\cos 2\theta = \dfrac{-527}{625}$
14. $\sin \dfrac{\theta}{2} = \dfrac{\sqrt{2}}{10}$; $\cos \dfrac{\theta}{2} = \dfrac{7\sqrt{2}}{10}$
15. $\sin \dfrac{\theta}{2} = \dfrac{7\sqrt{2}}{10}$; $\cos \dfrac{\theta}{2} = \dfrac{\sqrt{2}}{10}$

16. $\sin \dfrac{\theta}{2} = \dfrac{5\sqrt{34}}{34}$; $\cos \dfrac{\theta}{2} = \dfrac{-3\sqrt{34}}{34}$

17. $\sin \dfrac{\theta}{2} = \dfrac{3\sqrt{34}}{34}$; $\cos \dfrac{\theta}{2} = \dfrac{-5\sqrt{34}}{34}$

Lesson 14.5
Level C

1. $\tan \theta$
2. $-\left|3\sin \dfrac{3\theta}{2}\right|$
3. 2
4. $\cos \theta - \sin \theta$
5. $2\cos 10\theta$
6. $3\cos 6\theta$
7. $\dfrac{\sqrt{2+\sqrt{2}}}{2}$
8. $\dfrac{-\sqrt{2-\sqrt{2}}}{2}$
9. $\dfrac{\sqrt{2-\sqrt{3}}}{2}$
10. $\sin 2\theta = \dfrac{4\sqrt{21}}{25}$; $\cos 2\theta = \dfrac{17}{25}$
11. $\sin 2\theta = \dfrac{-5\sqrt{39}}{32}$; $\cos 2\theta = \dfrac{-7}{32}$
12. $\sin 2\theta = \dfrac{4\sqrt{2}}{9}$; $\cos 2\theta = \dfrac{7}{9}$
13. $\sin 2\theta = \dfrac{-\sqrt{695}}{72}$; $\cos 2\theta = \dfrac{-67}{72}$
14. $\sin \dfrac{\theta}{2} = \dfrac{\sqrt{10}}{6}$; $\cos \dfrac{\theta}{2} = \dfrac{\sqrt{26}}{6}$
15. $\sin \dfrac{\theta}{2} = \dfrac{\sqrt{8+2\sqrt{15}}}{4}$; $\cos \dfrac{\theta}{2} = \dfrac{\sqrt{8-2\sqrt{15}}}{4}$
16. $\sin \dfrac{\theta}{2} = \dfrac{-\sqrt{140}}{14}$; $\cos \dfrac{\theta}{2} = \dfrac{-2\sqrt{14}}{14}$
17. $\sin \dfrac{\theta}{2} = \dfrac{\sqrt{242-88\sqrt{7}}}{22}$; $\cos \dfrac{\theta}{2} = \dfrac{-\sqrt{242+88\sqrt{7}}}{22}$

Answers

Lesson 14.6
Level A

1. $\theta = 330° + 260n°; \theta = 210° + 360n°$
2. $\theta = 60° + 360n°; \theta = 300° + 360n$
3. $\theta = 60° + 360n°; \theta = 120° + 360n°$
4. $\theta = 30° + 180n°$
5. $\theta = 150° + 360n°; \theta = 210° + 360n°$
6. $\theta = 45° + 180n°$
7. $\theta = 60°, 120°, 240°, 300°$
8. $\theta = 180°$
9. $\theta = 30°, 60°, 210°, 240°$
10. $\theta = 0, 180°, 360°$
11. $\theta = \dfrac{\pi}{3}, \dfrac{4\pi}{3}$
12. $\theta = \dfrac{2\pi}{3}, \dfrac{4\pi}{3}$
13. $\theta = \dfrac{\pi}{4}, \dfrac{3\pi}{4}, \dfrac{5\pi}{4}, \dfrac{7\pi}{4}$
14. $\theta = \dfrac{\pi}{6}, \dfrac{5\pi}{6}, \dfrac{7\pi}{6}, \dfrac{11\pi}{6}$
15. $\theta = 48.6°, 131.4°$
16. $\theta = 200.4°, 270°$
17. $0.369 + 2n$ and $1.63 + 2n$

Lesson 14.6
Level B

1. $\theta = 180°n$
2. $\theta = 270° + 360n°$
3. $\theta = 60° + 360n°; \theta = 300° + 360n°$
4. $\theta = 30° + 180n°$
5. $\theta = 90°$
6. $\theta = 0°, 60°, 300°$
7. $\theta = 30°, 90°, 150°$
8. $\theta = 30°, 90°, 270°, 330°$
9. $\theta = 0, 2\pi$
10. $\theta = 0, \pi, 2\pi, \dfrac{\pi}{3}, \dfrac{5\pi}{3}$
11. $\theta = \dfrac{7\pi}{6}, \dfrac{11\pi}{6}$
12. $\theta = \dfrac{\pi}{6}, \dfrac{5\pi}{6}, \dfrac{7\pi}{6}, \dfrac{11\pi}{6}$
13. $\theta = 17.6°, 162.4°$
14. $\theta = 35.3°, 144.7°, 215.3°, 324.7°$
15. $\theta = 24.5°$

Lesson 14.6
Level C

1. $\theta = 45° + 180n°$
2. $\theta = 90° + 180n°$
3. $\theta = 45° + 180n°$
4. $\theta = 360n°$
5. $\theta = 0°, 60°, 300°$
6. $\theta = 0°, 180°, 210°, 330°, 360°$
7. $\theta = 60°, 120°, 240°, 300°$
8. $\theta = 60°, 120°, 240°, 300°$
9. $\theta = 0, \pi, 2\pi$
10. $\theta = 0, \dfrac{2\pi}{3}, \dfrac{4\pi}{3}, 2\pi$
11. $\theta = 0, 2\pi$
12. $\theta = 0, \dfrac{\pi}{2}, \dfrac{5\pi}{4}, 2\pi$
13. $\theta = 150.8°, 234.7°$
14. $\theta = 90°, 153.4°, 270°, 333.4°$
15. a. $32.1°$
    b. $27.1°$
    c. $24.3°$
    d. $32.1°$